Introduction to Thermo-Fluids Systems Design

Introduction to Thermo-Fluids Systems Design

André G. McDonald, Ph.D., P.ENG.

University of Alberta, Canada

Hugh L. Magande, M.B.A., M.S.E.M.

Rinnai America Corporation, USA

A John Wiley & Sons, Ltd., Publication

DISCLAIMER

The contents of this textbook are meant to supply information on the design of thermo-fluids systems.
The book is not meant to be the sole resource used in any design project. The examples and solutions
presented are not to be construed as complete engineered design solutions for any particular problem or
project. The authors and publisher are not attempting to render any type of engineering or other
professional services. Should these services be required, an appropriate professional engineer should be
consulted. The authors and publisher assume no liability or responsibility for any uses made of the
material contained and described herein.

Other Wiley Editorial Offices

John Wiley & Sons Ltd, The Atrium, Southern Gate, Chichester, West Sussex PO19 8SQ, UK

John Wiley & Sons, Inc., 111 River Street, Hoboken, NJ 07030, USA

Jossey-Bass, 989 Market Street, San Francisco, CA 94103-1741, USA

Wiley-VCH Verlag GmbH, Boschstr. 12, D-69469 Weinheim, Germany

John Wiley & Sons Australia Ltd, 42 McDougall Street, Milton, Queensland 4064, Australia

John Wiley & Sons Canada Ltd, 6045 Freemont Blvd, Mississauga, ONT, L5R 4J3, Canada

Wiley also publishes its books in a variety of electronic formats. Some content that appears in print may
not be available in electronic books.

A catalogue record for this book is available from the British Library.

ISBN: 9781118313633

Set in 10/12.5pt Palatino by Aptara Inc., New Delhi, India.
Printed in Malaysia by Ho Printing (M) Sdn Bhd

Contents

Preface   xi

List of Figures   xv

List of Tables   xix

List of Practical Notes   xxi

List of Conversion Factors   xxiii

1   Design of Thermo-Fluids Systems   1
    1.1   Engineering Design—Definition   1
    1.2   Types of Design in Thermo-Fluid Science   1
    1.3   Difference between Design and Analysis   2
    1.4   Classification of Design   2
    1.5   General Steps in Design   2
    1.6   Abridged Steps in the Design Process   2

2   Air Distribution Systems   5
    2.1   Fluid Mechanics—A Brief Review   5
          2.1.1   Internal Flow   5
    2.2   Air Duct Sizing—Special Design Considerations   12
          2.2.1   General Considerations   12
          2.2.2   Sizing Straight Rectangular Air Ducts   13
          2.2.3   Use of an Air Duct Calculator to Size Rectangular Air Ducts   18
    2.3   Minor Head Loss in a Run of Pipe or Duct   18
    2.4   Minor Losses in the Design of Air Duct Systems—Equal Friction
          Method   20

2.5   Fans—Brief Overview and Selection Procedures   44
    2.5.1   Classification and Terminology   44
    2.5.2   Types of Fans   44
    2.5.3   Fan Performance   46
    2.5.4   Fan Selection from Manufacturer's Data or Performance
          Curves   48
    2.5.5   Fan Laws   51
2.6   Design for Advanced Technology—Small Duct High-Velocity
     (SDHV) Air Distribution Systems   54
     Problems   66
     References and Further Reading   72

3   Liquid Piping Systems   **73**
3.1   Liquid Piping Systems   73
3.2   Minor Losses: Fittings and Valves in Liquid Piping Systems   73
    3.2.1   Fittings   73
    3.2.2   Valves   73
    3.2.3   A Typical Piping System—A Closed-Loop
          Fuel Oil Piping System   75
3.3   Sizing Liquid Piping Systems   75
    3.3.1   General Design Considerations   75
    3.3.2   Pipe Data for Building Water Systems   77
3.4   Fluid Machines (Pumps) and Pump–Pipe Matching   83
    3.4.1   Classifications and Terminology   83
    3.4.2   Types of Pumps   83
    3.4.3   Pump Fundamentals   83
    3.4.4   Pump Performance and System Curves   86
    3.4.5   Pump Performance Curves for a Family of Pumps   88
    3.4.6   A Manufacturer's Performance Plot for a Family of
          Centrifugal Pumps   89
    3.4.7   Cavitation and Net Positive Suction Head   92
    3.4.8   Pump Scaling Laws: Nondimensional Pump Parameters   97
    3.4.9   Application of the Nondimensional Pump
          Parameters—Affinity Laws   98
    3.4.10   Nondimensional Form of the Pump Efficiency   99
3.5   Design of Piping Systems Complete with In-Line or Base-Mounted
     Pumps   103
    3.5.1   Open-Loop Piping System   103
    3.5.2   Closed-Loop Piping System   111
     Problems   121
     References and Further Reading   126

4   Fundamentals of Heat Exchanger Design   **127**
4.1   Definition and Requirements   127

	4.2	Types of Heat Exchangers	127
		4.2.1 Double-Pipe Heat Exchangers	127
		4.2.2 Compact Heat Exchangers	129
		4.2.3 Shell-and-Tube Heat Exchangers	129
	4.3	The Overall Heat Transfer Coefficient	130
		4.3.1 The Thermal Resistance Network for Plane Walls—Brief Review	132
		4.3.2 Thermal Resistance from Fouling—The Fouling Factor	136
	4.4	The Convection Heat Transfer Coefficients—Forced Convection	138
		4.4.1 Nusselt Number—Fully Developed Internal Laminar Flows	139
		4.4.2 Nusselt Number—Developing Internal Laminar Flows—Correlation Equation	139
		4.4.3 Nusselt Number—Turbulent Flows in Smooth Tubes: Dittus–Boelter Equation	141
		4.4.4 Nusselt Number—Turbulent Flows in Smooth Tubes: Gnielinski's Equation	141
	4.5	Heat Exchanger Analysis	142
		4.5.1 Preliminary Considerations	142
		4.5.2 Axial Temperature Variation in the Working Fluids—Single Phase Flow	143
	4.6	Heat Exchanger Design and Performance Analysis: Part 1	147
		4.6.1 The Log-Mean Temperature Difference Method	147
		4.6.2 The Effectiveness-Number of Transfer Units Method: Introduction	148
		4.6.3 The Effectiveness-Number of Transfer Units Method: ε-NTU Relations	149
		4.6.4 Comments on the Number of Transfer Units and the Capacity Ratio (c)	151
		4.6.5 Procedures for the ε-NTU Method	156
		4.6.6 Heat Exchanger Design Considerations	157
	4.7	Heat Exchanger Design and Performance Analysis: Part 2	157
		4.7.1 External Flow over Bare Tubes in Cross Flow—Equations and Charts	157
		4.7.2 External Flow over Tube Banks—Pressure Drop	162
		4.7.3 External Flow over Finned-Tubes in Cross Flow—Equations and Charts	175
	4.8	Manufacturer's Catalog Sheets for Heat Exchanger Selection	202
		Problems	208
		References and Further Reading	211
5	**Applications of Heat Exchangers in Systems**		**213**
	5.1	Operation of a Heat Exchanger in a Plasma Spraying System	213
	5.2	Components and General Operation of a Hot Water Heating System	216

5.3   Boilers for Water   217
      5.3.1   Types of Boilers   217
      5.3.2   Operation and Components of a Typical Boiler   218
      5.3.3   Water Boiler Sizing   220
      5.3.4   Boiler Capacity Ratings   224
      5.3.5   Burner Fuels   226
5.4   Design of Hydronic Heating Systems c/w Baseboards
      or Finned-Tube Heaters   227
      5.4.1   Zoning and Types of Systems   227
      5.4.2   One-Pipe Series Loop System   227
      5.4.3   Two-Pipe Systems   229
      5.4.4   Baseboard and Finned-Tube Heaters   233
5.5   Design Considerations for Hot Water Heating Systems   236
      Problems   258
      References and Further Reading   265

6 Performance Analysis of Power Plant Systems   267
6.1   Thermodynamic Cycles for Power Generation—Brief Review   267
      6.1.1   Types of Power Cycles   267
      6.1.2   Vapor Power Cycles—Ideal Carnot Cycle   268
      6.1.3   Vapor Power Cycles—Ideal Rankine Cycle for Steam
              Power Plants   268
      6.1.4   Vapor Power Cycles—Ideal Regenerative Rankine Cycle for
              Steam Power Plants   269
6.2   Real Steam Power Plants—General Considerations   271
6.3   Steam-Turbine Internal Efficiency and Expansion Lines   272
6.4   Closed Feedwater Heaters (Surface Heaters)   280
6.5   The Steam Turbine   282
      6.5.1   Steam-Turbine Internal Efficiency and Exhaust End Losses   282
      6.5.2   Casing and Shaft Arrangements of Large Steam Turbines   284
6.6   Turbine-Cycle Heat Balance and Heat and Mass Balance Diagrams   286
6.7   Steam-Turbine Power Plant System Performance Analysis
      Considerations   288
6.8   Second-Law Analysis of Steam-Turbine Power Plants   300
6.9   Gas-Turbine Power Plant Systems   307
      6.9.1   The Ideal Brayton Cycle for Gas-Turbine Power Plant
              Systems   307
      6.9.2   Real Gas-Turbine Power Plant Systems   309
      6.9.3   Regenerative Gas-Turbine Power Plant Systems   312
      6.9.4   Operation and Performance of Gas-Turbine Power
              Plants—Practical Considerations   313
6.10 Combined-Cycle Power Plant Systems   324
      6.10.1   The Waste Heat Recovery Boiler   325
      Problems   332
      References and Further Reading   338

Appendix A: Pipe and Duct Systems   339

Appendix B: Symbols for Drawings   365

Appendix C: Heat Exchanger Design   373

Appendix D: Design Project— Possible Solution   383
D.1   Fuel Oil Piping System Design   383

Appendix E: Applicable Standards and Codes   413

Appendix F: Equipment Manufacturers   415

Appendix G: General Design Checklists   417
G.1   Air and Exhaust Duct Systems   417
G.2   Liquid Piping Systems   418
G.3   Heat Exchangers, Boilers, and Water Heaters   419

Index   421

Preface

Design courses and projects in contemporary undergraduate curricula have focused mainly on topics in solid mechanics. This has left graduating junior engineers with limited knowledge and experience in the design of components and systems in the thermo-fluids sciences. ABB Automation in their handbook on *Energy Efficient Design of Auxiliary Systems in Fossil-Fuel Power Plants* has mentioned that this lack of training in thermo-fluids systems design will limit our ability to produce high-performance systems. This deficiency in contemporary undergraduate curricula has resulted in an urgent need for course materials that underline the application of fundamental concepts in the design of thermo-fluids components and systems.

Owing to the urgent need for course materials in this area, this textbook has been developed to bridge the gap between the fundamental concepts of fluid mechanics, heat transfer, and thermodynamics and the practical design of thermo-fluids components and systems. To achieve this goal, this textbook is focused on the design of internal fluid flow systems, coiled heat exchangers, and performance analysis of power plant systems. This requires prerequisite knowledge of internal fluid flow, conduction heat transfer, convection heat transfer with emphasis on forced convection in tubes and over cylinders, analysis of constant area fins, and thermodynamic power cycles, in particular, the Rankine and Brayton cycles. The fundamental concepts are used as tools in an exhaustive design process to solve various practical problems presented in the examples. For junior design engineers with limited practical experience, use of fundamental concepts of which they have previous knowledge will help them to increase their confidence and decision-making capabilities.

The complete design or modification of modern equipment and systems will require knowledge of current industry practices. While relying on and demonstrating the application of fundamental principles, this textbook highlights the use of manufacturers' catalogs to select equipment and practical rules to guide decision-making in the design process. Some of these practical rules are included in the text as **Practical Notes**, to underline their importance in current practice and provide additional information. While great emphasis is placed upon the use of these rules, an effort was made to ensure that the reader understands the fundamental concepts that support these guidelines. It is strongly believed that this will also enable the design engineer to make quick and accurate decisions in situations where the guidelines may not be applicable.

The topics covered in the text are arranged so that each topic builds on the previous concepts. It is important to convey to the reader that, in the design process, topics are not stand-alone items and they must come together to produce a successful design. There are three main topical areas, arranged in six chapters.

Introductory material on the design process is presented in *Chapter 1*. Since the book focuses on the detailed, technical design of thermo-fluids components and systems, the chapter ends with an abridged version of the full design process.

Chapters 2 and 3 deal with the design of air duct and liquid piping systems, respectively. It is in these initial chapters that a brief review of internal fluid flow is presented. System layout, component sizing, and equipment selection are also covered.

An introduction to heat exchanger design and analysis is presented in *Chapter 4*. This chapter presents the most fundamental material in the textbook. Extensive charts are used to design and analyze the performance of bare-tube and finned-tube coiled heat exchangers. The chapter ends with a description of excerpts from a manufacturer's catalog used to select heating coil models that are used in high-velocity duct systems.

Chapter 5 continues the discussion of heat exchangers by focusing on the sizing and selection of various heat exchangers such as boilers, water heaters, and finned-tube baseboard heaters. Various rules and data are presented to guide the selection and design process.

Chapter 6 focuses on the analysis of power plant systems. Here, the reader is introduced to a review of thermodynamic power cycles and various practical considerations in the analysis of steam-turbine and gas-turbine power generation systems. Combined-cycle systems and waste heat recovery boilers are also presented.

There are seven *Appendices* at the end of this book. They contain a wide variety of charts, tables, and catalog sheets that the design engineer will find useful during practice. Also included in the appendices are: a possible solution of a design project, the names of organizations that provide applicable codes and standards, and the names of some manufacturers and suppliers of equipment used in thermo-fluids systems.

The writing of this textbook was inspired, in part, by the difficulty to find appropriate textbooks that presented a detailed practical approach to the design of thermo-fluids components and systems in industrial environments. It is hoped that the readers and design engineers, in particular, will find it useful in practice as a reference during design projects and analysis.

The authors have made no effort to claim complete originality of the text. We have been motivated by the work of many others that have been appropriately referenced throughout the textbook.

While we feel that this textbook will be a valuable resource for design engineers in industry, it is offered as a guide, and as such, judgement is required when using the text to design systems or for application to specific installations. The authors and the publisher are not responsible for any uses made of this text.

We express our deepest gratitude to and acknowledge the advice, critiques, and suggestions that we received from, our advisory committee of professors, professional engineers, and students. These individuals include Dr. Roger Toogood, P. Eng.; Mr. Mark Ackerman, P. Eng.; Mr. Curt Stout, P. Eng.; Dr. Larry Kostiuk, P. Eng.; Mr. Dave DeJong, P. Eng.; Mr. Michael Ross; and Mr. David Therrien.

A.G. McDonald
H.L. Magande

List of Figures

1.1	General steps in the design process	3
2.1	Duct shapes and aspect ratios	13
2.2	Photo of a typical air duct calculator	19
2.3	A ductwork system to transport air (*ASHRAE Handbook*, Fundamentals Volume, 2005; reprinted with permission)	21
2.4	Axial fans	45
2.5	Centrifugal fans	45
2.6	Classification of centrifugal fans based on blade types	46
2.7	Typical performance curves of centrifugal fans	47
2.8	Forward-curved centrifugal fan performance curves (Morrison Products, Inc.; reprinted with permission)	49
3.1	Some typical industrial valves	74
3.2	A typical fuel oil piping system complete with a pump set (*ASHRAE Handbook*, Fundamentals Volume, 2005; reprinted with permission)	75
3.3	Plastic pipe (Schedule 80) friction loss chart (*ASHRAE Handbook*, Fundamentals Volume, 2005; reprinted with permission)	79
3.4	Pipes supported on hangers	79
3.5	Pipes and an in-line pump mounted on brackets	81
3.6	Types of industrial pumps: (a) three-lobe rotary pump; (b) two-screw pump; (c) in-line centrifugal pump; (d) vertical mutistage submersible pump (Hydraulic Institute, Parsippany, NJ, www.pumps.org; reprinted with permission)	84
3.7	Schematic of a H_{pump} versus $\dot{V}$ curve for a centrifugal pump	86
3.8	Schematic of a η_{pump} versus $\dot{V}$ curve	87
3.9	Schematic of a system curve intersecting a pump performance curve	88
3.10	Performance curves for a family of geometrically similar pumps	89
3.11	Pump performance plot (Taco, Inc.; reprinted with permission)	89
3.12	A typical open-loop condenser piping system for water	104
3.13	Diagrams of closed-loop piping systems	112

4.1   Temperature profiles and schematics of (a) parallel and (b) counter
      flow double-pipe heat exchangers   128
4.2   Cross-flow heat exchangers   129
4.3   Picture of a continuous plate-fin-tube type cross-flow heat exchanger 130
4.4   Schematics of shell-and-tube heat exchangers   131
4.5   Temperature distribution around and through a 1D plane wall   132
4.6   Thermal resistance network around a plane wall   135
4.7   Axial temperature variation in parallel flow heat exchanger   144
4.8   Axial temperature variation in counter flow heat exchanger   145
4.9   Axial temperature variation in a balanced heat exchanger   145
4.10 Axial temperature variation in a heat exchanger with condensation   146
4.11 Axial temperature variation in a heat exchanger with boiling   146
4.12 Effectiveness charts for some heat exchangers (Kays and London [2])   153
4.13 (a) Finned tube and (b) bare tube bank bundles   158
4.14 Flow pattern for an in-line tube bank (Çengel [3], reprinted with
      permission)   159
4.15 Data for flow normal to an in-line tube bank (Kays and London [2])   160
4.16 Flow pattern for a staggered tube bank (Çengel [3], reprinted with
      permission)   161
4.17 Data for flow normal to a staggered tube bank (Kays and London [2])   162
4.18 Schematic drawing of tube bank showing the total length, L_{total} 163
4.19 Examples of finned heat exchangers   176
4.20 General constant area, straight fins attached to a surface   177
4.21 Staggered tube bank with a hexangular finned-tube array   178
4.22 Data for flow normal to a finned staggered tube bank (*ASHRAE
      Transactions*, Vol. 79, Part II, 1973; reprinted with permission)   179
4.23 Data for flow normal to staggered tube banks: multiple tube rows
      (*ASHRAE Transactions*, Vol. 81, Part I, 1975; reprinted with permission) 180
4.24 M series heating coil from Unico, Inc. (a) Page 1 of the M series heating
      coil from Unico, Inc. (Unico, Inc., reprinted with permission) (b) Page 2
      of the M series heating coil from Unico, Inc. (Unico, Inc.; reprinted with
      permission) (c) Page 3 of the M series heating coil from Unico, Inc.
      (Unico, Inc., reprinted with permission) Page 4 of the M series heating
      coil from Unico, Inc. (Unico, Inc.; reprinted with permission)   203
5.1   A Praxair SG-100 plasma spray torch in operation   214
5.2   The Sulzer Metco Climet-HE^{TM}-200 heat exchanger (Sulzer Metco,
      Product Manual MAN 41292 EN 05; reprinted with permission)   214
5.3   Functional diagram for the Sulzer Metco Climet-HE^{TM}-200 (Sulzer
      Metco, Product Manual MAN 41292 EN 05; reprinted with permission)   215
5.4   Flow diagram for cooling a typical plasma torch (modified from Sulzer
      Metco, Product Manual MAN 41292 EN 05; reprinted with permission)   216
5.5   Schematic of a closed-loop hydronic heating system c/w a boiler   217
5.6   A typical gas-fired hot water boiler   218
5.7   Schematic of the internal section of typical water heaters   220

5.8   (a) A Rinnai noncondensing tankless water heater. (b) Schematic of
        Rinnai noncondensing tankless water heater (reprinted with
        permission)   221
5.9   Brochure showing specifications for a line of gas-fired boilers (Smith
        Cast Iron Boilers, GB100 series technical brochure; reprinted with
        permission)   225
5.10   Schematic diagram of a one-pipe series loop system   227
5.11   Schematic diagram of a split series loop system   228
5.12   Schematic of a one-pipe "monoflow" series loop system   229
5.13   Schematic diagram of a multizone system of one-pipe series loops   230
5.14   Schematic of a two-pipe direct return system   230
5.15   Schematic of a two-pipe reverse return system   231
5.16   Unbalanced flow in a two-pipe direct return system   232
5.17   Improved balance in a two-pipe direct return system   232
5.18   Diagrams of baseboard heaters. (a) 1-tiered baseboard heater; (b)
        2-tiered finned-tube heater   233
6.1   Ideal Carnot cycle   268
6.2   Ideal Rankine cycle   269
6.3   Ideal regenerative Rankine cycles. (a) Single-stage feedwater heating;
        (b) four-stage feedwater heating   270
6.4   Mollier diagram for water   273
6.5   Mollier diagram for water showing an expansion line   274
6.6   Drain disposals for closed feedwater heaters (surface heaters)   281
6.7   Turbine operation   283
6.8   Exhaust diffuser of a LP turbine   284
6.9   Casing and shaft arrangements for large condensing turbines. (a)
        Tandem-compound 2 flows from 150 to 400 MW; (b)
        Tandem-compound 4 flows from 300 to 800 MW; (c) Cross-compound 2
        flows from 300 to 800 MW; (d) Cross-compound 4 flows from 800 to
        1200 MW   285
6.10   Heat-and-mass balance diagram for a fossil-fuel power plant (Li and
        Priddy [1]; reprinted with permission)   287
6.11   Ideal Brayton cycle   308
6.12   Real Brayton cycle   309
6.13   Regenerative Brayton cycle   313
6.14   Regenerative Brayton cycle with intercooling   313
6.15   Schematic of a combined-cycle power plant   324
6.16   Piping schematic of a single-pressure waste heat recovery boiler   325
6.17   Temperature profile in a single-pressure waste heat recovery boiler   326
A.1   Friction Loss in Round (Straight) Ducts. *Source: System Design Manual,
        Part 2: Air Distribution*, Carrier Air Conditioning Co., Syracuse, NY,
        1974 (Reprinted with permission)   351
A.2   Schematics elbows in ducts   352

A.3   Copper tubing friction loss (open and closed piping systems) (Carrier
      Corp.; reprinted with permission)   353
A.4   Commercial steel pipe (Schedule 40) friction loss. (a) *Open piping*
      systems (Carrier Corp.; reprinted with permission); (b) *closed piping*
      systems (Carrier Corp.; reprinted with permission)   354
A.5   Bell & Gosset pump catalog (ITT Bell & Gossett; reprinted with
      permission)   356
C.1   *j*-factor versus Re_G charts for in-line tube banks. Transient tests
      (2 charts): (a) For $X_t = 1.50$ and $X_L = 1.25$; (b) For $X_t = 1.25$ and
      $X_L = 1.25$.
      (Kays, W. and London, A. (1964) *Compact Heat Exchangers*, 2nd edn,
      McGraw-Hill, Inc., New York)   375
C.2   *j*-factor versus Re_G charts for staggered tube banks. Transient tests
      (6 charts): (a) For $X_t = 1.50$ and $X_L = 1.25$; (b) For $X_t = 1.25$ and
      $X_L = 1.25$; (c) For $X_t = 1.50$ and $X_L = 1.0$; (d) For $X_t = 1.5$ and $X_L = 1.5$;
      (e) For $X_t = 2$ and $X_L = 1$; (f) For $X_t = 2.5$ and $X_L = 0.75$.
      (Kays, W. and London, A. (1964) *Compact Heat Exchangers*, 2nd edn,
      McGraw-Hill, Inc., New York)   376
C.3   *j*-factor versus Re_{x_L} charts for staggered tube banks (finned tubes):
      (a) five rows of tubes (*ASHRAE Transactions*, vol. 79, Part II, 1973;
      reprinted with permission); (b) multiple rows of tubes (*ASHRAE*
      Transactions, vol. 81, Part I, 1975; reprinted with permission)   380
C.4   *j*-factor versus Re_G charts for staggered tube banks (finned tubes).
      (a) Tube outer diameter $= 0.402$ in.; (b) tube outer diameter $= 0.676$ in.
      (Kays, W. and London, A. (1964) *Compact Heat Exchangers*, 2nd edn,
      McGraw-Hill, Inc., New York)   381

List of Tables

2.1	Maximum duct velocities	14
2.2	Typical values of component pressure losses [9]	21
2.3	Maximum supply duct velocities	54
2.4	Sound data during airflow through a rectangular elbow	55
2.5	Maximum main duct air velocities for acoustic design criteria	56
2.6	Acoustic design criteria for unoccupied spaces [21]	57
3.1	Typical average velocities for selected pipe flows	76
3.2	Pipe data for copper and steel	78
3.3	Hanger spacing for straight stationary pipes and tubes [1]	80
3.4	Minimum hanger rod size for straight stationary pipes and tubes [1]	80
4.1	Values of the overall heat transfer coefficient (US)	136
4.2	Values of the overall heat transfer coefficient (SI)	137
4.3	Representative fouling factors in heat exchangers	138
4.4	Nusselt numbers and friction factors for fully developed laminar flow in tubes of various cross sections: constant surface temperature and surface heat flux [3]	140
4.5	Effectiveness relations for heat exchangers	152
5.1	Minimum recovery rates and minimum usable storage capacities	224
5.2	Approximate heating value of fuels	226
5.3	Baseboard heater rated outputs at 1 gpm water flow rate	233
5.4	"Front outlet" finned-tube heater ratings for Trane heaters	234
5.5	Flow rate correction factors for water velocities less than 3 fps	235
5.6	Temperature correction factors for hot water ratings	236
6.1	Pressure drops at the gas-turbine plant inlet and exhaust [1]	315
6.2	Common steam conditions for waste heat recovery boilers [1]	327
A.1	Average roughness of commercial pipes	339
A.2	Correlation equations for friction factors	340
A.3	Circular equivalents of rectangular ducts for equal friction and capacity	341
A.4	Approximate equivalent lengths for selected fittings in circular Ducts	342
A.5	Approximate equivalent lengths for elbows in ducts	342

A.6   Data for copper pipes   343
A.7   Data for schedule 40 steel pipes   344
A.8   Data for schedule 80 steel pipes   345
A.9   Data for class 150 cast iron pipes   346
A.10   Data for glass pipes   346
A.11   Data for PVC plastic pipes   347
A.12   Typical average velocities for selected pipe flows[a]   348
A.13   Erosion limits: maximum design fluid velocities for water flow in
       small tubes   348
A.14   Loss coefficients for pipe fittings   349
A.15   Typical pipe data format   350
A.16   Typical pump schedule format   350
B.1   Airmoving devices and ductwork symbols   365
B.2   Piping symbols   367
B.3   Symbols for piping specialities   368
B.4   Additional/alternate valve symbols   369
B.5   Fittings   370
B.6   Radiant Panel Symbols   372
C.1   Representative values of the overall heat transfer coefficients (US)   373
C.2   Representative values of the overall heat transfer coefficients (SI)   374
C.3   Representative fouling factors in heat exchangers   374

List of Practical Notes

2.1	Total Static Pressure Available at a Plenum or Produced by a Fan	20
2.2	Diffuser Discharge Air Volume Flow Rates in SDHV Systems	56
3.1	Link Seals	75
3.2	Piping Systems Containing Air	76
3.3	Higher Pipe Friction Losses and Velocities	77
3.4	Piping System Supported by Brackets	81
3.5	Manufacturers' Pump Performance Curves	88
3.6	"To-the-point" Design	90
3.7	Oversizing Pumps	90
3.8	NPSH	93
3.9	Bypass Lines	104
3.10	Regulation and Control of Flow Rate across a Pump	104
3.11	In-Line and Base-Mounted Pumps	105
3.12	Flanged or Screwed Pipe Fittings?	113
4.1	Industrial Flows	142
4.2	Flow in Rough Pipes	142
4.3	Condensers and Boilers	147
4.4	Real Heat Exchangers	149
4.5	Heat Transfer from Staggered Tube Banks	161
4.6	Coil Arrangement in Air-to-Water Heat Exchangers	164
4.7	Pressure Drop Over Tube Banks	164
4.8	L and M values	179
5.1	Condensing Boilers	219
5.2	Typical OSF Values	222
5.3	Domestic Water Data for Edmonton, Alberta, Canada	223
5.4	Hot Water Temperatures from Faucets	223
5.5	Temperature Data for Sizing Finned-Tube Heaters	235

6.1   Optimizing the Number of Feedwater Heaters   271
6.2   *DCA* and *TTD* Values   281
6.3   Stages of a Steam Turbine   282
6.4   Exhaust End Loss   284
6.5   Units of the Net Heat Rate (NHR)   288
6.6   How Does One Initiate Operation of a Power Plant System?   289
6.7   Reference Pressure and Temperature for Availability Analysis   302
6.8   Combustion Air and Cracking in a Burner   309

List of Conversion Factors

Dimension	Conversion
Energy	1 Btu $= 778.28$ lbf ft 1 kWh $= 3412.14$ Btu 1 hp h $= 2545$ Btu 1 therm $= 10^5$ Btu (natural gas)
Force	1 lbf $= 32.2$ lbm ft/s^2 $= 16$ ozf 1 dyne $= 2.248 \times 10^{-6}$ lbf
Length	1 ft $= 12$ in. 1 yard $= 3$ ft 1 in. $= 25.4$ mm 1 mile $= 5280$ ft
Mass	1 slug $= 32.2$ lbm 1 lbm $= 16$ ounces (oz) 1 ton mass $= 2000$ lbm
Power	1 kW $= 3412.14$ Btu/h 1 hp $= 550$ lbf ft/s 1 hp (boiler) $= 33475$ Btu/h 1 ton refrigeration $= 12000$ Btu/h
Pressure	1 atm $= 14.7$ psia 1 psia $= 2.0$ in Hg at 32°F
Temperature	$T(\text{R}) = T(°\text{F}) + 460$ $T(°\text{F}) = 1.8T(°\text{C}) + 32$
Viscosity (dynamic)	1 lbm/(ft s) $= 1488$ centipoises (cp)
Viscosity (kinematic)	1 ft^2/s $= 929$ stokes (St)
Volume	1 British gallon $= 1.2$ US gallon 1 ft^3 $= 7.48$ US gallons 1 US gallon $= 128$ fluid ounces
Volume Flow Rate	35.315 ft^3/s $= 15850$ gal/min (gpm) $= 2118.9$ ft^3/min (cfm)

1

Design of Thermo-Fluids Systems

1.1 Engineering Design—Definition

Process of devising a system, subsystem, component, or process to meet desired needs.

1.2 Types of Design in Thermo-Fluid Science

(i) *Process Design:* The manipulation of physical and/or chemical processes to meet desired needs.

   Example: (a) Introduce boiling or condensation to increase heat transfer rates.

(ii) *System Design:* The process of defining the components and their assembly to function to meet a specified requirement.

   Examples: (a) Steam turbine power plant system consisting of turbines, pumps, pipes, and heat exchangers.

   (b) Hot water heating system, complete with boilers.

(iii) *Subsystem Design:* The process of defining and assembling a small group of components to do a specified function.

   Example: Pump/piping system of a large power plant. The pump/piping system is a subsystem of the larger power plant system used to transport water to and from the boiler or steam generator.

(iv) *Component Design:* Development of a piece of equipment or device.

Introduction to Thermo-Fluids Systems Design, First Edition. André G. McDonald and Hugh L. Magande.
© 2012 André G. McDonald and Hugh L. Magande. Published 2012 by John Wiley & Sons, Ltd.

1.3 Difference between Design and Analysis

Analysis: Application of fundamental principles to a well-defined problem. All supporting information is normally provided, and one closed-ended solution is possible.

Design: Application of fundamental principles to an undefined, open problem. All supporting information may not be available and assumptions may need to be made. Several alternatives may be possible. No single correct answer exists.

1.4 Classification of Design

  (i) Modification of an existing device for
    (a) cost reduction;
    (b) improved performance and/or efficiency;
    (c) reduced mean time between "breakdowns";
    (d) satisfy government codes and standards;
    (e) satisfy customer/client preferences.
 (ii) Selection of existing components for the design of a subsystem or a complete system.
(iii) Creation of a new device or system.

1.5 General Steps in Design

The general steps in the design process are shown schematically in Fig. 1.1.

1.6 Abridged Steps in the Design Process

1. *Project Definition*: One or two sentences describing the system or component to be designed. Check the problem statement for information.
2. *Preliminary Specifications and Constraints*: List the requirements that the design should satisfy. Requirements could come from the problem statement provided by the client or from the end users' preferences.

   At this point, develop detailed, quantifiable specifications. For example, the client wants a fan-duct system that is quiet. What does "quiet" mean? What are the maximum and minimum noise levels for this "quiet" range? 60 dB may be satisfactory. Could the maximum noise level be 70 dB?

   Detailed specifications or requirements could originate from the client ("client desired"), could be internally imposed by the designer to proceed with the design, or could be externally imposed by international/federal/provincial/municipal/industry standards or codes.

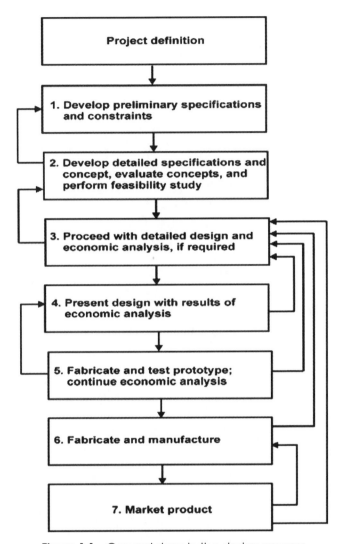

Figure 1.1 General steps in the design process

3. *Detailed Design and Calculations*
    (i) Objective
    (ii) Data Given or Known
    (iii) Assumptions/Limitations/Constraints
    (iv) Sketches (where appropriate)
    (v) Analysis
    (vi) Drawings (where appropriate) or other documentation such as manufac-
        turer's catalog sheets and Specifications.
    (vii) Conclusions

2

Air Distribution Systems

2.1 Fluid Mechanics—A Brief Review

2.1.1 Internal Flow

Flow is **laminar**: smooth streamlines; highly ordered motion.
   Or
Flow is **turbulent**: velocity fluctuates with time; highly disordered motion.
   Use the **Reynolds number** to characterize the flow regime:

$$\text{Re}_D = \frac{\rho V_{\text{ave}} D}{\mu} = \frac{V_{\text{ave}} D}{\nu} = \frac{\text{inertial forces}}{\text{viscous forces}}. \tag{2.1}$$

Note: For noncircular pipes or ducts, Re_D is based on the hydraulic diameter, D_h:

$$D_h = \frac{4A_c}{p}, \tag{2.2}$$

where A_c is the cross-sectional area and p is the perimeter wetted by the fluid.
   For square ducts,

$$D_h = \frac{4(L \times L)}{L + L + L + L} = \frac{4L^2}{4L} = L. \tag{2.3}$$

For rectangular ducts,

$$D_h = \frac{4(L \times w)}{2L + 2w} = \frac{2Lw}{L + w}. \tag{2.4}$$

It is important to note that, for volume flow rate calculations, D_h should not be used to find the cross-sectional area. Use the true cross-sectional area.

Introduction to Thermo-Fluids Systems Design, First Edition. André G. McDonald and Hugh L. Magande.
© 2012 André G. McDonald and Hugh L. Magande. Published 2012 by John Wiley & Sons, Ltd.

Therefore, for a rectangular duct,

$$\dot{V} = V_{\text{ave}} A_{\text{c}} = V_{\text{ave}}(Lw). \tag{2.5}$$

But

$$\dot{V} = V_{\text{ave}} A_{\text{c}} \neq V_{\text{ave}}\left(\frac{\pi D_{\text{h}}^2}{4}\right). \tag{2.6}$$

Criteria for Flow Characterization

$\text{Re} \leq 2300$	Laminar flow	(2.7)
$2300 \leq \text{Re} \leq 4000$	Transitional flow: laminar to turbulent flow	(2.8)
$\text{Re} \geq 4000$	Fully turbulent flow.	(2.9)

For engineering design analysis, use a critical Reynolds number, Re_{cr}:

$\text{Re}_{\text{cr}} = 2300$		(2.10)
$\text{Re} < \text{Re}_{\text{cr}}$	For laminar flow	(2.11)
$\text{Re} > \text{Re}_{\text{cr}}$	For turbulent flow.	(2.12)

2.1.2 Frictional Losses in Internal Flow—Head Losses

For fully developed laminar flow, the volume flow rate is related to the pressure drop via Poiseuille's law:

$$\dot{V} = \frac{\pi D^4 \Delta P}{128 \, \mu L}. \tag{2.13}$$

So,

$$\dot{V} \propto \Delta P \quad \text{and} \quad V_{\text{ave}} \propto \Delta P. \tag{2.14}$$

From these relationships, it can be seen that an increase in the average velocity within the duct/pipe system will result in an increased pressure drop within the duct/pipe owing to the higher frictional losses.

Head losses are the frictional losses that occur in ducts/pipes due to flow. There are two types of head losses:

Major head losses, H_l: These are due to viscous effects in fully developed flow in constant area pipes or ducts.

Minor head losses, H_{lm}: These are due to entrances, fittings, valves, and area changes. In addition, for ductwork this could be caused by filters, cooling or heating coils, and volume dampers, to name a few.

Given the point above, the total head loss is determined by

$$H_{lT} = H_l + H_{lm}. \tag{2.15}$$

Head losses are expressed in units of meter, feet, or inches **of fluid**. Head loss expressed in terms of units of length of water is preferred by practicing engineers in industry.

Under some conditions, the total head loss in a pipe/duct system is directly related to the pressure drop in the length of pipe/duct. Consider the energy equation (without fluid machines included in the pipe/duct section):

$$\left(\frac{p_1}{\rho g} + \frac{V_1^2}{2g} + z_1 \right) - \left(\frac{p_2}{\rho g} + \frac{V_2^2}{2g} + z_2 \right) = H_{lT}, \tag{2.16}$$

where points 1 and 2 are points selected at the beginning and end points of the length of pipe/duct. The average pipe/duct velocity is V.

For a constant area pipe/duct, $V_1 = V_2$. For a horizontal pipe/duct, $z_1 = z_2$. Hence,

$$\frac{p_1}{\rho g} - \frac{p_2}{\rho g} = \frac{\Delta p}{\rho g} = H_{lT}, \tag{2.17}$$

where Δp is the pressure drop across the length of pipe/duct.

2.1.3 Major Head Loss in a Run of Pipe or Duct—Pipe/Duct Sizing

In general, the following expression for head loss applies:

$$H_l = f \frac{L}{D} \frac{V^2}{2g}, \tag{2.18}$$

where f is Darcy friction factor.

For laminar flows,

$$f_{\text{laminar}} = \frac{64}{\text{Re}_D}. \tag{2.19}$$

So, f_{laminar} depends on the Reynolds number only.

For turbulent flows, $f_{turbulent}$ depends on the Reynolds number and the pipe/duct roughness, e.

So,

$$f_{turbulent} = \text{function}\left(\text{Re}_D, \frac{e}{D}\right),\tag{2.20}$$

where $\frac{e}{D}$ is the relative roughness and $f_{turbulent}$ must be determined experimentally.

To find $f_{turbulent}$,

(i) use the Moody chart, or
(ii) use appropriate correlation equations.

Procedure for Using the Moody Chart

(i)   Calculate Re_D.
(ii) Find $\frac{e}{D}$. Refer to Table A.1 for representative values of roughness for a variety of pipe and duct materials.
(iii) Use the Re_D and $\frac{e}{D}$ values to find estimates of $f_{turbulent}$ from the Moody chart.
(iv) Calculate the major head loss from $H_l = f \frac{L}{D} \frac{V^2}{2g}$.

Mathematical Formula (Empirical Correlation Equations) to Find Friction Factors for Fully Developed Turbulent Flows

Several correlation equations are available to find the friction factor in fully developed turbulent flow in pipes/ducts. These equations were developed after fitting curves to experimental data. Note the restrictions on the Reynolds number that may apply to some of the equations.

(i) *Colebrook Equation* [1]

$$\frac{1}{\sqrt{f}} = -2.0 \, \log\left(\frac{\varepsilon/D}{3.7} + \frac{2.51}{\text{Re}_D\sqrt{f}}\right).\tag{2.21}$$

The friction factor cannot be found directly. While this equation is very well established and widely used, its use will require guessing and iterations to determine f.

(ii) *Swamee–Jain Formulae* [2]

$$f_0 = 0.25\left[\log\left(\frac{\varepsilon/D}{3.7} + \frac{5.74}{\text{Re}_D^{0.9}}\right)\right]^{-2}.\tag{2.22}$$

This equation may be used to get a first estimate of f to within 1% of the true initial guess of f for use in the Colebrook equation [3]. Start with this equation, and continue with the Colebrook equation until sufficient convergence in f occurs.

"Sufficient convergence" depends on the degree of accuracy required by the design engineer.

(iii) *Haaland's Equation* [4]

$$\frac{1}{\sqrt{f}} = -1.8 \log \left[\frac{6.9}{\text{Re}_D} + \left(\frac{\varepsilon/D}{3.7} \right)^{1.11} \right]. \tag{2.23}$$

No iteration is required. It has been shown that f is within 2% of the values obtained from the Colebrook equation for a given pipe/duct roughness and Reynolds number. The value obtained from Haaland's equation may be used as a first estimate or guess in the Colebrook equation, if greater accuracy is desired.

(iv) *Blasius Correlation* [5]

$$f = \frac{0.3164}{\text{Re}_D^{0.25}}, \text{for } 3 \times 10^3 < \text{Re}_D < 2 \times 10^5. \tag{2.24}$$

This correlation applies to turbulent flow in smooth pipes or ducts. The Reynolds number is restricted as shown.

(v) *Churchill's Equation* [6]

$$f = 8 \left[\left(\frac{8}{\text{Re}_D} \right)^{12} + (A+B)^{-3/2} \right]^{1/12}, \tag{2.25}$$

where $A = \left[2.457 \ln \left(\frac{1}{C} \right) \right]^{16}$, $B = \left(\frac{37530}{\text{Re}_D} \right)^{16}$, $C = \left(\frac{7}{\text{Re}_D} \right)^{0.9} + 0.27 \left(\frac{\varepsilon}{D} \right)$.

(vi) *First Petukhov Equation* [7]

$$f = [0.79 \ln \text{Re}_D - 1.64]^{-2}, \text{for } 3000 < \text{Re}_D < 5 \times 10^6. \tag{2.26}$$

This equation applies to turbulent flow in smooth ducts or pipes.

Example 2.1   Determining the Size of an Air Duct

Heated air at 1 atm, 100°F, and 23% relative humidity is to be transported in a 490 ft long circular plastic duct at a rate of 740 cfm (ft^3/min). If the head loss in the duct is not to exceed 790 in. of air, determine the minimum diameter of the duct.

Solution. The fundamental assumption in the solution of this problem is that the head loss occurs in a constant area duct. There are no transitions or fittings in the run of duct.

The major loss (in units of length) is

$$H_l = f \frac{L}{D} \frac{V^2}{2g}.$$

Focus on the terms in the head loss equation.

The velocity may be written in terms of the volume flow rate: $V = \dfrac{\dot{V}}{A} = \dfrac{4\dot{V}}{\pi D^2}$.

As a result,

$$H_l = f \frac{L}{D} \left(\frac{4\dot{V}}{\pi D^2} \right)^2 \frac{1}{2g} = f \frac{8 L \dot{V}^2}{\pi^2 D^5 g}.$$

Assume that the flow through the duct is turbulent. The Reynolds number will be verified after determination of the duct diameter. Therefore, the friction factor, f, can be determined from the Colebrook equation or another appropriate correlation equation. The Colebrook equation is

$$\frac{1}{\sqrt{f}} = -2.0 \log \left[\frac{\varepsilon/D}{3.7} + \frac{2.51}{\mathrm{Re}_D \sqrt{f}} \right].$$

It is observed that the solution for f will be difficult to determine with the Colebrook equation since the diameter is unknown. The Haaland's approximation could be used instead

$$\frac{1}{\sqrt{f}} = -1.8 \log \left[\left(\frac{\varepsilon/D}{3.7} \right)^{1.11} + \frac{6.9}{\mathrm{Re}_D} \right].$$

For plastic, $\varepsilon = 0$. So,

$$\frac{1}{\sqrt{f}} = -1.8 \log \left[\frac{6.9}{\mathrm{Re}_D} \right]$$

$$f = \left[-1.8 \log \left[\frac{6.9}{\mathrm{Re}_D} \right] \right]^{-2}.$$

The Reynolds number is $\mathrm{Re}_D = \dfrac{\rho V D}{\mu} = \dfrac{4\rho \dot{V}}{\pi \mu D}$.

Hence,

$$H_l = \left[-1.8 \log \left[\frac{6.9}{\mathrm{Re}_D} \right] \right]^{-2} \frac{8 L \dot{V}}{\pi^2 D^5 g}$$

$$H_l = \left[-1.8 \log \left[\frac{1.7\pi \mu D}{\rho \dot{V}} \right] \right]^{-2} \frac{8 L \dot{V}^2}{\pi^2 D^5 g}.$$

The diameter is the only unknown parameter in the major head loss equation. An iterative process will be needed to find the diameter. An initial value of the diameter will be guessed.

Guess #1: $D = 8$ in. $= 0.667$ ft.

For air at 1 atm and 100°F, $\rho = 0.07088$ lb/ft^3 and $\mu = 1.281 \times 10^{-5}$ lb/(ft s) $= 0.0007686$ lb/(ft min).

Therefore,

$$H_l = \left[-1.8 \log \left[\frac{1.7\pi\ (0.0007686\ \text{lb/ft/min})\ (0.667\ \text{ft})}{\left(0.07088\ \text{lb/ft}^3\right)\left(740\ \text{ft}^3/\text{min}\right)}\right]\right]^{-2} \frac{8\,(490\ \text{ft})\left(740\ \text{ft}^3/\text{min}\right)^2}{\pi^2\,(0.667\ \text{ft})^5\left(32.2\ \text{ft/s}^2\right)}$$

$$\times \left(\frac{1\ \text{min}}{60\ \text{s}}\right)^2$$

$H_l = 239$ ft of air $= 2870$ in. of air.

The head loss calculated for the 8 in. diameter duct is much greater than the 790 in. of air constraint. Increasing the duct diameter will reduce the head loss, if all the other parameters are held constant.

Guess #2: $D = 10$ in. $= 0.833$ ft.

For air at 1 atm and 100°F, $\rho = 0.07088$ lb/ft^3 and $\mu = 1.281 \times 10^{-5}$ lb/(ft s) $= 0.0007686$ lb/(ft min).

As a result,

$$H_l = \left[-1.8 \log \left[\frac{1.7\pi\ (0.0007686\ \text{lb/ft/min})\ (0.833\ \text{ft})}{\left(0.07088\ \text{lb/ft}^3\right)\left(740\ \text{ft}^3/\text{min}\right)}\right]\right]^{-2} \frac{8\,(490\ \text{ft})\left(740\ \text{ft}^3/\text{min}\right)^2}{\pi^2\,(0.833\ \text{ft})^5\left(32.2\ \text{ft/s}^2\right)}$$

$$\times \left(\frac{1\ \text{min}}{60\ \text{s}}\right)^2$$

$H_l = 82.4$ ft of air $= 988$ in. of air.

The head loss decreased significantly with an increase in duct diameter from 8 to 10 in.

Guess #3: $D = 11$ in. $= 0.917$ ft.

For air at 1 atm and 100°F, $\rho = 0.07088$ lb/ft^3 and $\mu = 1.281 \times 10^{-5}$ lb/(ft s) $= 0.0007686$ lb/(ft min).

Thus,

$$H_l = \left[-1.8 \log \left[\frac{1.7\pi\ (0.0007686\ \text{lb/ft/min})\ (0.917\ \text{ft})}{\left(0.07088\ \text{lb/ft}^3\right)\left(740\ \text{ft}^3/\text{min}\right)}\right]\right]^{-2} \frac{8\,(490\ \text{ft})\left(740\ \text{ft}^3/\text{min}\right)^2}{\pi^2\,(0.917\ \text{ft})^5\left(32.2\ \text{ft/s}^2\right)}$$

$$\times \left(\frac{1\ \text{min}}{60\ \text{s}}\right)^2$$

$H_l = 52.0$ ft of air $= 624$ in. of air.

The head loss is lower than 790 in. of air if the duct diameter is 11 in. Therefore, the minimum diameter is between 10 and 11 in.

Further iterations will show that a duct diameter of 10.5 in. produces a head loss of approximately 777 in. of air, which deviates about 1.6% from the maximum head loss value (790 in.). This error would be acceptable in engineering practice.
So,

$$D_{minimum} = 10.5 \text{ in.}$$

Check the Reynolds number:

$$Re_D = \frac{4(0.07088 \text{ lb/ft}^3)(740 \text{ ft}^3/\text{min})}{\pi (0.0007686 \text{ lb/ft/min})(0.875 \text{ ft})} = 99\ 188.$$

Since $Re_D \gg 4000$, the flow is fully turbulent.

2.2 Air Duct Sizing—Special Design Considerations

2.2.1 General Considerations

The following points should be considered when sizing a duct to transport air:

1. Friction loss (head loss) **must** be determined in order to design duct and fan systems. Note that the smaller the duct perimeter, the lower the friction losses.
2. Several duct shapes for a given cross-sectional area are possible, and are shown in Figure 2.1.
3. Circular ducts are good choices because
    (i) lower perimeters result in less material required for fabrication;
    (ii) lower perimeters result in lower head loss;
    (iii) they can be purchased prefabricated, resulting in lower labor costs.
4. Circular ducts may be impractical due to
    (i) clearance restrictions;
    (ii) need for easy transitions in one dimension.
5. *Rectangular ducts:* Rectangular ducts with low perimeters may be used instead of circular ducts in cases where circular ducts are impractical. In this case, the **aspect ratio** of the duct must also be considered. Aspect ratio is

$$\text{Aspect ratio} = \frac{\text{length}}{\text{width}} = \frac{L}{w}. \tag{2.27}$$

Typically, aspect ratios for rectangular ducts should be **less than 4**.
For aspect ratios greater than 4:

(i) Expensive to fabricate and install due to larger perimeters.
(ii) High friction losses.

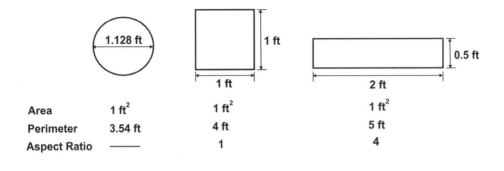

Area	1 ft^2	1 ft^2	1 ft^2
Perimeter	3.54 ft	4 ft	5 ft
Aspect Ratio	——	1	4

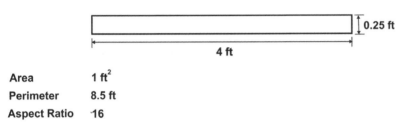

Area	1 ft^2
Perimeter	8.5 ft
Aspect Ratio	16

Figure 2.1   Duct shapes and aspect ratios

2.2.2 Sizing Straight Rectangular Air Ducts

The **Circular Equivalent Method** presents a practical approach to size rectangular ducts. First, determine the diameter of a round duct that satisfies the airflow requirement at an acceptable velocity and frictional loss. Charts may be used to facilitate the selection of an appropriate round duct diameter, rather than using the correlation equations presented in Example 2.1. For applications requiring low noise, velocities lower than about 1200 ft/min are desired (low-velocity systems). Table 2.1 presents an exhaustive list of maximum duct velocities for low-velocity systems [9]. Tables or charts may then be used to select equivalent rectangular ducts, based on the round duct diameter. Since many configurations of equivalent rectangular duct dimensions will be available for a given round duct diameter, the final choice of rectangular duct dimensions will depend on the following:

(i) Architectural barriers and limitations.
(ii) Location of structural members.
(iii) *Noise constraint*: Higher flow velocities produce more noise in the ductwork (typical).
(iv) *Aspect ratio*: Lower aspect ratio ducts require less material for fabrication and will produce less head loss (typical).

Table 2.1   Maximum duct velocities

	Low-Velocity Systems		
Designation	Private Residences	Schools, Theaters, Public Buildings	Industrial Buildings
	Maximum Velocities (fpm)		
Main ducts	800–1200	1100–1600	1300–2200
Branch ducts	700–1000	800–1300	1000–1800
Branch risers	650–800	800–1200	1000–1600
	Typical Velocities (fpm)[a]		
Throwaway filter	200–800		
Heating coil	400–500 (200 min, 1500 max)		
Cooling coil	500–600		

Source: Howell, Sauer, and Coad [9].
[a]*Ductulator*, Trane US Inc., La Crosse, Wisconsin.

Example 2.2   Sizing a Rectangular Air Duct

Size an appropriate rectangular air duct under the following conditions from Example 2.1:

(a) Heated air at 1 atm, 100°F, and 23% RH
(b) Flow rate of 740 cfm
(c) Head loss of 790 in. of air over 490 ft of duct
(d) Duct material is plastic tubing ($\varepsilon \approx 0$)

Solution. Figure A.1 can be used to find the diameter of the circular duct equivalent based on the air volume flow rate and friction loss. Figure A.1 applies to clean, round, smooth, galvanized metal duct.

This problem provides the friction loss in units of in of air. However, the friction loss chart of Figure A.1 uses friction loss as **inch water gage (in. wg)** per 100 ft of equivalent length of duct. Therefore, the head loss of 790 in. of air (H_a) should be converted to head loss in terms of inches water gage (H_w) for use with the friction loss chart. From fluid statics,

$$\rho_{air} g H_a = \rho_{water} g H_w.$$

Thus,

$$H_w = \frac{\rho_{air}}{\rho_{water}} H_a = SG_{air} H_a.$$

The specific gravity of air (SG_{air}) is at $100°F$.
On a per unit 100 ft of equivalent length of duct basis,

$$h_w = \frac{H_w}{L_{duct}} \times 100 \text{ ft} = \frac{SG_{air} H_a}{L_{duct}} \times 100 \text{ ft}.$$

Therefore, at $100°F$,

$$h_w = \left(\frac{0.07088 \text{ lb/ft}^3}{62.00 \text{ lb/ft}^3} \right) \frac{790 \text{ in. of air}}{490 \text{ ft}} \times 100 \text{ ft}$$

$h_w = 0.184$ in. wg per 100 ft of equivalent length of duct.

The circular duct is then sized with the aid of the friction loss chart for an airflow rate of 740 cfm and a friction loss of 0.184 in. wg per 100 ft of equivalent length of duct. Since the roughness of the drawn plastic is approximately zero, the chart of Figure A.1 will be applicable to drawn plastic tubes. Assume that the drawn plastic tube is clean.

From the chart:

$$V \approx 1180 \text{ fpm}$$

$$D \approx 11 \text{ in.}$$

With the circular diameter known, an appropriate rectangular duct with equivalent friction losses can be selected. Refer to Table A.3 to find possible choices. Some possible duct size choices are

10 in. × 10 in. ($D_{equivalent} = 10.9$ in.), aspect ratio $= 1.0$;
12 in. × 8 in. ($D_{equivalent} = 10.7$ in.), aspect ratio $= 1.5$;
14 in. × 8 in. ($D_{equivalent} = 11.5$ in.), aspect ratio $= 1.75$;
18 in. × 6 in. ($D_{equivalent} = 11.0$ in.), aspect ratio $= 3.0$.

The best choice is the **10 in. × 10 in.** rectangular duct equivalent.
This choice is most appropriate because

(a) the equivalent circular diameter from the chart is close to the calculated value of Example 2.1;
(b) the aspect ratio is 1;
(c) the duct is small, resulting in low friction losses;
(d) the duct will be easier to fabricate and install as compared to larger duct sizes.

While the 10 in. × 10 in. rectangular duct is the best option based on aspect ratio and ease of fabrication and installation, there may be other constraints to consider when selecting the final rectangular duct geometry. For example, structural barriers in the building may force the selection of the 12 in. × 8 in. duct as in the case of an opening in the structural wall that is 14 in. × 10 in. wide through which the duct must penetrate.

Example 2.3   Sizing a Rectangular Air Duct to Transport Other Gases

A researcher has developed a process that requires the transport of dry carbon dioxide (CO_2) at 1 atm and 50°F through a well-sealed rectangular plastic duct. The required flow rate of gas is 740 cfm and the equivalent length of duct is 490 ft. When air was transported through the duct under the same conditions and volume flow rate (see Examples 2.1 and 2.2), the friction loss was restricted to 0.184 in. wg per 100 ft of equivalent length of duct. Will the minimum duct size be the same if cold CO_2 is transported instead of heated air? Conduct an analysis to justify your response.

Solution. The friction loss chart of Figure A.1 applies to clean, round, smooth, galvanized metal ducts that transport air. Therefore, direct use of the chart in this problem for CO_2 may not produce an accurate result. The fundamental head loss expression should be used to aid in sizing the duct. Similar to Examples 2.1 and 2.2, the head loss occurs in a constant area duct and the roughness of the plastic duct is approximately zero ($\varepsilon = 0$).

The major loss (in units of length) is

$$H_l = f \frac{L}{D} \frac{V^2}{2g} = f \frac{8L\dot{V}^2}{\pi^2 D^5 g}.$$

Assume that the flow through the duct is turbulent. The Reynolds number will be verified after determination of the duct diameter. Thus, the friction factor, f, can be determined from Haaland's approximation with $\varepsilon = 0$:

$$f = \left[-1.8 \log \left[\frac{6.9}{Re_D} \right] \right]^{-2},$$

where the Reynolds number is $Re_D = \dfrac{\rho V D}{\mu} = \dfrac{4\rho \dot{V}}{\pi \mu D}$.

Therefore,

$$H_l = \left[-1.8 \log \left[\frac{1.7\pi \mu D}{\rho \dot{V}} \right] \right]^{-2} \frac{8L\dot{V}^2}{\pi^2 D^5 g}.$$

In general industrial practice, head loss is presented in inches water gage to represent pressure drop in a flowing fluid. However, use of the major head loss equations will require that the head loss have units of inches of CO_2. Therefore,

$$H_{CO_2} = \frac{\rho_{water}}{\rho_{CO_2}} H_w$$

$$H_{CO_2} = \frac{\left(62.41 \text{ lb/ft}^3 \right)}{\left(0.11825 \text{ lb/ft}^3 \right)} \times \frac{0.184 \text{ in.}}{100 \text{ ft}} \times 490 \text{ ft}$$

$$H_{CO_2} = 476 \text{ in. of } CO_2.$$

The diameter is the only unknown parameter in the major head loss equation. An iterative process will be needed to find the diameter. An initial value of the diameter will be guessed. Begin with a duct diameter of 10.5 in.

Guess #1: $D = 10.5$ in. $= 0.875$ ft.

For CO_2 at 1 atm and 50°F, $\rho = 0.11825$ lb/ft^3 and $\mu = 9.564 \times 10^{-6}$ lb/(ft s) $= 0.0005738$ lb/(ft min).
   So,

$$H_l = \left[-1.8 \log \left[\frac{1.7\pi \, (0.0005738 \text{ lb/ft/min}) \, (0.875 \text{ ft})}{\left(0.11825 \text{ lb/ft}^3\right) \left(740 \text{ ft}^3/\text{min}\right)} \right] \right]^{-2} \frac{8\,(490 \text{ ft}) \left(740 \text{ ft}^3/\text{min}\right)^2}{\pi^2 \, (0.875 \text{ ft})^5 \left(32.2 \text{ ft/s}^2\right)}$$
$$\times \left(\frac{1 \text{ min}}{60 \text{ s}} \right)^2$$

$H_l = 55$ ft of $CO_2 = 665$ in. of CO_2.
   The head loss calculated for the 10.5 in. diameter duct is greater than the 476 in. of CO_2 constraint. Increasing the duct diameter will reduce the head loss, if all the other parameters are held constant.

Guess #2: $D = 11$ in. $= 0.917$ ft.

$$H_l = \left[-1.8 \log \left[\frac{1.7\pi \, (0.0005738 \text{ lb/ft/min}) \, (0.917 \text{ ft})}{\left(0.11825 \text{ lb/ft}^3\right) \left(740 \text{ ft}^3/\text{min}\right)} \right] \right]^{-2} \frac{8\,(490 \text{ ft}) \left(740 \text{ ft}^3/\text{min}\right)^2}{\pi^2 \, (0.917 \text{ ft})^5 \left(32.2 \text{ ft/s}^2\right)}$$
$$\times \left(\frac{1 \text{ min}}{60 \text{ s}} \right)^2$$

$H_l = 44$ ft of air $= 531$ in. of CO_2.
   The head loss decreased further with an increase in duct diameter.

Guess #3: $D = 11.5$ in. $= 0.958$ ft.

$$H_l = \left[-1.8 \log \left[\frac{1.7\pi \, (0.0005738 \text{ lb/ft/min}) \, (0.958 \text{ ft})}{\left(0.11825 \text{ lb/ft}^3\right) \left(740 \text{ ft}^3/\text{min}\right)} \right] \right]^{-2} \frac{8\,(490 \text{ ft}) \left(740 \text{ ft}^3/\text{min}\right)^2}{\pi^2 \, (0.958 \text{ ft})^5 \left(32.2 \text{ ft/s}^2\right)}$$
$$\times \left(\frac{1 \text{ min}}{60 \text{ s}} \right)^2$$

$H_l = 36$ ft of $CO_2 = 430$ in. of CO_2.
   The head loss is lower than 476 in. of CO_2 if the duct diameter is 11.5 in. Therefore, the minimum diameter is between 11 and 11.5 in.
   Take the circular duct diameter to be 11.5 in., which produces a head loss of approximately 430 in. of CO_2, which deviates about 10% from the maximum head loss value (476 in.). This error would be acceptable in engineering practice.
   Therefore,

$$D_{minimum} = 11.5 \text{ in.}$$

Check the Reynolds number:

$$\text{Re}_D = \frac{4(0.11825 \text{ lb/ft}^3)(740 \text{ ft}^3/\text{min})}{\pi \, (0.0005738 \text{ lb/ft/min}) \, (0.958 \text{ ft})} = 202683.$$

Since $\text{Re}_D \gg 4000$, the flow is fully turbulent.

The diameter of the circular duct estimated for the transport of CO_2 is no more than 1 in. larger than the diameter of the duct used to transport air. This represents a difference of no more than 10%. Further, this error would be lower because the actual duct size would be slightly lower than 11.5 in. Under these circumstances, the 11 in. circular duct selected from the friction loss chart of Figure A.1 in Example 2.2 could be used in this application without the generation of significant head loss to undermine the performance of the system. The actual head loss, in inches water gage, would be 0.205 in. wg per 100 ft of equivalent length of duct, which corresponds to 531 in. of CO_2 in 490 ft of duct.

Table A.3 and the circular equivalent method can be used to find the rectangular duct size. So, a possible choice could be 14 in. × 8 in. ($D_{\text{equivalent}} = 11.5$ in.). In this case, the aspect ratio is 1.75, which is lower than 4. This duct size and geometry was one of the possible choices that were presented in Example 2.2.

This exercise demonstrates that the error generated through the use of the friction loss charts that are based on air for gases may be small. Provided that these small errors are acceptable, the chart may be used to size ducts that transport a variety of gases, as well as air at various temperatures.

2.2.3 Use of an Air Duct Calculator to Size Rectangular Air Ducts

The air duct calculator is a device used to size circular and rectangular ducts. Figure 2.2 shows a typical air duct calculator. With any one of the following four parameters known, the other three are found from a single setting of the calculator:

(i) Friction loss (head loss)
(ii) Duct velocity (for round duct)
(iii) Round duct diameter
(iv) Dimensions for rectangular duct

2.3 Minor Head Loss in a Run of Pipe or Duct

Minor losses occur when a fluid passes through (i) fittings, (ii) bends, (iii) valves, (iv) abrupt area changes, (v) other devices or components in the flow path that add resistance to flow (filters, strainers, cooling/heating coils, louvers, dampers, flowmeters), and (vi) entrances and exits.

In **long** air duct or liquid piping systems, these minor losses may be small compared to the major losses in the constant area straight runs of duct or pipe.

In **shorter** air duct or liquid piping systems, these losses are significant.

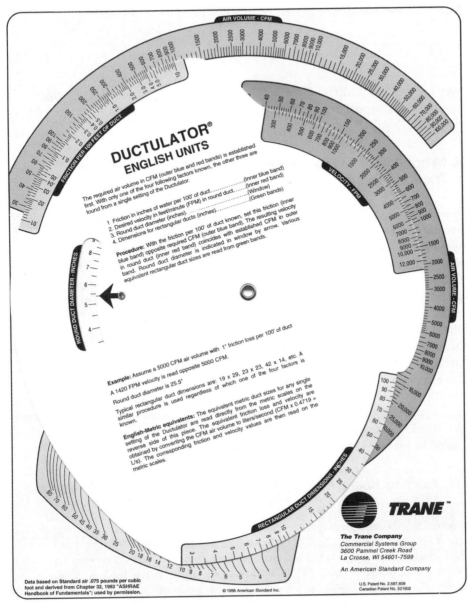

Figure 2.2 Photo of a typical air duct calculator

Minor losses are defined as

$$H_{\text{lm}} = K \frac{V^2}{2g} \text{ (in units of length)},$$ (2.28)

where K is the loss coefficient.

Minor losses may also be presented in terms of equivalent lengths:

$$H_{lm} = f \frac{L_{equiv}}{D} \frac{V^2}{2g},$$ (2.29)

where L_{equiv} is the additional equivalent length of straight pipe/duct, which corresponds to the component (i.e., source of the minor loss). Data for L_{equiv} are tabulated in Tables A.4 and A.5.

2.4 Minor Losses in the Design of Air Duct Systems—Equal Friction Method

The **equal friction method** is typically used to size small duct systems. In this method, the major head loss in the constant area straight run of duct is added to the equivalent lengths for the sources of minor losses (fittings, bends, transitions). This is used to calculate the total head loss per 100 ft of equivalent length of straight duct.

For duct design and sizing purposes, choose a loss of approximately **0.1 in. of H_2O gage (in. wg) per 100 equivalent ft** of straight ductwork. With this design parameter, the friction loss in the longest continuous duct branch can be found. Other branches of the ductwork **must** be designed to increase their friction loss to match this higher value. This equal friction will help to balance the flow in the entire duct network system.

An alternate procedure could be to select an absolute 0.1 in. of H_2O loss for the longest branch of the ductwork system (applicable to small duct systems, only). In this case, the design engineer would be constraining the friction loss in the system. A check would be required to ensure that the total pressure was available at a plenum or could be produced by a fan.

Practical Note 2.1   Total Static Pressure Available at a Plenum or Produced by a Fan

In some cases, the design of the ductwork system is constrained by the total pressure available at a plenum or produced by a fan. In addition, the pressure losses across components in the ductwork such as filters and coils must be substrated from the total pressure available at the plenum or from the fan. Therefore, the pressure loss per 100 ft of the longest branch duct is dependent on the available pressure that remains (see Example 2.4). When calculated, the pressure loss per 100 ft should then be used to size the remaining duct sections. The design engineer must always verify that the pressure loss in each section of the ductwork system, which may be complete with components, does not exceed the total pressure available at a plenum or produced by a fan (balancing).

Figure 2.3 shows a partial ductwork system that includes several branches and fittings. The schematic drawing shows dampers (labeled 2, 6, 7, and 13), a centrifugal fan, and fittings such as wyes (labeled 3 and 8) and bends (labeled 5, 12, and 15). The

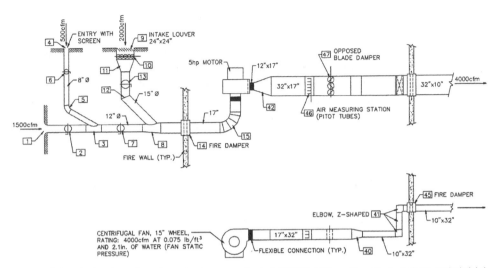

Figure 2.3 A ductwork system to transport air (*ASHRAE Handbook*, Fundamentals Volume, 2005; reprinted with permission)

flow rates and sizes of each duct section are also shown. Fire dampers are shown in areas where the ducts penetrate a firewall. This is required by Section 705.11 of the International Building Code [10], Section 607 of the International Mechanical Code [11], and the National Fire Protection Association (NFPA) Standard 90A [12].

As mentioned in Practical Note 2.1, components in the ductwork system will generate air pressure losses during transport. These losses must be considered in order to size the fan or plenum and balance the pressure losses in the system. Table 2.2 presents some typical pressure loss values across several components. The list is not extensive or exhaustive. Therefore, the design engineer may need to consult the manufacturer to obtain detailed information for specific components.

Table 2.2 Typical values of component pressure losses [9]

Component	Static Pressure Loss (in. wg)
Supply plenum	0.50
Supply grille (diffuser)	0.05
Raised floor perforations	0.05
High-efficiency particulate air (HEPA) filter	0.70
30/30 prefilters	0.30
Bag filters	0.80
Return plenum	0.05
Cooling coil	0.35

Source: Howell, Sauer, and Coad [9].

It is important to note that the construction and installation of fittings and components into a ductwork system will affect the total static pressure that is generated within the system. Duct construction standards provided by organizations such as the Sheet Metal and Air Conditioning Contractors National Association, Inc. (SMACNA) [13] are useful in that they allow contractors to construct the ductwork in accordance with the design documents and specifications. However, the onus rests on the design engineer to present clear guidance regarding the types of fittings and components that are required.

Example 2.4   Sizing a Simple Air Duct System

The layout of a simple low-velocity (maximum velocity = 1000 ft/min) duct system is presented below. Size an appropriate rectangular air duct for this system, if the plenum can only provide 0.18 in. wg of static pressure. The client will install a diffuser that is rated for 0.03 in. wg at the exit of the duct system.

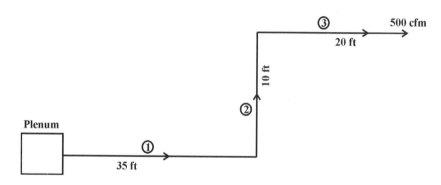

Possible Solution

Definition

Size an appropriate rectangular air duct for the given system. Select a suitable material.

Preliminary Specifications and Constraints

(i) The working fluid will be air.
(ii) The duct system must be attached to an air plenum and have two 90° elbows.
(iii) The air exits the system at 500 cfm.
(iv) The plenum can only provide 0.18 in. wg of static pressure.

Detailed Design

Objective

To size a rectangular air duct. The size and material of the duct will be determined.

Data Given or Known

(i) The length of each duct section is given: $L_1 = 35$ ft; $L_2 = 10$ ft; $L_3 = 20$ ft.
(ii) The total flow rate of air in the system is 500 cfm.
(iii) The duct system is connected to an air plenum.
(iv) The plenum can provide 0.18 in. wg of static pressure.
(v) The diffuser is rated for 0.03 in. wg.

Assumptions/Limitations/Constraints

(i) Maximum air velocity is 1000 fpm.
(ii) An attempt will be made to keep the total friction losses close to 0.1 in. of water per 100 ft of ductwork, as per common industry practice.
(iii) Typically, rectangular duct aspect ratios should be less than 4.
(iv) Galvanized steel is typically used to fabricate air duct systems. It will be chosen as the material.
(v) Choose 90° pleated elbows since they have lower equivalent length and lower losses.

Sketch

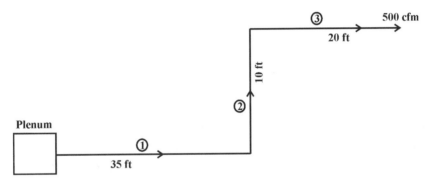

Analysis

Designers of duct systems usually use charts and tables to size the ducts. That approach will be taken in this design problem.

System Pressure (Friction) Loss

The plenum can provide a total of 0.18 in. wg of static pressure to move air through the system. The presence of the diffuser at the exit of the system will reduce the amount of static pressure available to move air through the constant area duct, only. So, the total losses due to friction in the duct, only is

$$\Delta P_{\text{duct}} = (0.18 - 0.03) \text{ in. wg} = 0.15 \text{ in. wg}.$$

The equivalent length of the ductwork is the sum of the actual total length (straight runs) of the duct, plus the equivalent lengths for the 90° abrupt entrance from the large plenum and the two 90° bends:

$$L_{e,\text{total}} = L_1 + L_2 + L_3 + L_{e,\text{entrance}} + 2L_{e,90° \text{ elbow}}.$$

To choose the appropriate equivalent length for the fittings, the circular duct diameter is required. However, the purpose of this design is to determine the duct diameter. So, assume that the duct diameter is 10 in., and verify this assumption at the end of the design. Some iteration may be required.

The equivalent lengths are found from Table A.4 for a 10 in. diameter circular galvanized duct: $L_{e,\text{entrance}} = 25$ ft and $L_{e,90° \text{ elbow}} = 13$ ft.

Therefore,

$$L_{e,\text{total}} = 35 \text{ ft} + 10 \text{ ft} + 20 \text{ ft} + 25 \text{ ft} + 2(13 \text{ ft}) = 116 \text{ ft}.$$

The total friction loss in the main run of the ductwork is restricted to 0.15 in. wg. Hence, the pressure loss in the ductwork is

$$\Delta P = \frac{0.15 \text{ in. of water}}{116 \text{ ft}} \times 100 \text{ ft of ductwork} = 0.13 \text{ in. wg per 100 ft of duct}.$$

Note that the loss is close to 0.1 in. wg per 100 ft of ductwork.

Duct Size

With the friction loss and the airflow rate (500 cfm) known, the chart for friction in round, straight galvanized steel ducts can be used to select the appropriate duct diameter, and specify the flow velocity (Figure A.1).

Thus, $D_{\text{circular}} = 10.2$ in. and $V = 890$ fpm. Note that $V = 890$ fpm < 1000 fpm.

Since the circular diameter is known, an appropriate rectangular duct with equivalent friction losses can be chosen from Table A.3.

Possible Choices and Their Aspect Ratios. Rectangular Size (in. × in.)	Circular Diameter (in.)	Aspect Ratio
10 × 9	10.4	1.1
11 × 8	10.2	1.4
13 × 7	10.3	1.9
15 × 6	10.1	2.5
16 × 6	10.4	2.7

Drawings

No additional drawings are required.

Conclusions

A rectangular galvanized sheet metal duct with dimensions of 10 in. × 9 in. is chosen. The size is chosen because of the following reasons:

(i) $D_{circular} \sim 10.2$ in.
(ii) Aspect ratio is close to 1.
(iii) Duct is small. So, friction losses will be low.
(iv) Ductwork will be easier to fabricate compared to the other choices.

Other constraints (e.g., structural or architectural), if provided, would need to be considered when making the final selection of duct geometry.

Note that, though more expensive, the bellmouth entrance could have been chosen to reduce losses at the entrance of the duct system at the plenum.

Example 2.5   System Design: Designing a Simple Air Duct System

Miss Cherry is the proprietor of a local massage-beauty-relaxation spa. She has solicited the services of an engineer to size a rectangular duct system to deliver air to three private rooms in her establishment. It is expected that the air will be delivered from the dropped ceiling height directly from openings in the ductwork. Miss Cherry has some knowledge of mechanical design, and has provided the following sketch, complete with air requirements per room.

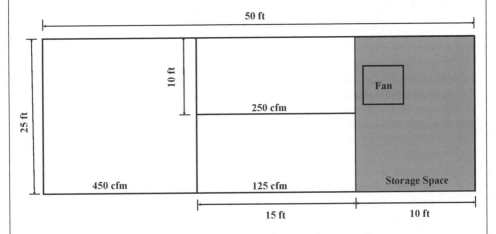

A new Greenheck fan will be located on top of the roof, and the new duct system will be connected to this fan. The roof is 3 ft above the dropped ceiling. The wall partitions shown in the sketch do not extend beyond the dropped ceiling. Layout an appropriate rectangular duct system to deliver the air to the private rooms.

Possible Solution

Definition

Size an appropriate rectangular air duct system for the given space.

Preliminary Specifications and Constraints

(i) The working fluid will be air.
(ii) The three rooms in the establishment have the following air requirements: 250, 125, and 450 cfm.
(iii) The duct system must be attached to an existing rooftop fan.
(iv) The length of the ductwork is constrained by the dimensions of the room.

Detailed Design

Objective

To design a rectangular air duct system. The size and material of the duct will be determined.

Data Given or Known

(i) Air will be delivered from the dropped ceiling height directly from openings in the ductwork. This implies that no duct elbows will be needed to deliver the air to the space. No diffusers will be attached at the duct exit.
(ii) All the dimensions of the rooms were provided by the client.
(iii) The air requirements for the three rooms were given as 250, 125, and 450 cfm. The total air requirement is 825 cfm.
(iv) The roof is 3 ft above the dropped ceiling.
(v) The wall partitions do not extend beyond the dropped ceiling.

Assumptions/Limitations/Constraints

(i) Total friction losses available for the ductwork should be about 0.1 in. of water per 100 ft of ductwork, as per industry standard. In this case, it is assumed that the fan or plenum will be sized after the ductwork system since no technical information on the Greenheck fan was provided.
(ii) Typically, rectangular duct aspect ratios should be less than 4, as per industry standard.
(iii) Galvanized steel is typically used to fabricate air duct systems. It will be chosen as the material.
(iv) Where appropriate, branch fittings will be 45° wyes or mitered 90° elbows with turning vanes. This will reduce the minor loss equivalent lengths.
(v) Losses due to duct size reductions will be ignored since they are small compared to other losses.
(vi) The entrance to the duct system from the fan will be a bellmouth to reduce frictional losses.
(vii) This is a massage-beauty-relaxation spa. Therefore, a low-velocity ductwork system would probably be required by the client. The air velocity in the duct should not exceed 1200 fpm (assuming a private residence).

Sketch

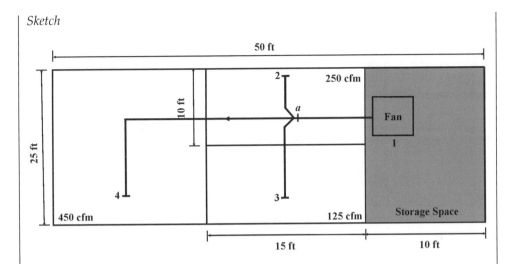

This sketch will be used to provide *estimates* of the duct lengths, and will serve as justification of these estimates. All the fittings are shown (except bellmouth at the entrance to the duct system). All attempts were made to center the ductwork in the rooms to promote uniform distribution of air in the spaces.

Analysis

Designers of duct systems usually use charts and tables to size the ducts. That approach will be taken in this design problem.

Estimates of the Lengths of the Sections of the Ductwork

Section 1-*a*: 7.5 ft
Section *a*-2: 7.5 ft
Section *a*-3: 10 ft
Section *a*-4: 32.5 ft

Flow Rates in the Sections of the Ductwork

Section 1-*a*: 825 cfm
Section *a*-2: 250 cfm
Section *a*-3: 125 cfm
Section *a*-4: 450 cfm

Sizing the Circular Duct Equivalent

The flow rate through each section of the ductwork system is known. It was assumed that the pressure loss in the system will be constrained to 0.1 in. wg per 100 ft of ductwork. The diameter of an equivalent circular duct will be found by using the flow rates, pressure losses, and the appropriate friction loss chart for round, straight galvanized steel ducts (Figure A.1).

Therefore,

Section 1-*a*: 13 in. diameter, 900 fpm
Section *a*-2: 8.25 in. diameter, 690 fpm
Section *a*-3: 6.50 in. diameter, 580 fpm
Section *a*-4: 10.25 in. diameter, 800 fpm

Note that the velocities of the air in each section of the ductwork system are less than the 1200 fpm maximum that is allowed for low-velocity duct systems.

Sizing the Rectangular Duct Equivalent

The equal friction and capacity chart (Table A.3) will be used to select an appropriate rectangular duct equivalent for the circular ducts. The aspect ratio will be 4 or lower.
Accordingly,

Section 1-*a*: 12 in. × 12 in. (13.1 in. diameter; aspect ratio = 1)
Section *a*-2: 8 in. × 7 in. (8.2 in. diameter; aspect ratio = 1.1)
Section *a*-3: 6 in. × 6 in. (6.6 in. diameter; aspect ratio = 1)
Section *a*-4: 10 in. × 9 in. (10.4 in. diameter; aspect ratio = 1.1)

Drawings

The final drawing shows the layout and size of the ducts.

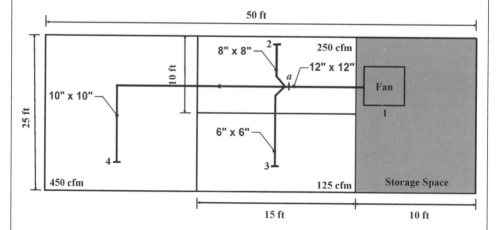

Conclusions

The rectangular ductwork system has been sized. Galvanized sheet metal will be used to fabricate the system. The aspect ratios are low (1.1 or lower) to reduce friction losses and facilitate fabrication. The air velocities in each section of the duct are less than 1200 fpm. If the total available pressure from the fan was known, then the pressure drops through the sections of the system would need to be calculated to ensure that the total pressure would not be exceeded. If the fan or plenum was to be sized, then the pressure drop in the longest (main) branch would be needed. Dampers will probably be needed to balance the airflow in the system. The following final duct sizes are recommended to facilitate further the fabrication of the system and reduce the aspect ratios.

Section 1-*a*: 12 in. × 12 in.
Section *a*-2: 8 in. × 8 in.
Section *a*-3: 6 in. × 6 in.
Section *a*-4: 10 in. × 10 in.

Example 2.6   System Design: Sizing an Air Duct System

Golash Professional Engineers have struck a contract with the Alberta Department of Motor Vehicles (DMV) to design an air distribution system. A design engineer has used AUTOCAD to prepare the following schematic of the system based on an architectural drawing that was provided by Basian Architecture, Interior Design, and Planning, Ltd. All that remains is the determination of the sizes of the rectangular ducts that will be installed below a concrete slab and the total static pressure required by the fan. No information was provided on the diffusers. A low-noise, low-vibration system is desired.

Further Information: All design engineers know that the diffusers cannot be ignored when sizing ducts and fans.

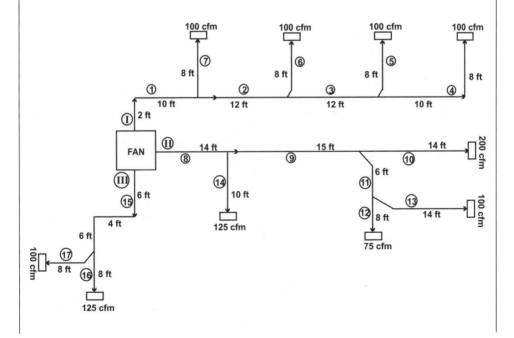

Possible Solution

Definition

Size an appropriate rectangular air duct system for the air distribution system.

Preliminary Specifications and Constraints

  (i) The working fluid will be air.
 (ii) A low-noise, low-vibration system is required.
(iii) The duct lengths and airflow rates through the sections are constrained to the values shown in the drawing.

Detailed Design

Objective

To design a rectangular air duct system. The size and material of the duct will be determined.

Data Given or Known

  (i) An architectural drawing has been provided.
 (ii) All the duct lengths were provided by the architect.
(iii) All the airflow rates were provided by the architect.
 (iv) The system will be installed below a concrete slab.

Assumptions/Limitations/Constraints

  (i) Friction losses within the ductwork should be about 0.1 in. of water per 100 ft of ductwork, as per industry standard. In this case, the fan will be sized after the ductwork system has been sized since no technical information was provided.
 (ii) Typically, rectangular duct aspect ratios should be less than 4, as per industry standard.
(iii) Galvanized steel is typically used to fabricate air duct systems. However, since the client wishes to install the ducts below a concrete slab, smooth stainless steel will be chosen to provide strength and corrosion resistance.
 (iv) Where appropriate, branch fittings will be 45° wyes or mitered 90° elbows with turning vanes. This will reduce the minor loss equivalent lengths.
  (v) The entrance to the duct system from the fan will be a bellmouth to reduce frictional losses.
 (vi) A low-noise, low-vibration ductwork system is required by the client. Therefore, the air velocity in the duct should not exceed 1200 fpm. If the DMV wishes to install the ducts in an industrial setting, then the velocity limit could be as high as 2200 fpm. However, since limited information was provided, a conservative maximum velocity will be chosen to ensure that the system will meet the client's requirement of low-noise, low-vibration under most circumstances.
(vii) Assume that the diffuser pressure loss is on the order of 0.05 in. wg. This is a typical order of magnitude value for diffusers. In practice, the pressure drop across the diffuser will depend on the airflow rate (in cfm) and the air velocity (in. fpm).

Sketch

A complete architectural drawing was provided. The sections of the duct have also been labeled for the engineer.

Analysis

Designers of duct systems usually use charts and tables to size the ducts. That approach will be taken in this design problem.

Flow Rates in the Sections of the Ductwork

Section 1: 400 cfm
Section 2: 300 cfm
Section 3: 200 cfm
Section 4: 100 cfm
Section 5: 100 cfm
Section 6: 100 cfm
Section 7: 100 cfm
Section 8: 500 cfm
Section 9: 375 cfm
Section 10: 200 cfm
Section 11: 175 cfm
Section 12: 75 cfm
Section 13: 100 cfm
Section 14: 125 cfm
Section 15: 225 cfm
Section 16: 125 cfm
Section 17: 100 cfm

Sizing the Circular Duct Equivalent

The flow rate through each section of the ductwork system is known. It was assumed that the pressure loss in the system will be constrained to 0.1 in. wg per 100 ft of duct-work. Note that since no information was given regarding the total pressure available from the fan, the design engineer will opt to use this industry standard. The approximate diameter of an equivalent circular duct will be found by using the flow rates, pressure losses, and the appropriate friction loss chart for round, straight galvanized steel ducts (Figure A.1).
  So,

Section 1: 9.9 in. diameter, 780 fpm
Section 2: 8.9 in. diameter, 710 fpm
Section 3: 7.5 in. diameter, 650 fpm
Section 4: 5.9 in. diameter, 540 fpm
Section 5: 5.9 in. diameter, 540 fpm
Section 6: 5.9 in. diameter, 540 fpm

Section 7: 5.9 in. diameter, 540 fpm
Section 8: 10.3 in. diameter, 810 fpm
Section 9: 9.5 in. diameter, 760 fpm
Section 10: 7.5 in. diameter, 650 fpm
Section 11: 7.2 in. diameter, 620 fpm
Section 12: 5.2 in. diameter, 500 fpm
Section 13: 5.9 in. diameter, 540 fpm
Section 14: 6.3 in. diameter, 560 fpm
Section 15: 7.9 in. diameter, 670 fpm
Section 16: 6.3 in. diameter, 560 fpm
Section 17: 5.9 in. diameter, 540 fpm

Note that the velocities of the air in each section of the ductwork system are less than 1200 fpm, as required for a low-velocity, low-noise, low-vibration duct system.

Sizing the Rectangular Duct Equivalent

The equal friction and capacity chart (Table A.3) will be used to select an appropriate rectangular duct equivalent for the circular ducts. The aspect ratio will be 4 or lower.
So,

Section 1: 10 in. × 8 in. (aspect ratio: 1.25)
Section 2: 8 in. × 8 in. (aspect ratio: 1.00)
Section 3: 8 in. × 6 in. (aspect ratio: 1.33)
Section 4: 6 in. × 6 in. (aspect ratio: 1.00)*
Section 5: 6 in. × 6 in. (aspect ratio: 1.00)*
Section 6: 6 in. × 6 in. (aspect ratio: 1.00)*
Section 7: 6 in. × 6 in. (aspect ratio: 1.00)*
Section 8: 10 in. × 10 in. (aspect ratio: 1.00)**
Section 9: 10 in. × 8 in. (aspect ratio: 1.25)**
Section 10: 8 in. × 6 in. (aspect ratio: 1.00)
Section 11: 8 in. × 6 in. (aspect ratio: 1.00)**
Section 12: 6 in. × 6 in. (aspect ratio: 1.00)*
Section 13: 6 in. × 6 in. (aspect ratio: 1.00)*
Section 14: 6 in. × 6 in. (aspect ratio: 1.00)*
Section 15: 10 in. × 6 in. (aspect ratio: 1.67)*
Section 16: 6 in. × 6 in. (aspect ratio: 1.00)*
Section 17: 6 in. × 6 in. (aspect ratio: 1.00)*

*Note: The smallest size available from Table A.3 is 6 in. × 6 in. This is gives an equivalent diameter of 6.6 in. At this diameter, the pressure loss will be lower than the design equivalent diameters of 5.9 in. and 6.3 in., resulting in a smaller fan. The increase in material is small. This should be acceptable from a cost and installation perspective.
**Note: The equivalent diameter is exceeded slightly. This decision was made to keep the aspect ratio low.

Total Static Pressure Required from the Fan

For this problem, all the circular duct sizes were determined with a friction loss of 0.1 in. wg per 100 ft of ductwork. This friction loss can be used to estimate the total static pressure required by the fan. The total equivalent length (straight duct + equivalent lengths of the fittings) of the longest branch should be used.

From visual inspection of the drawing supplied by the design engineer, it appears as though either branch I-1-2-3-4 or branch II-8-9-11-13 will be the longest branch in this system. The lengths should be calculated.

For branch I-1-2-3-4,

$$L_{e,total} = L_{straight} + L_{ent} + L_{90} + L_{tee,thru} + L_{contract} + L_{wye,thru} + L_{contract} + L_{wye,thru}$$
$$+ L_{contract} + L_{90}$$

$$L_{e,total} = 54 \text{ ft} + 10 \text{ ft} + 8 \text{ ft} + 7 \text{ ft} + 3 \text{ ft} + 6 \text{ ft} + 3 \text{ ft} + 5 \text{ ft} + 3 \text{ ft} + 5 \text{ ft}$$

$$L_{e,total} = 104 \text{ ft.}$$

For branch II-8-9-11-13,

$$L_{e,total} = L_{straight} + L_{ent} + L_{tee,thru} + L_{contract} + L_{wye,branch} + L_{wye,branch}$$

$$L_{e,total} = 49 \text{ ft} + 10 \text{ ft} + 7 \text{ ft} + 3 \text{ ft} + 12 \text{ ft} + 10 \text{ ft}$$

$$L_{e,total} = 91 \text{ ft.}$$

Therefore, branch I-1-2-3-4 is the longest branch in this air distribution system.

It should be noted that in some cases the L_e/D ratio (Table A.4) was used to estimate the equivalent lengths. The duct diameter must be converted to feet to use this ratio.

The pressure drop in the duct is

$$\Delta P_{duct} = \frac{0.1 \text{ in. of water}}{100 \text{ ft}} \times 104 \text{ ft of ductwork} = 0.104 \text{ in. wg.}$$

The total static pressure required from the fan is

$$\Delta P_{fan} = \Delta P_{duct} + \Delta P_{diffuser} = (0.104 + 0.05) \text{ in. wg}$$

$$\Delta P_{fan} = 0.16 \text{ in. wg.}$$

If the fan is able to move air through the longest duct branch with this total static pressure, then it will be able to move air through all the other smaller branches in the air distribution system.

Drawings

The duct sizes determined by the design engineer can be included on the architectural drawing.

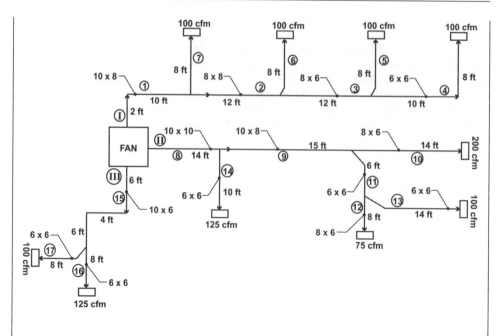

Conclusions

The rectangular ductwork system has been sized. Stainless steel will be used to fabricate the system. The aspect ratios are small (1.67 or lower) to reduce friction losses and facilitate fabrication. Some sections were slightly oversized to maintain a low aspect ratio. This will reduce slightly the static pressure required from the fan. The air velocities in each section of the duct are less than 1200 fpm. With the total available pressure from the fan known, the pressure drops through the other branches of the system can be calculated to ensure that the total pressure would not be exceeded. Dampers will probably be needed to balance the airflow in the system.

Example 2.7 Design of a Kitchen Exhaust System—Application of Codes

La Ronde Jambalaya Grill and Restaurant has decided to install a variety of cooking appliances, including a deep fat fryer for chicken, grills for beef burger patties, and gas-fired oven ranges. The appliances will be installed along a wall in a one-story building and will require approximately 14 ft of wall space. The depth and height of the appliances range from about 2 to 4 ft. The NFPA Standard 96 requires that all cooking equipment used in processes that generate smoke and vapors containing grease be equipped with an exhaust system [14]. To that end, the management team of the restaurant has hired a mechanical engineer to design and layout the required exhaust system. There is a fire-rated roof-ceiling assembly in the building, which, according to NFPA Standard 96, requires special consideration during installation of the exhaust system. Of all the components that will be required, including a hood, ductwork, and exhaust fan, the client has clearly stated that a wall-mounted canopy type hood would fit within the architectural esthetic of the kitchen. Through consultation with NFPA Standard 96, the International Mechanical Code, and the American Society of Heating, Refrigerating, and Air-Conditioning Engineers (ASHRAE) Standard 154 [15], design the exhaust system for the appliances based on the drawings of the building sections provided by the client.

Further Information: It is expected that the mechanical engineer will specify a hood and exhaust fan from manufacturers' catalog(s) and will present excerpts for review, if available.

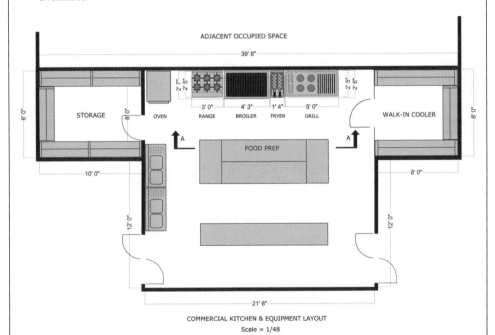

COMMERCIAL KITCHEN & EQUIPMENT LAYOUT
Scale = 1/48

COMMERCIAL KITCHEN & EQUIPMENT ELEVATION "A-A"
Scale = 1/48

Possible Solution

Definition

Design an appropriate exhaust duct system for this kitchen application.

Preliminary Specifications and Constraints

(i) The working fluid will be an exhaust gas mixture from the appliances.
(ii) The building has one story and a fire-rated roof-ceiling assembly.
(iii) NFPA Standard 96 requires the use of an exhaust system in this type of kitchen application.
(iv) The International Mechanical Code and ASHRAE Standard 154 should be consulted.
(v) A wall-mounted canopy type hood, ductwork, and an exhaust fan will be required.

Detailed Design

Objective

To design a duct system to remove odors, latent heat, and combustion by-products. The size and material of the duct will be determined. The fan and hood will be selected.

Data Given or Known

(i) An architectural drawing has been provided, complete with dimensions.
(ii) The types and sizes of the appliances have been provided by the client.
(iii) A 14 ft wall space has been allocated for the appliances that will be located under the hood.

Assumptions/Limitations/Constraints

(i) The friction losses in the ductwork should be about 0.1 in. of water per 100 ft of ductwork to mitigate noise and vibration. In this case, the fan will be sized after the ductwork system has been designed.
(ii) Circular ducts will be used. This will facilitate installation and reduce pressure loss and noise.
(iii) Type I exhaust ducts and hoods are those that are installed where cooking appliances produce grease or smoke, such as with fryers and oven ranges. According to Section 506.3.1.1 of the International Mechanical Code, for Type I exhaust ducts, steel of not less than No. 16 gage (0.0575 in. thick) or stainless steel of not less than No. 18 gage (0.044 in. thick) must be used to construct the duct system. Since excessive amounts of water will likely not be present to induce corrosion or other chemical reactions, No. 16 galvanized steel shall be used as the duct material. This is permitted by Sections 7.5.1.1 and A.7.5.1 of NFPA Standard 96.
(iv) Where appropriate, 90° elbows will possess long radii. This will reduce the minor loss equivalent lengths.
(v) The exit from the duct system to the fan will be a bellmouth to reduce frictional losses.
(vi) Assume that the kitchen is an industrial setting.

(vii) Section 8.2.1.1 of NFPA Standard 96 and Section 506.3.4 of the International Mechanical Code require that the exhaust gas velocity in the ductwork be at least 500 fpm. In order to maintain a low-noise, low-vibration ductwork system, the gas velocity should not exceed 2200 fpm in the main duct and 1800 fpm in the branch duct. Therefore, a target maximum velocity of 1800 fpm will be chosen for the duct system for a conservative design approach. A note to Section 5.3.1 of ASHRAE Standard 154 also suggests general design velocities up to 1800 fpm for commercial exhaust systems for cooking operations.

(viii) Additional pressure losses will occur when the exhaust gas enters the duct from the hood, passes through the grease filters, and is ejected from the fan against external prevailing winds. Assume that the losses at the entrance of the duct are similar to that encountered in a bellmouth entrance ($L_e/D = 12$). The losses across the filters will be assumed to be the same as those across 30/30 filters (0.30 in. wg). The losses due to wind currents may be between 0.1 and 0.5 in. wg. For a conservative design and equipment sizing, assume the loss to be 0.5 in. wg.

Sketch

A complete architectural drawing was provided. A sketch of the ductwork is shown in the drawing.

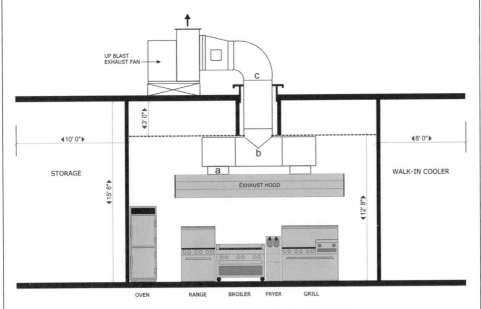

COMMERCIAL KITCHEN & EQUIPMENT ELEVATION "A-A"
Scale = 1/48

Analysis

The ductwork and fan will need to be sized. Reference will be made to the sketch of the ductwork.

Size of the Hood

Section 507.12 of the International Mechanical Code and Section 4.3 of ASHRAE Standard 154 provide clear guidance regarding the minimum size of the exhaust hood relative to the size of the appliances. These codes require at least 6 in. of end overhang and 12 in. of front overhang for wall-mounted canopy hoods. Therefore, for this design problem and for the appliances selected, the minimum size of the hood should be 14 ft 7 in. long and 3 ft 7 in. deep to cover all the appliances. Section 507.15 of the International Mechanical Code requires that each exhaust outlet on a hood service no more than a 12 ft section of hood. Therefore, two exhaust outlets will be required in this hood.

A box exhaust canopy from American Hood Systems, Inc. will be chosen. A drawing of the custom-made hood in plan and elevation views is shown. **The depth of this hood will be 3 ft 7 in.**, which meets the minimum requirement established by the code. **The length of the hood will be specified as 15 ft**, which satisfies Section 507.15 of the International Mechanical Code. The features of the hoods include compliance with NFPA Standard 96 and 16 and 18 gage steel construction. The baffle-type filters are each 20 in. wide and the stainless steel cup tray traverses the entire length of the hood.

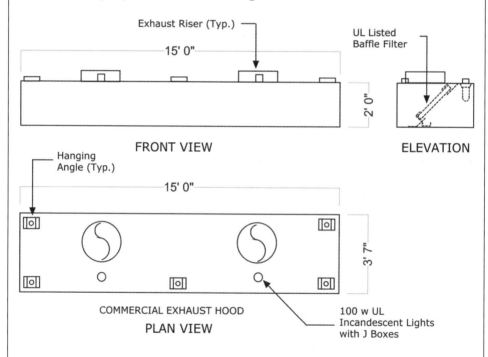

Flow Rates in the Sections of the Ductwork

The volume flow rates of the gases that need to be exhausted through the system are usually specified by international or municipal codes. Section 8.2.2.1 of NFPA Standard 96 states that the exhaust hood and system should move a sufficient volume of gas to ensure capture and removal of grease-laden cooking vapors. However, Section 507.13 of the International Mechanical Code provides specific guidelines on the capacity (volume flow rate) of hoods for appliances that are defined as heavy duty, medium duty, or light duty. Section 3 of ASHRAE Standard 154 provides full definitions of the duty levels of the appliances. For this design problem, Section 4.2 of ASHRAE Standard 154 defines the gas-fired ranges as heavy-duty appliances and the deep fat fryer and grills as medium-duty appliances. According to Section 507.13 of the International Mechanical Code, type I wall-mounted canopy hoods should exhaust at least 400 cfm per linear ft of hood for heavy-duty appliances and 300 cfm per linear ft of hood for medium-duty appliances. The same code section requires that the minimum exhaust flow rate required by the heaviest duty appliance covered by the hood should apply to the entire hood. Therefore, 500 cfm per linear ft of hood will be used to design the exhaust duct system and select the fan.

Therefore, for the hood in question, the exhaust gas volume flow rate will be

$$\dot{V} = \frac{500 \text{ cfm}}{1 \text{ ft hood}} \times 15 \text{ ft hood} = 7500 \text{ cfm}.$$

For each section of the ductwork system, the volume flow rates are

Section a-b: 3750 cfm (for each side of the symmetrical system),
Section b-c: 7500 cfm.

Sizing the Circular Duct

The flow rate through each section of the ductwork system is known. It was assumed that the pressure loss in the system will be guided by a value of 0.1 in. wg per 100 ft of ductwork and the target maximum velocity will be 1800 fpm. Note that since the designer will size and select a fan as a part of this problem solution, this industry standard will be used. The approximate diameter of the circular ducts will be found by using the flow rates, pressure losses, and the appropriate friction loss chart for round, straight galvanized steel ducts (Figure A.1).

Thus,

Section a-b: 22 in. diameter, 1500 fpm, 0.14 in. wg per 100 ft
Section b-c: 30 in. diameter, 1550 fpm, 0.10 in. wg per 100 ft

Note that the velocities of the exhaust gas in each section of the ductwork system are less than 1800 fpm, as required for a low-velocity, low-noise, low-vibration duct system.

It is likely that the horizontal portion of Section a-b will be 30 in. to accommodate installation and attachment to the 30 in. duct of Section b-c.

Total External Static Pressure Required from the Fan

The friction loss for each section can be used to estimate the pressure loss in the duct. The equivalent length (straight duct + equivalent lengths of the fittings) of the longest branch should be used. The longest branch is section a-b-c.

For section a-b,

$$L_{e,a-b} = L_{\text{straight}} + L_{\text{ent}} + L_{\text{tee,branch}}.$$

The exhaust outlet is located about 3 ft from the edge of the hood to facilitate centering the outlets. Therefore, the horizontal length of straight duct is approximately 4 ft. Therefore, $L_{straight} = 4$ ft.
So,

$$L_{e,a-b} = 4 \text{ ft} + 22 \text{ ft} + 73 \text{ ft}$$

$$L_{e,a-b} = 99 \text{ ft}.$$

It should be noted that the use of a 90° bend instead of a tee fitting would have reduced the losses significantly. If a 5-piece or 3-piece 90° bend was chosen, the radius of curvature would need to be 1.5 times the duct diameter. For a 22 in. diameter duct, the radius of curvature would need to be 33 in. For the 30 in. duct, it would need to be 45 in. Given that a 30 in. diameter duct may be used for the horizontal portion of section *a-b*, the 90° bend would likely not be feasible or easy to install given the 40 in. space that is available between the top of the hood and the finished ceiling (see final drawing). Therefore, the tee fitting might be the most likely fitting that would be installed.
For section *b-c*,

$$L_{e,b-c} = L_{straight} + L_{tee,branch}.$$

Section 4.2.1 of NFPA Standard 96 requires a minimum 18 in. clearance from the fan to the roof if the roof were made of combustible material. To facilitate installation, a 2 ft clearance will be provided, which will be added to the 3 ft length over which section *b-c* penetrates the attic space for connection with section *a-b*. Therefore, the length of section *b-c* is approximately 5 ft.
So,

$$L_{e,b-c} = 5 \text{ ft} + 100 \text{ ft}$$

$$L_{e,b-c} = 105 \text{ ft}.$$

It should be noted that in some cases the L_e/D ratio (Table A.4) was used to estimate the equivalent lengths. The duct diameter must be converted to feet to use this ratio.
The pressure drop in the duct is

$$\Delta P_{duct} = \frac{0.14 \text{ in. of water}}{100 \text{ ft}} \times 99 \text{ ft of ductwork}$$

$$+ \frac{0.10 \text{ in. of water}}{100 \text{ ft}} \times 105 \text{ ft of ductwork} = 0.244 \text{ in. wg.}$$

The total static pressure required from the fan is

$$\Delta P_{fan} = \Delta P_{duct} + \Delta P_{filter} + \Delta P_{wind} = (0.244 + 0.30 + 0.5) \text{ in. wg}$$

$$\Delta P_{fan} = 1.1 \text{ in. wg.}$$

A Greenheck tubular centrifugal belt drive roof upblast fan will be selected for this application. The specifications are marked in the catalog sheet. The fan should be able to move 7500 cfm of exhaust gas over 1.1 in. wg of external static pressure. The fan that can produce 1.5 in. wg of external static pressure will meet the design requirements. From the selection table for the **Greenheck TCBRU-2-22 line of fans, a 1542 rpm speed and a 4.64 hp motor is selected**.

TCB-2-22
Level 2 Construction

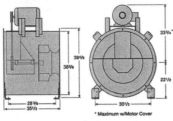

Dimensions shown in inches
Maximum rpm — 1722
Outlet Velocity = cfm X 0.139
Tip Speed = rpm X 5.83
Maximum bhp at a given rpm — (rpm/910)^3
Max Motor Frame Size - 213T

Each cell below shows **RPM | BHP** with **Sones** on the second line.

CFM	OV	0.25	0.5	1	1.5	2	2.5	3	3.5	4
2000	277	484 | 0.15 — 4.7	598 | 0.27 — 5.5							
2500	346	549 | 0.22 — 6.4	648 | 0.36 — 6.5	823 | 0.67 — 9.4						
3000	415	620 | 0.30 — 7.6	705 | 0.47 — 7.8	863 | 0.82 — 10.0	1004 | 1.21 — 12.2	1127 | 1.61 — 14.6				
3500	484	695 | 0.42 — 9.0	771 | 0.60 — 9.3	913 | 1.00 — 11.0	1040 | 1.43 — 12.9	1162 | 1.88 — 15.3	1270 | 2.35 — 17.1			
4000	554	773 | 0.56 — 10.8	841 | 0.77 — 11.0	969 | 1.21 — 12.4	1089 | 1.68 — 14.2	1197 | 2.18 — 16.0	1305 | 2.69 — 17.7	1403 | 3.22 — 19.7	1493 | 3.76 — 22	
4500	623	853 | 0.74 — 12.8	915 | 0.97 — 12.9	1030 | 1.45 — 14.0	1141 | 1.96 — 15.8	1246 | 2.50 — 17.1	1341 | 3.07 — 18.4	1438 | 3.64 — 20	1528 | 4.23 — 22	1612 | 4.83 — 25
5000	692	933 | 0.96 — 14.6	991 | 1.21 — 14.8	1098 | 1.73 — 16.0	1199 | 2.28 — 17.5	1297 | 2.87 — 18.4	1390 | 3.47 — 19.5	1476 | 4.10 — 21	1563 | 4.73 — 23	1646 | 5.38 — 25
5500	761	1015 | 1.22 — 16.6	1069 | 1.49 — 17.0	1167 | 2.05 — 18.4	1261 | 2.65 — 19.0	1353 | 3.27 — 19.7	1440 | 3.92 — 21	1525 | 4.59 — 22	1604 | 5.27 — 24	1682 | 5.97 — 26
6000	831	1098 | 1.52 — 18.7	1148 | 1.82 — 19.3	1241 | 2.43 — 20	1328 | 3.06 — 20	1411 | 3.73 — 21	1495 | 4.42 — 22	1575 | 5.13 — 24	1653 | 5.85 — 25	
6500	900	1181 | 1.88 — 21	1228 | 2.20 — 21	1315 | 2.85 — 22	1396 | 3.53 — 22	1476 | 4.24 — 23	1553 | 4.97 — 24	1630 | 5.72 — 25	1704 | 6.49 — 27	
7000	969	1265 | 2.29 — 23	1309 | 2.64 — 23	1391 | 3.34 — 24	1468 | 4.06 — 24	1543 | 4.80 — 25	1614 | 5.57 — 26	1687 | 6.36 — 27		
7500	1038	1349 | 2.76 — 25	1390 | 3.14 — 25	1468 | 3.88 — 26	1542 | 4.64 — 26	1611 | 5.43 — 27	1681 | 6.24 — 28			
8000	1108	1433 | 3.30 — 27	1472 | 3.70 — 28	1546 | 4.49 — 28	1617 | 5.30 — 29	1683 | 6.13 — 29				
8500	1177	1518 | 3.90 — 30	1555 | 4.33 — 30	1626 | 5.17 — 31	1692 | 6.02 — 32					
9000	1246	1603 | 4.57 — 33	1638 | 5.03 — 34	1706 | 5.92 — 34						

Performance certified is for installation type B: Free inlet, Ducted outlet. Power rating (bhp) does not include transmission losses. Performance ratings do not include the effects of appurtenances (accessories) in the airstream.
The sound ratings shown are loudness values in fan sones at 5 ft. (1.5m) in a hemispherical free field calculated per AMCA Standard 301. Values shown are for installation type B: free inlet fan sone levels.

Building Value in Air.

Source: Greenheck Fan, Corp. (reprinted with permission)

Drawings

The duct sizes determined by the design engineer are included on the architectural drawing that was supplied by the client. Further details on the required clearances are also shown.

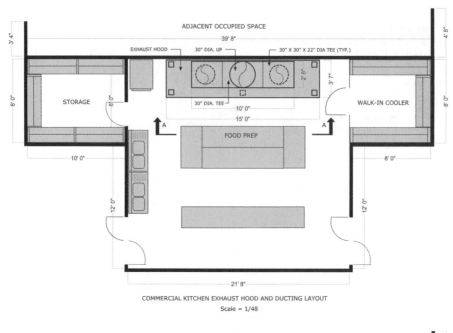

COMMERCIAL KITCHEN EXHAUST HOOD AND DUCTING LAYOUT
Scale = 1/48

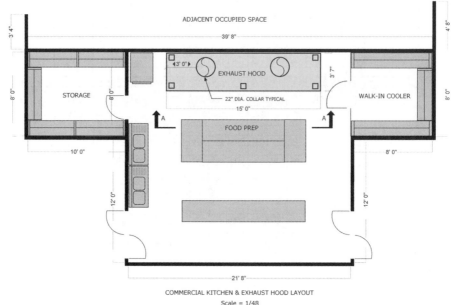

COMMERCIAL KITCHEN & EXHAUST HOOD LAYOUT
Scale = 1/48

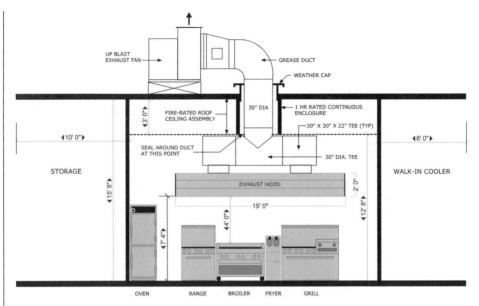

COMMERCIAL KITCHEN & EQUIPMENT ELEVATION "A-A"
Scale = 1/48

Conclusions

The exhaust system has been sized, taking into consideration the appropriate codes and standards. Galvanized steel will be used to fabricate the exhaust system based on circular ducts. The air velocities in each section of the duct are less than 1800 fpm, which will maintain a low-velocity, low-noise, low-vibration duct system. The pressure drops through the ductwork, fittings, and components were used to size and select a fan from a manufacturer's catalog.

The pressure loss in components such as the filters was estimated based on generic data that is available. The losses across 30/30 filters were used; but, baffle-type filters were actually specified with the hood. Baffle-type filters have higher losses than 30/30 filters, and the manufacturer's catalog should be consulted for actual pressure loss values across the filters. The loss specified due to wind currents was likely larger than that which will be encountered during operation. A loss of 0.5 in. wg is likely if the fan exhausts directly in the direction opposite to the prevailing wind. With the use of a roof upblast fan, the prevailing wind is likely to move perpendicular to the exhaust gas flow direction.

In this exhaust system, the main duct branch that is connected to the fan (section b-c) was designed such that the pressure loss in that section was lower than that in section a-b. A check will show that $\Delta P_{\text{duct},b-c} = 0.103$ in. wg $< \Delta P_{\text{duct},a-b} = 0.141$ in. wg. The lower pressure loss will facilitate exhaust gas flow through section b-c and final ejection from system. Given that 30 in. duct sections will be used in section a-b, pressure losses will be slightly lowered and balancing issues will be mitigated.

Since this is already a custom-made hood because of the total length and the International Mechanical Code requirement to have two exhaust outlets, it may be possible to request

---

the fabrication of circular exhaust outlets on the hood to avoid the use of transitions from square outlets to round ducts.

Codes and standards have placed other requirements on the system, some of which have been addressed in this design. Section 8.1.2.1 of NFPA Standard 96 requires that the fan motor be located outside of the exhaust gas stream. The catalog sheet from the Greenheck Fan, Corp. shows that the in-line upblast exhaust fan selected meets this requirement. As exhaust air is moved from the kitchen, the air must be replaced by another fan or appropriate makeup air unit, in accordance with Section 508.1 of the International Mechanical Code. Section 8.3.1 of NFPA Standard 96 requires that an adequate quantity of air be supplied such that negative pressures in the cooking area do not exceed 0.02 in. wg. Section 6 of ASHRAE Standard 154 provides details on delivery of the replacement or makeup air. While outside the scope of this present design problem, consideration of the makeup air system would be required since it is required by code. Finally, Section 5.7 of ASHRAE Standard 154 requires that the exhaust system become immediately operational whenever the appliances are on through use of direct or indirect interlocks. Heat-sensitive switches could be used for automatic operation of the exhaust system. A detailed control system design would be required.

2.5 Fans—Brief Overview and Selection Procedures

2.5.1 Classification and Terminology

A fan is a fluid machine that is used to move and induce flow of a gas (gas pumps). Rotational mechanical energy is imparted upon the gas, causing an increase in gas pressure. The increased pressure and energy of the gas is used to overcome frictional and component losses in the system to which the fan is connected. The pressure difference of the gas that is generated across the inlet and discharge of the fan will usually be reported in inches water gage. Fans that generate pressure differences in excess of 30 in. wg are known as **compressors**. Most fans that are used in other systems such as those found in commercial or residential buildings tend to generate pressure differences less than 15 in. wg. These types of fans are introduced in Section 2.5.2. The *Fan Application Manual* published by Carrier Corporation [16] is a resource that may be consulted for detailed information on fans.

2.5.2 Types of Fans

(a) *Axial Fans*: In an axial fan, gas flow enters and leaves the fan in a straight line. The fluid flows through the impeller and parallel to the driveshaft, which is used to rotate the fan blade (impeller). Motors may be connected directly to the impeller for **direct-drive motor** arrangements. Alternatively, the impeller shaft may be connected to the motor via a drive pulley to a **belt drive** arrangement. Several types of axial fans are available, including propeller, tubeaxial, and vaneaxial fans. These types of fans are shown pictorially in Figure 2.4. A propeller fan is used to propel air into an open ambient, and is devoid of a housing

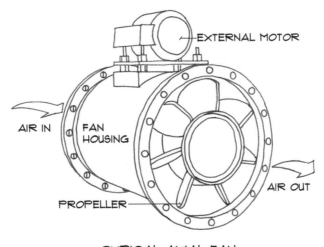

TYPICAL AXIAL FAN

Figure 2.4   Axial fans

enclosure. A tubeaxial or vaneaxial fan is a propeller fan that is enclosed within a
ducted system.

(b) *Centrifugal Fans*: Centrifugal fans are fabricated such that an impeller wheel, that
is complete with blades, rotates within an enclosure or **fan housing** as illustrated
in Figure 2.5. Air enters the fan axially, through one or both sides (through the
"eye"), and is propelled radially through the impeller and discharge outlet. The
fan may be connected directly to a motor (direct-drive) or connected to the motor

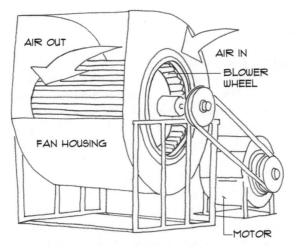

TYPICAL CENTRIFUGAL FAN

Figure 2.5   Centrifugal fans

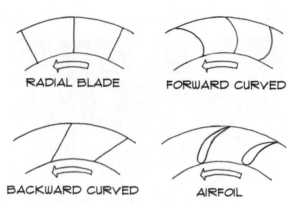

Figure 2.6   Classification of centrifugal fans based on blade types

via a belt drive (belt-driven). Centrifugal fans may be classified based on the design of the fan blades as **forward-curved** (curved in the direction of rotation of the fan) or **backward-curved** (curved in the direction opposite to the rotation of fan) as shown in Figure 2.6. Some fans with backward-curved blades may have those blades in the shape of airfoils (shape of a wing). These **airfoil** fans increase operational efficiency due to the streamlined flow of the gas over the blades. However, they are more expensive than typical backward-curved fans that possess simpler blades. The orientation of the blades in centrifugal fans will have an impact on the performance of the fan. In particular, forward-curved fans are usually used in low-pressure (less than 5 in. wg) systems found in residential and light commercial applications. For applications that require high pressures and high efficiency, backward-curved fans may be the more suitable choice, though they could also be used in low-to-medium-pressure applications.

2.5.3 Fan Performance

The design, installation, and inclusion of components in a duct system will produce pressure losses that the fan must overcome in order to move the gas through the system. Once the design of the ductwork system is complete, the design engineer will determine the maximum amount of flow (in cfm) and the total pressure loss (in in. wg) that will be experienced by the moving gas. That total volume flow rate of gas and static pressure must be supplied by the fan. Given that the static pressure required by the ductwork system is external to fan and housing, the pressure difference required is usually referred to as **external static pressure**.

Manufacturers will present catalogs that specify the performance of their line of fans based in the total volume flow rate of gas that can be delivered (usually air) and the external static pressure that the fan can provide. This data may be provided in either tabular or graphical forms. When in graphical form, curves of the external static pressure versus the volume flow rate for a fan operating at different speeds are known as **performance curves**. Figure 2.7 shows typical performance curves for

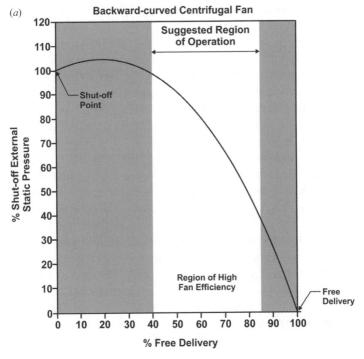

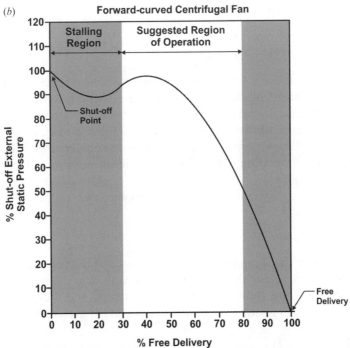

Figure 2.7 Typical performance curves of centrifugal fans

backward-curved and forward-curved centrifugal fans. The figure shows that no flow occurs when the external static pressure is maximum (i.e., maximum resistance to flow). This point on the curve is known as the **shut-off point**. The point on the curves at which the external static pressure is a minimum and the volume flow rate is a maximum (i.e., no resistance to flow) is known as **free delivery**. Shut-off and free delivery will be explored in greater detail in Chapter 3.

The shapes of the performance curves shown in Figure 2.7 are specific to the type of centrifugal fan. Backward-curved and airfoil fans usually have a curve with a parabolic shape, with external static pressure increasing from the shut-off point to a maximum pressure and decreasing steeply to free delivery. Forward-curved fans usually have the performance curve shown in Figure 2.7, in which the static pressure decreases, then increases as the volume flow increases shortly after the shut-off point. Operation of the fan in this "dip" section of the curve will result in unsteady pressure and flow, lower efficiency, and high noise. Operation of the fan in this **stalling** region should be avoided. Close to free delivery, pulsations in flow, pressure, and air speed will occur. Therefore, operation of the fan close to free delivery should also be avoided. Figure 2.7 shows the suggested region of operation of the fans as a percentage of the free delivery volume flow rate. This range was selected after conducting a survey of several performance curves from different manufacturers. The design engineer should consult the manufacturer for details on the fan that is selected for use in the design system.

2.5.4 Fan Selection from Manufacturer's Data or Performance Curves

Figure 2.8 shows a typical set of manufacturer's performance curves for fans operating at different speeds (in revolutions per minute, rpm). The torque provided by the motor, which is directly proportional to the motor brake horsepower, is also shown. The combination of maximum volume flow rate and external static pressure required by the ductwork system will produce a point on the appropriate performance curve known as the **system operating point**. This point and the performance curve are used to make the fan selection.

The following points pertaining to the performance curves shown in Figure 2.8 should be noted:

(a) *Fan Housing Size*: The dimensions and a schematic of the fan housing are shown in the insert at the upper right corner of the graph. The diameter of the inlet to the eye is 8.81 in. Not shown are the actual diameter and width of the impeller wheel. The 10–7 designation indicates that the wheel diameter is 10 in. and the width is 7 in.

(b) *Motor Arrangement*: This is a direct-drive fan motor arrangement.

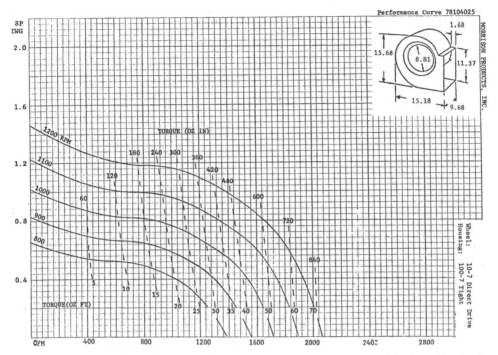

Figure 2.8 Forward-curved centrifugal fan performance curves (Morrison Products, Inc.; reprinted with permission)

(c) *Fan Speed*: The speed of the fan ranges from 800 to 1200 rpm, producing different curves for each speed.

(d) *Motor Torque*: The torque provided by the motor is shown in ounce feet (oz ft) and ounce-inch (oz in). The torque and speed can be used to determine the motor brake horsepower. The minimum motor torque available is 5 oz ft and the maximum torque is 70 oz ft.

Example 2.8   Selecting a Fan for a Designed System

Select a centrifugal fan from a manufacturer's performance curves if the total air requirement is 1400 cfm and the minimum external static pressure required by the system is 0.75 in. wg at 70°F. Specify the fan motor size, that is, the brake horsepower (bhp).

Solution. A forward-curved centrifugal fan would be suitable for this application since the external static pressure required is low and less than 5 in. wg. The operating point in this problem would be located on the performance curve at an air volume flow rate of 1400 cfm and an external static pressure of 0.75 in. wg. The manufacturer's performance curve shown shows the operating point located on the performance curve for the fan that rotates at 1100 rpm.

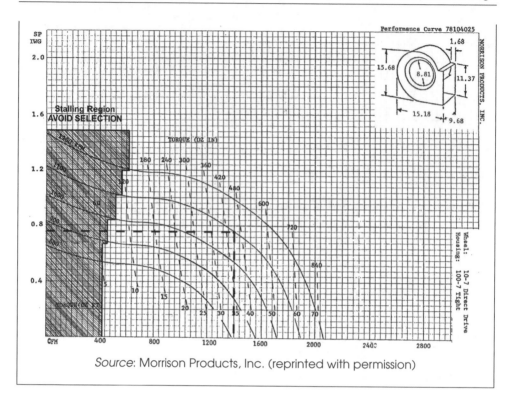

Source: Morrison Products, Inc. (reprinted with permission)

The details of the fan housing and wheel are provided with the curves. The fan speed must be **1100 rpm**. It is important to note that the operating point falls within the recommended region of operation shown in Figure 2.7. Operating points that fall below 570 cfm (30% of free delivery) or above 1520 cfm (80% of free delivery) should not be matched with this fan operating at 1100 rpm.

The operating point falls between the 35 and 40 oz ft torque curves. Since the 35 oz ft torque curve is to the left of the operating point, it should not be chosen. The exact torque requirement is 37 oz ft. However, to avoid **motor overloading** and damage, 40 oz ft of torque is chosen. Motor overloading will be explored further in Chapter 3. In Imperial units, the ounce (oz) presented in the unit of torque is ounce-force (ozf), and 16 ozf is equal to 1 lbf (pound-force). In terms of brake horsepower, the motor size, based on 40 ozf ft of torque at 1100 rpm, is

$$bhp = T\omega$$

$$bhp = (40 \text{ ozf ft}) \left(1100 \, \frac{\text{rev}}{\text{min}}\right) \left(\frac{2\pi \text{ rad}}{1 \text{ rev}}\right) \left(\frac{1 \text{ lbf}}{16 \text{ ozf}}\right) \left(\frac{1 \text{ min}}{60 \text{ s}}\right) \left(\frac{1 \text{ hp}}{550 \text{ lbf-ft/s}}\right)$$

$$bhp = 0.53 \text{ hp}.$$

2.5.5 Fan Laws

In practice, and after a fan has been selected and installed, it may be necessary to change the operating parameters of the fan to meet new design and operating criteria. **Fan laws** may be used to specify the new conditions under which the fan should operate. The **fan laws** are relationships among the various performance parameters (volume flow rate, external static pressure, fan speed, gas density, wheel diameter, brake horsepower) and they can be used to determine new values of selected parameters as a result of changes in any of the other remaining parameters. Derivation of the laws will not be detailed here. The laws were derived by utilizing the Buckingham Pi theory and applying the method of repeating variables [17, 18] to develop nondimensional ratios of the performance parameters. Provided that the fan type does not change to ensure geometric, kinematic, and dynamic similarity, the laws may be applied for cases of changes in gas volume flow rate ($\dot{V}$), external static pressure (P_s), fan speed (N), gas density (ρ), type of gas and wheel diameter (D), and brake horsepower (bhp).

Consider two fans (fan A and fan B) or consider a fan at two different states of volume flow rate, fan speed, gas density, wheel diameter, and brake horsepower. The general forms of the fans laws that are most frequently used in industry are

$$(a) \quad \frac{\dot{V}_B}{\dot{V}_A} = \frac{N_B}{N_A} \left(\frac{D_B}{D_A} \right)^3 ; \tag{2.30}$$

$$(b) \quad \frac{P_{sB}}{P_{sA}} = \frac{\rho_B}{\rho_A} \left(\frac{N_B}{N_A} \right)^2 \left(\frac{D_B}{D_A} \right)^2 ; \tag{2.31}$$

$$(c) \quad \frac{bhp_B}{bhp_A} = \frac{\rho_B}{\rho_A} \left(\frac{N_B}{N_A} \right)^3 \left(\frac{D_B}{D_A} \right)^5 . \tag{2.32}$$

Example 2.9   Application of the Fan Laws in System Design

After selecting the centrifugal fan described in Example 2.8, the client has decided to use the same ductwork system and fan under the following conditions, each at different periods each month:

(a) for dry air at 20°F;
(b) for carbon dioxide at 0°F.

For both these cases, the client requires that the fan operate at 1100 rpm, as initially selected for air at 70°F. Will the fan selected in Example 2.8 still be able to deliver 1400 cfm of gas under these two updated conditions? If this is not possible, recommend a course of action that could be taken with the selected fan. The client and contractor understand fully that strict adherence to the guidelines and standards of SMACNA [13] will be required to avoid gas leakage from the ducts.

Solution. The fan laws can be used to determine the performance of the fan under the new operating conditions of the gas. Let state A be governed by the performance parameters

---

found in Example 2.8. So, $\dot{V}_A = 1400$ cfm, $P_{sA} = 0.75$ in. wg, $N_A = N_B = 1100$ rpm, $D_A = D_B = 10$ in., and $\rho_{A,air} = 0.07489$ lbm/ft^3. The exact torque requirement is 37 oz ft. So, the exact brake horsepower requirement is

$$\text{bhp}_A = (37 \text{ ozf ft}) \left(1100 \frac{\text{rev}}{\text{min}}\right) \left(\frac{2\pi \text{ rad}}{1 \text{ rev}}\right) \left(\frac{1 \text{ lbf}}{16 \text{ ozf}}\right) \left(\frac{1 \text{ min}}{60 \text{ s}}\right) \left(\frac{1 \text{ hp}}{550 \text{ lbf-ft/s}}\right) = 0.48 \text{ hp}.$$

Based on Equation (2.30), the volume flow rate at the new conditions is

$$\frac{\dot{V}_B}{\dot{V}_A} = \frac{1100 \text{ rpm}}{1100 \text{ rpm}} \left(\frac{10 \text{ in.}}{10 \text{ in.}}\right)^3 = 1$$

$$\dot{V}_B = \dot{V}_A.$$

(a) For the case of dry air at 20°F, $\rho_{B,air} = 0.08270$ lbm/ft^3. Under this condition, the static pressure is

$$\frac{P_{sB}}{0.75 \text{ in. wg}} = \frac{0.08270 \text{ lb/ft}^3}{0.07489 \text{ lb/ft}^3} \left(\frac{1100 \text{ rpm}}{1100 \text{ rpm}}\right)^2 \left(\frac{10 \text{ in.}}{10 \text{ in.}}\right)^2$$

$$P_{sB} = 0.83 \text{ in. wg}$$

$$P_{sB} = 0.83 \text{ in. wg.}$$

The brake horsepower would be

$$\frac{\text{bhp}_B}{0.48 \text{ hp}} = \frac{0.08270 \text{ lb/ft}^3}{0.07489 \text{ lb/ft}^3} \left(\frac{1100 \text{ rpm}}{1100 \text{ rpm}}\right)^3 \left(\frac{10 \text{ in.}}{10 \text{ in.}}\right)^5$$

$$\text{bhp}_B = 0.53 \text{ hp.}$$

The brake horsepower is equivalent to approximately 40 oz ft of torque at 1100 rpm. The performance plot obtained from the manufacturer shows that for an external static pressure of 0.83 in. wg, the operating point is now above the performance curve for the fan that operates at 1100 rpm. The brake horsepower of the motor that is required under these new conditions is nearly equal to that of the existing motor, increasing the risk of motor overload. Therefore, **the client would not be advised to use the fan at this speed in this application to provide 1400 cfm of air at 20°F.**

One course of action could be to move the operating point over to the performance curve (1100 rpm) for the fan at the external static pressure of 0.83 in. wg, that is, to the left. This would result in a decrease in the volume flow rate of the air to 1280 cfm. The motor torque would also decrease to an exact value of 33 ozf ft. However, a motor torque of 35 ozf ft would likely be selected to avoid motor overload and damage. The current motor is rated for 40 ozf ft of torque, as shown in Example 2.8. Therefore, if the reduction in volume flow rate of air is acceptable to the client, this proposed solution would be sufficient and would avoid motor overload. The design engineer would need to use fundamental principles of fluid mechanics to determine the system static pressure loss that would occur due to the flow of the denser air through the duct system. Major and minor losses would need to be considered. This would serve as a useful comparison to the fan static pressure of 0.83 in. wg.

If the reduction in volume flow rate of air is unacceptable to the client, an alternate course of action could be to increase the fan speed to 1200 rpm. As would be the case for the fan operating at 1100 rpm, knowledge of the new system static pressure loss would refine the selection of the fan to operate at 1200 rpm.

(b) For the case of carbon dioxide at 0°F, $\rho_{B,CO_2} = 0.1311$ lb/ft³. Under this condition, the static pressure is

$$\frac{P_{sB}}{0.75 \text{ in. wg}} = \frac{0.13111 \text{ lb/ft}^3}{0.07489 \text{ lb/ft}^3} \left(\frac{1100 \text{ rpm}}{1100 \text{ rpm}}\right)^2 \left(\frac{10 \text{ in.}}{10 \text{ in.}}\right)^2$$

$$P_{sB} = 1.31 \text{ in. wg.}$$

The brake horsepower would be

$$\frac{bhp_B}{0.48 \text{ hp}} = \frac{0.13111 \text{ lb/ft}^3}{0.07489 \text{ lb/ft}^3} \left(\frac{1100 \text{ rpm}}{1100 \text{ rpm}}\right)^3 \left(\frac{10 \text{ in.}}{10 \text{ in.}}\right)^5$$

$$bhp_B = 0.84 \text{ hp.}$$

The brake horsepower is equivalent to approximately 65 oz ft of torque at 1100 rpm.

The operating point is now well above all the performance curves of this series of fans. Therefore, **the client would not be advised to use this fan at any speed in this application to provide 1400 cfm of carbon dioxide at 0°F.** The client would be advised to replace the fan.

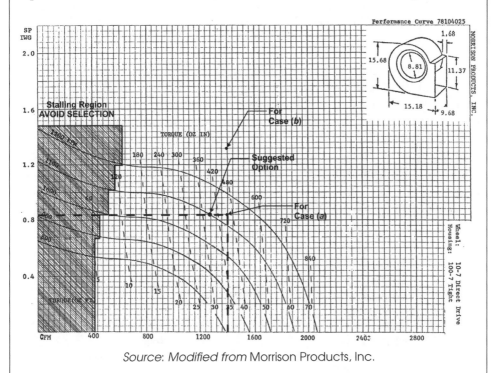

Source: Modified from Morrison Products, Inc.

2.6 Design for Advanced Technology—Small Duct High-Velocity (SDHV) Air Distribution Systems

Small duct, high-velocity (SDHV) air distribution systems are those in which small-sized duct branches and high velocities are used to transport air through the system and to the spaces. This type of system can be used when space restrictions are present in a building such that large ducts cannot be accommodated. The SDHV systems may be used during the retrofit of older residential buildings that were not constructed with conventional duct systems. Due to the small sizes of the ducts required, extensive demolition may not be required and the interior and exterior appearance of the building may be preserved. The SDHV air distribution system is not a new technology. However, the use of this type of system is becoming popular in residential building applications.

In SDHV air distribution systems, duct sizes can be as low as 2 in. in diameter, in which flexible ductwork would be used. Table 2.3 shows the supply air velocities that typify SDHV air distribution systems based on the air volume flow rate. The **supply ductwork** provides air from the fan or blower to the spaces or the applications. A comparison of the velocities presented in Tables 2.1 and 2.3 show that for residential applications with main duct airflow rates between 1000 and 3000 cfm, an SDHV air distribution system may be defined as a system in which the supply air velocity is between approximately 1200 and 2500 fpm. For industrial (commercial) buildings, the velocity range would be 2200–2500 fpm. Any velocity greater than 2500 fpm would force classification of the system as a high-velocity duct system. It should be noted that as the air returns from the space back to the fan or blower, the return air ducts are low-velocity ducts.

The small ducts and high velocities required by SDHV air distribution systems will produce higher external static pressures. The higher velocities will produce higher pressure losses through the ductwork system and across components such as

Table 2.3   Maximum supply duct velocities

High-Velocity Systems (Main Branches)	
Airflow Rate in Duct (cfm)	Maximum Velocity (fpm)
1000–3000	2500
3000–6000	3000
6000–10000	3500
10000–15000	4000
15000–25000	4500
25000–40000	5000
40000–60000	6000

Source: Howell, Sauer, and Coad [9].

heating/cooling coils. This will result in larger power (bhp and torque) requirements by the fan or blower that is used to move the air. The design engineer needs to ensure that the increased requirements of the fan or blower are met since undersized fans or blowers will produce lower air volume flows rates, which may undermine the performance of the installed components and the design criteria may not be satisfied. To that end, Section 3.11 of the Air-Conditioning, Heating, and Refrigeration Institute (AHRI) Standard 210/240-2005 [19] requires that a cooling product contain a blower that produces at least 1.2 in. wg of external static pressure when operating at the certified airflow rate of 220–350 cfm per rated ton of cooling. One ton of cooling is equivalent to 12000 Btu/h of energy. In this case, the minimum external static pressure requirement of the fan or blower is specified by the standard, and it should be utilized in the design of the ductwork and selection of equipment.

Noise may be an issue to address during operation of SDHV air distribution systems. ASHRAE have led the development of criteria that are used to stipulate acceptable noise levels for spaces, equipment, and air duct systems. These criteria are specific to the application and use of the space, and are based on values of the **noise criteria (NC)**. NC values are established after conducting acoustic tests at different sound frequencies. The maximum sound pressure (power) level in decibels (dB) that is achieved during the tests will represent the value for the NC rating. Table 2.4 shows results from a sound test of an 8 in. × 8 in. rectangular elbow with 7 circular arc turning vanes. The air velocity was 3000 fpm. Since the maximum pressure level is approximately 55 dB (at 500 Hz), the NC rating for the elbow is NC-55. Higher NC values indicate higher levels of noise.

NC can be used to guide and justify decisions during the design of air distribution systems. Further constraints on air velocities, similar to those provided in Tables 2.1 and 2.3, can be established based on acceptable NC for spaces. Table 2.5 presents recommended maximum velocities that are required to achieve specified acoustic design criteria for main ducts. In order to achieve the same NC values listed in the

Table 2.4 Sound data during airflow through a rectangular elbow

Frequency (Hz)	Sound Pressure Level (dB)
125	50
250	53
500	55
1000	52
2000	51
4000	46
8000	40

Source: ASHRAE [20].

Table 2.5 Maximum main duct air velocities for acoustic design criteria

Duct Location	Noise Criteria (NC)	Maximum Velocity (fpm)	
		Rectangular Duct	Circular Duct
In shaft or above	45	3500	5000
drywall ceiling	35	2500	3500
	25	1700	2500
Above suspended	45	2500	4500
acoustic ceiling	35	1750	3000
	25	1200	2000
In occupied space	45	2000	3900
	35	1450	2600
	25	950	1700

Source: ASHRAE [20].

table, the velocities in branch ducts should be 80% of the values listed [20]. Where applicable, for duct runouts from the branch ducts to outlets such as diffuser boots, the velocities should be 50% or less of those in the main duct. Table 2.6 follows with the recommended design guidelines for sound in unoccupied spaces. The tables may be used in conjunction with other design criteria to make and justify duct system design decisions. The design engineer should consult with an acoustical engineer regarding acoustical critical spaces such as concert and recital halls or TV and radio studios.

There will be cases where modifications of the design of the ductwork will not eliminate excess noise. In such cases, the design engineer will need to take additional steps. Noise control protocols may include the use of acoustical liners in the ducts, nonmetallic insulated flexible duct, and/or duct silencers (sound attenuators, sound traps, or mufflers). These solutions may add further pressure loss to the duct system and they should be considered when sizing and selecting appropriate fans.

Practical Note 2.2   Diffuser Discharge Air Volume Flow Rates in SDHV Systems

The velocities at air supply outlets such as diffusers in SDHV air distribution systems may be in excess of the constraints established by Table 2.5 for duct runouts. This becomes amplified when the air volume flow rates through the diffusers are high and the duct size is small. To mitigate this problem, designers and contractors will constrain the air volume flow rate through the duct runouts and diffusers at approximately 40–70 cfm. While this constraint will establish the need for multiple diffusers in a space, the constraints on noise levels and velocities will be met successfully.

Table 2.6 Acoustic design criteria for unoccupied spaces [21]

Type of Space	Noise Criteria (NC)
Private residences, apartments, condominia	25–35
Hotels and motels	
Individual rooms or suites; meeting and banquet rooms	25–35
Halls, corridors, lobbies, service and support areas	35–45
Office buildings	
Executive and private offices, conference rooms	25–35
Teleconferencing rooms	25 (max)
Open-plan offices	30–40
Circulation and public lobbies	40–45
Hospitals and clinics	
Private rooms	25–35
Wards	30–40
Operating rooms	25–35
Corridors, public areas	30–40
Laboratories (with fume hoods)	
Testing or research, minimal speech communication	45–55
Research, extensive telephone usage, speech communication	40–50
Group teaching	35–45
Places of worship	25–35
Schools	
Classrooms, less than 750 ft^2	40 (max)
Classrooms, more than 750 ft^2	35 (max)
Libraries	30–40
Courtrooms	
Unamplified speech	25–35
Amplified speech	30–40
Performing arts spaces	
Drama theaters and music teaching studios	25 (max)
Music practicing rooms	25–35
School and university gymnasia and natatoria	40–50

Example 2.10   Designing a Simple SDHV Air Distribution System

Miss Cherry wishes to modify the air duct system that was designed in Example 2.5 to that of an SDHV air distribution system. In this case, she has requested that all the ducts be circular to deliver air to three private rooms in her establishment. Instead of delivering air from the dropped ceiling height directly from openings in the ductwork, she has requested that diffusers be used. The original sketch of the floor plan, complete with air requirements per room, is shown below.

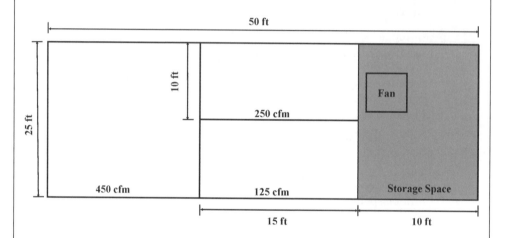

Miss Cherry has also decided to include a finned-tube heating coil and a Camfil Farr 30/30 medium efficiency pleated air filter in the main duct to heat the air that will be delivered to her rooms. It is expected that the pressure drop across the coil will be approximately 0.25 in. wg. A new fan, which will be located on top of the roof, will be required. In compliance with Section 1209 of the 2006 International Building Code, the roof is 3 ft above the dropped ceiling. The wall partitions shown on the sketch do not extend beyond the dropped ceiling. Layout an appropriate SDHV air distribution system based on circular ducts. Size and select a fan.

Possible Solution

Definition

Size and layout an appropriate circular SDHV air distribution system for the given space. Select a fan.

Preliminary Specifications and Constraints

   (i) The working fluid will be air.
   (ii) The system should be based on small circular ducts and high air velocities.

(iii) The three rooms in the establishment have the following air requirements: 250, 125, and 450 cfm.

(iv) A finned-tube heating coil is required by the client, which will induce air pressure drop in then system.

(v) A Camfil Farr 30/30 medium efficiency pleated air filter will be installed.

(vi) A new rooftop fan is required.

(vii) The length of the ductwork is constrained by the dimensions of the room.

(viii) The attic space between the roof and the dropped ceiling is 3 ft.

Detailed Design

Objective

To design a circular SDHV air distribution system that is complete with a selected fan.

Data Given or Known

(i) Air will now be delivered from diffusers connected to ductwork.

(ii) All the dimensions of the rooms were provided by the client.

(iii) The air requirements for the three rooms were given as 250, 125, and 450 cfm. The total air requirement is 825 cfm.

(iv) The finished ceiling is 3 ft above the dropped ceiling.

(v) The wall partitions do not extend beyond the dropped ceiling.

(vi) A finned-tube heating coil with an air pressure drop of 0.25 in. wg will be installed.

(vii) A Camfil Farr 30/30 medium efficiency pleated air filter will be installed.

Assumptions/Limitations/Constraints

(i) Section 3.11 of the AHRI Standard 210/240-2005 requires that a cooling product contain a blower that produces at least 1.2 in. wg of external static pressure when operating at the certified airflow rate of 220–350 cfm per rated ton of cooling. In this design problem, the client (Miss Cherry) is heating the air. Also, the design problem preamble statement does not specify if cooling will be required and the tonnage of cooling that would be needed. This presents some uncertainty in the possible design solution. However, it is expected that Miss Cherry may need cooling during the summer months and she may wish to use the same duct-fan system that is proposed here. Therefore, a conservative design approach would be to assume that the fan or blower will provide at least 1.2 in. wg of external static pressure to provide the client with some flexibility.

(ii) Galvanized steel is typically used to fabricate air duct systems. It will be chosen as the material.

(iii) Where appropriate, branch fittings will be 45° wyes or 90° elbows with long smooth radii. This will reduce the minor loss equivalent lengths.

(iv) Converging transitions (duct size reductions) will be at 20° angles to reduce losses.

(v) The entrance to the duct system from the fan will be a bellmouth to reduce frictional losses.

(vi) Assume that the dropped ceiling is a suspended acoustic ceiling.

(vii) This is a massage-beauty-relaxation spa. Therefore, it will be assumed that the space will be similar to a private residence, apartment, or condominia. So, based on Table 2.6, the noise criterion will be restricted to 35.

(viii) Based on the NC restriction and Table 2.5, the maximum air velocity in the main ducts will be 3000 fpm since the ducts will be located above a suspended acoustic ceiling.

(ix) Restrict the flow rate through each diffuser to 70 cfm or less.

Sketch

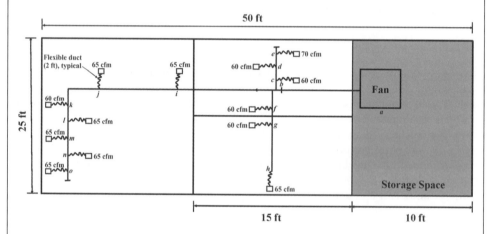

This sketch will be used to provide *estimates* of the duct lengths, and will serve as justification of these estimates. All the fittings are shown (except the bellmouth fitting at the entrance to the duct system). Typically, high-velocity diffusers are placed at the perimeter of the structure (close to the walls) to minimize the direct discharge of high-speed air on the occupants. Therefore, every attempt was made to center the ductwork and locate the diffusers close to the walls. An attempt was also made to promote uniform distribution of air in the spaces.

Flexible duct was also used to connect the galvanized smooth duct to the diffusers. Given the corrugated nature of flexible duct, Section 3.23 of the SMACNA HVAC duct construction standards requires that a minimum length be used. This will minimize both pressure losses in the duct system and fan sizes. As a practical guide, flexible duct lengths should not exceed 3 ft whenever possible. In this solution, the lengths were constrained to 2 ft.

Analysis

Designers of duct systems usually use charts and tables to size the ducts. That approach will be taken in this design problem.

Estimates of the Lengths of the Main and Branch Sections of the Ductwork

Section *a-b*: 7.5 ft
Section *b-e*: 7.5 ft
Section *b-h*: 10 ft
Section *b-o*: 40 ft

Flow Rates in the Sections of the Ductwork

Section *a-b*: 825 cfm
Section *b-c*: 190 cfm
Section *c-d*: 130 cfm
Section *d-e*: 70 cfm
Section *b-f*: 185 cfm
Section *f-g*: 125 cfm
Section *g-h*: 65 cfm
Section *b-i*: 450 cfm
Section *i-j*: 385 cfm
Section *j-k*: 320 cfm
Section *k-l*: 260 cfm
Section *l-m*: 195 cfm
Section *m-n*: 130 cfm
Section *n-o*: 65 cfm

Sizing the Circular Duct

The flow rate through each section of the ductwork system is known. Given that the pressure drop in this SDHV air distribution system is expected to be large, the maximum velocities that are expected for each section will be used to guide the sizing of the duct, in lieu of the friction loss. The diameter of the circular ducts will be found by using the flow rates, air velocities, and the appropriate friction loss chart for round, straight galvanized steel ducts (Figure A.1). The NC will be used to confirm the final duct sizes based on the expected performance of the system.

For this design problem, the following maximum velocities are expected to maintain a maximum noise criterion of 35:

1. Main duct: 3000 fpm
2. Branch duct: 2400 fpm
3. Duct runouts (flexible ducts): 1500 fpm

Based on the sketch, the 2-ft-long flexible ducts should deliver 60, 65, or 70 cfm of air to the diffusers. So, based on the airflow rates,

Flexible duct: 60 cfm, 3 in. diameter, 1210 fpm, friction loss ≈ 1.0 in. wg per 100 ft
Flexible duct: 65 cfm, 3 in. diameter, 1310 fpm, friction loss ≈ 1.2 in. wg per 100 ft
Flexible duct: 70 cfm, 3 in. diameter, 1450 fpm, friction loss ≈ 1.4 in. wg per 100 ft.

The friction loss chart for round, straight galvanized steel ducts shown in Figure A.1 applies to smooth, galvanized steel ducts. Corrugated flexible ducts will have higher friction loss than smooth ducts for a given airflow rate. However, the short lengths of the ducts will make the error in the static pressure loss negligible. The Carrier Corporation [22] has provided a friction loss chart similar to that shown in Figure A.1 for flexible ducts with 3 in. and 4 in. diameters.

For the main and branch ducts,

Section *a-b*: 8 in. diameter, 2350 fpm, friction loss = 1.0 in. wg per 100 ft (main duct)
Section *b-c*: 4 in. diameter, 2150 fpm, friction loss = 1.9 in. wg per 100 ft (branch duct)
Section *c-d*: 4 in. diameter, 1450 fpm, friction loss = 1.0 in. wg per 100 ft (branch duct)
Section *d-e*: 3 in. diameter, 1450 fpm, friction loss = 1.4 in. wg per 100 ft (branch duct)
Section *b-f*: 4 in. diameter, 2020 fpm, friction loss = 1.8 in. wg per 100 ft (branch duct)
Section *f-g*: 4 in. diameter, 1380 fpm, friction loss = 0.9 in. wg per 100 ft (branch duct)
Section *g-h*: 3 in. diameter, 1310 fpm, friction loss = 1.2 in. wg per 100 ft (branch duct)
Section *b-i*: 6 in. diameter, 2300 fpm, friction loss = 1.4 in. wg per 100 ft (main duct)
Section *i-j*: 6 in. diameter, 2000 fpm, friction loss = 1.1 in. wg per 100 ft (main duct)
Section *j-k*: 6 in. diameter, 1680 fpm, friction loss = 0.8 in. wg per 100 ft (main duct)
Section *k-l*: 5 in. diameter, 1850 fpm, friction loss = 1.2 in. wg per 100 ft (main duct)
Section *l-m*: 5 in. diameter, 1425 fpm, friction loss = 0.7 in. wg per 100 ft (main duct)
Section *m-n*: 4 in. diameter, 1450 fpm, friction loss = 1.0 in. wg per 100 ft (main duct)
Section *n-o*: 3 in. diameter, 1310 fpm, friction loss = 1.2 in. wg per 100 ft (main duct).

The velocities are lower than the maximum values prescribed for main and branch ducts and duct runouts in order to achieve NC less than 35. However, in comparison to low-velocity air distribution systems, where friction losses are approximately 0.1 in. wg per 100 ft, the friction losses are an order of magnitude higher at approximately 1.0 in. wg per 100 ft.

Sizing and Selecting a Fan

For this problem, the friction loss varies in each section of the ductwork system where flexible ducts are attached. These friction loss values should be used to estimate the total external static pressure required by the fan. The total equivalent length (straight duct + equivalent lengths of the fittings) of the longest branch should be used. The additional pressure losses across the heating coil and the filter should be included.

From visual inspection of the drawing, the longest branch in this system is *a-b-j-k-l-m-n-o*-diffuser (from the fan at *a* to the diffuser at *o*). The equivalent length is

$$L_{e,total} = L_{straight} + L_{ent} + 2L_{tee,thru,8''} + L_{contract,8''-6''} + L_{90} + 2L_{tee,thru,6''} + 3L_{tee,thru,5''} + L_{tee,thru,4''}$$

$$L_{tee,branch,3''} + L_{contract,6''-5''} + L_{contract,5''-4''} + L_{contract,4''-3''}$$

$$L_{e,total} = (7.5\ ft +\ 40\ ft +\ 2\ ft) + 8\ ft + 2\ (5\ ft) + 3\ ft + 4.5\ ft + 2\ (4\ ft) + 3\ (3.5\ ft) + 3\ ft$$

$$10\ ft + 2\ ft + 2\ ft + 1.5\ ft$$

$$L_{e,total} = 112\ ft.$$

It should be noted that in some cases the L_e/D ratio (Table A.4) was used to estimate the equivalent lengths. The duct diameter must be converted to feet to use this ratio.

The total equivalent length should be used with the friction loss to determine the total external static pressure loss due to the duct and fittings. As mentioned earlier, the friction loss at each section of the main duct where flexible ducts are attached will be different. The length of these sections will also differ. Specifying exact duct section lengths between the flexible duct connections may not be useful since field installation will usually require modifications to the exact connection points. Therefore, an average friction loss for the main duct will be used with the total equivalent length of duct to find the static pressure drop. The friction loss in the main duct varies from 0.7 to 1.4 in wg per 100 ft of equivalent length of duct. The average friction loss is 1.1 in. wg per 100 ft of equivalent length of duct.

The pressure drop in the duct is

$$\Delta P_{\text{duct}} = \frac{1.1 \text{ in. of water}}{100 \text{ ft}} \times 112 \text{ ft of ductwork} = 1.23 \text{ in. wg.}$$

The total static pressure required from the fan is

$$\Delta P_{\text{fan}} = \Delta P_{\text{duct}} + \Delta P_{\text{coils}} + \Delta P_{\text{filter}} + \Delta P_{\text{diffuser}},$$

$$\Delta P_{\text{fan}} = (1.23 + 0.25 + 0.30 + 0.05) \text{ in. wg,}$$

$$\Delta P_{\text{fan}} = 1.83 \text{ in. wg.}$$

Table 2.2 provided typical values of the static pressure losses through the filter and the diffuser. If the fan is able to move air through the longest duct branch with this total static pressure, then it will be able to move air through all the other smaller branches in the air distribution system. At this point, it is important to note that the total external static pressure of the fan is in excess of 1.2 in. wg, which is the minimum external static pressure required by Section 3.11 of the AHRI Standard 210/240-2005 when a cooling product is used in the SDHV air distribution system.

A Greenheck tubular centrifugal belt drive roof supply fan will be selected for this application. The performance curves and specifications are shown in the catalog sheet. The fan should provide 825 cfm of air over 1.83 in. wg of external static pressure. From the performance curves for the **Greenheck TCBRS-1-09 line of fans, a 2545 rpm speed and a $^3/_4$ hp motor is selected.**

TCB-1-09
Level 1 Construction

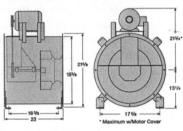

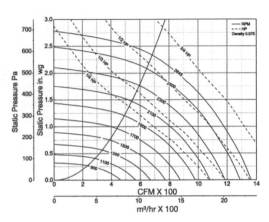

Dimensions shown in inches
Maximum rpm — 2645
Outlet Velocity = cfm X 0.543
Tip Speed = rpm X 2.928
Maximum bhp at a given rpm — $(rpm/3114)^3$
Max Motor Frame Size - 56

CFM	OV	0.125 RPM	0.125 BHP	0.25 RPM	0.25 BHP	0.5 RPM	0.5 BHP	0.75 RPM	0.75 BHP	1 RPM	1 BHP	1.25 RPM	1.25 BHP	1.5 RPM	1.5 BHP	1.75 RPM	1.75 BHP	2 RPM	2 BHP
360	195	846	0.02	999	0.03	1260	0.06												
	Sones	5.3		6.0		7.5													
420	228	941	0.03	1071	0.04	1314	0.07	1526	0.11										
	Sones	5.7		6.3		7.7		9.2											
480	260	1040	0.04	1155	0.05	1380	0.09	1576	0.12	1758	0.17								
	Sones	6.3		6.8		8.0		9.3		10.6									
540	293	1144	0.05	1245	0.06	1449	0.10	1635	0.14	1807	0.19	1968	0.23	2115	0.28				
	Sones	7.1		7.4		8.4		9.5		10.7		12.0		13.3					
600	326	1250	0.06	1341	0.08	1523	0.12	1704	0.16	1865	0.21	2017	0.26	2163	0.31	2298	0.36		
	Sones	7.8		8.1		8.8		9.9		11.0		12.2		13.5		14.7			
660	358	1357	0.07	1439	0.09	1606	0.14	1774	0.18	1930	0.23	2074	0.29	2212	0.34	2346	0.40	2472	0.46
	Sones	8.6		8.8		9.3		10.4		11.4		12.6		13.7		14.9		16.2	
720	391	1465	0.09	1542	0.11	1692	0.16	1845	0.21	1999	0.26	2137	0.32	2270	0.38	2395	0.43	2520	0.50
	Sones	9.4		9.6		10.0		10.9		12.0		13.1		14.0		15.0		16.3	
780	423	1576	0.12	1647	0.14	1786	0.18	1928	0.24	2069	0.29	2206	0.35	2331	0.41	2454	0.47	2570	0.54
	Sones	10.2		10.4		10.8		11.5		12.7		13.6		14.5		15.4		16.5	
840	456	1688	0.14	1752	0.16	1882	0.21	2012	0.27	2142	0.33	2276	0.39	2400	0.45	2515	0.52	2629	0.59
	Sones	11.0		11.2		11.7		12.3		13.3		14.2		15.0		16.0		17.0	
900	489	1801	0.17	1859	0.19	1980	0.25	2100	0.30	2225	0.36	2347	0.43	2469	0.50	2584	0.56		
	Sones	12.0		12.2		12.7		13.3		13.9		14.8		15.7		16.6			
960	521	1914	0.20	1966	0.23	2081	0.28	2195	0.34	2310	0.40	2424	0.47	2540	0.54				
	Sones	13.0		13.2		14.0		14.2		14.7		15.4		16.4					
1020	554	2028	0.24	2075	0.26	2184	0.32	2292	0.38	2395	0.44	2507	0.52	2612	0.59				
	Sones	14.4		14.9		15.0		15.1		15.5		16.3		17.2					
1080	586	2141	0.28	2186	0.31	2288	0.37	2389	0.43	2490	0.50	2592	0.57						
	Sones	15.9		15.9		16.0		16.2		16.6		17.2							
1140	619	2255	0.33	2298	0.36	2394	0.42	2487	0.49	2585	0.55								
	Sones	16.9		16.9		17.1		17.4		17.8									
1200	652	2370	0.38	2410	0.41	2500	0.47	2590	0.54										
	Sones	17.9		18.0		18.3		18.6											

Performance certified is for installation type B: Free inlet, Ducted outlet. Power rating (bhp) does not include drive losses. Performance ratings do not include the effects of appurtenances in the airstream.

The sound ratings shown are loudness values in fan sones at 5 ft. (1.5m) in a hemispherical free field calculated per AMCA Standard 301. Values shown are for installation type B: free inlet fan sone levels.

Source: Greenheck Fan, Corp. (reprinted with permission)

Drawings

The final drawing shows the layout and size of the ducts.

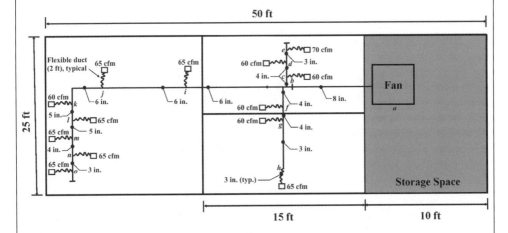

Conclusions

The SDHV air distribution system based circular ducts has been designed. Galvanized sheet metal will be used to fabricate the system. A suitable fan has also been selected. The air velocity in each section of the duct and the air volume flow rate was used to size the ducts. The NC used was a performance metric to support final duct size selections. Dampers will probably be needed to balance the airflow in the system.

Flow across the filter and heating coil will need to occur at a velocity that is much lower than 2350 fpm, which is the velocity in the main duct section that is connected to the fan. This will prevent damage to the filter and mitigate pressure losses across the heating coil. It may be common to find filters complete with the fan assembly, in which the filter static pressure loss would be included with the static pressure that is internal to the fan and its housing. The section of duct that houses the heating coil would likely be larger to reduce the flow velocity across the coil. This will likely be possible because the outlet orifice of the fan is $18^{3}/_{8}$ in., which is much larger than the diameter of the duct that is connected to the fan (8 in.).

Problems

2.1. A design engineer wishes to select an appropriate fan for the following galvanized steel duct system. Estimate the pressure loss for each branch of the duct system.

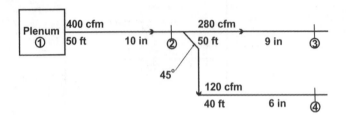

2.2. The duct system shown is one branch of a complete low-velocity air distribution system. The system is a perimeter type, located below the finished floor. The diffuser boots are shown, complete with the pressure losses. Design a round duct system, bearing in mind that a total pressure of 0.21 in. wg is available at the plenum.

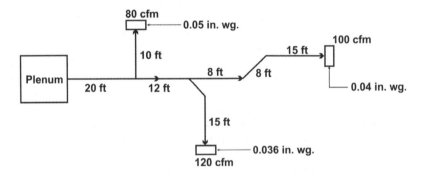

2.3. For most building design projects, the architectural trade tends to be the consultant (i.e., the lead consultant for the project) who hires the mechanical and the electrical trades as subconsultants on the project. In most cases, the mechanical engineering subconsultant has expertise in the design of ductwork to transport air for the purposes of heating and/or cooling an occupied space. The following section of a second floor tenant plan of an office building has been given by an architect.

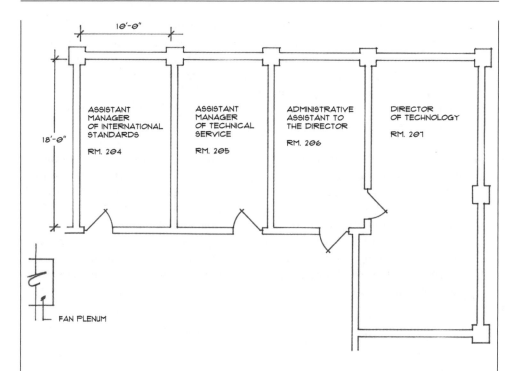

For the offices shown in the plan above (complete with the occupant and work function), the architect has requested the design of a ductwork system to provide air at 75°F to heat the occupied spaces. A HVAC engineer has determined the amount of air required to maintain the space temperature, and they are shown in the following table.

Office Space	Heating Air Requirement (cfm)
Office 204	310
Office 205	450
Office 206	170
Office 207	500

However, the engineer missed the fact that ASHRAE Standard 62 requires that 20 cfm per person of fresh outdoor air must be provided.

(a) To ensure an esthetically pleasing finish in the space, the architect has requested the design of a ductwork system based on round ducts. Because most of the occupants of this section of the floor are managers and/or directors in the complex hierarchy of the client's company, the architect would like to have a dedicated fan installed with the ductwork for this section of offices. The fan is to be located on the roof above the offices, and it will be fitted with a plenum section.

(b) Based on the design of the ductwork, specify the minimum operating condition of the fan.

2.4. A draw-through air-handling unit (AHU) will be used to supply conditioned air as shown in the schematic drawing below. Within the AHU assembly, the filter section has a pressure loss of 0.10 in. wg, the heating/cooling coil section has a pressure loss of 0.20 in. wg, and the casing has a miscellaneous loss of 0.05 in. wg. The AHU is a modular unit complete with a fan that can produce 0.60 in. wg of total pressure at the required design flows. Design a round ductwork system, ensuring that the location of and pressure drops across appropriate dampers for balancing the system is clear for the convenience of the mechanical contractor and the client.

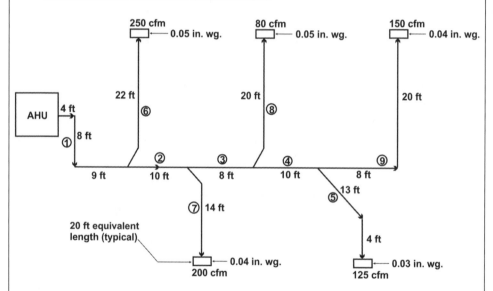

2.5. An *SDHV* system is to be developed to distribute conditioned air to a factory. This type of air distribution system results in smaller duct sizes, and is desired due to space limits and high construction costs. As a guide, Section 3.11 of the AHRI Standard 210/240-2005 [19] requires that a cooling product contain a blower that produces at least 1.2 in. wg of external static pressure when operating at the certified airflow rate of 220–350 cfm per rated ton of cooling. For high-velocity systems for a factory, a maximum velocity of 5000 fpm has been recommended [23]. *Terminal boxes* will be introduced at the duct exit to the space to throttle the air to a low velocity, control the airflow, and reduce noise. The terminal boxes are usually designed to operate at a minimum pressure loss of about 0.25–1.0 in. wg, that is, the branch pressure loss should be on the order of at least 0.25–1.0 in. wg. The cooling product is an AHU, capable of producing 5 tons of cooling. Based on the sketch provided below, design a round duct, high-velocity system to distribute air in the factory. Will the pressure drops across the terminal boxes be sufficient to balance the system? Make appropriate recommendations to the client.

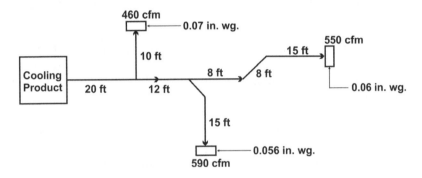

2.6. The National Research Council has decided to pursue research in the area of spray-dried agglomeration of nano-sized powder particles to produce micron-sized powder particles. Safety and health regulations permit only a limited amount of these particles to escape into the ambient air of the space. To that end, the Council has contracted the services of Alliance Engineering Corp. to design a *high-velocity* duct system for a Farr® Gold Series 10 dust collector. The dust collector will draw 4000 cfm of air in an effort to eliminate any powder particles from the space. The air will be drawn through a hood and a filter system, as shown. High efficiency, open-pleat style cartridge filters with flame-retardant media and average pressure loss of 2.7 in. wg were used. A commercial shop environment will be provided by the Council, in which the maximum duct velocity can be on the order of 2500–6000 fpm. Size and specify the round ductwork between the hood and the duct collector. The layout, accessories, and fittings that are chosen should be such that losses are kept at a minimum. Specify an appropriate fan or blower.
Further Information: Given that this is an industrial application, the designer may consider specification of a utility or industrial centrifugal fan.

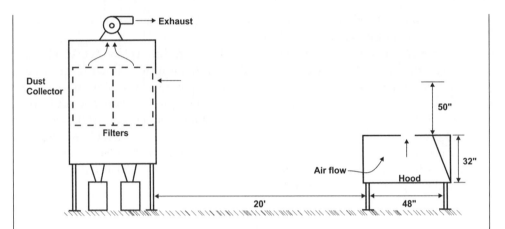

2.7. Refer to Problem 2.2 and redesign the system with rectangular ducts.

2.8. Refer to Problem 2.4 and redesign the system with rectangular ducts. Specify a fan from a manufacturer's catalog.

2.9. Hot combustion gases from a large burner are being considered to heat cold water in a heat exchanger. To facilitate operation and maintenance of the two units, they have been separated and installed individually. The client failed to provide specific information regarding the heat exchanger and conduct verification of the presence of electronics and electrical boards on the unit. As per the 2006 NFPA Standard 31, Section 4.3.6, oil-burning equipment must be installed so that a minimum 3 ft separation is maintained from any electrical panel-board. The design strategy will be to connect a duct to the burner and route it to the bottom of the heat exchanger. It is expected that 25000 lb/h of corrosive combustion gases at 600°F will be transported through the duct after combustion with a low air/fuel ratio. The duct will be routed through the concrete slab of the floor in a trench to provide insulation, support, and protection. Design and layout a low-velocity rectangular duct system.
Further Information: For the system designed, the maximum length of straight duct will depend on the fact that the burner blower cannot provide more than 0.35 in. wg of pressure.

2.10. Refer to Problem 2.5. The client has decided to upgrade the factory space that is serviced by the SDHV system such that it will be classified as a clean room space for use in fabrication of microelectronic devices. To that end, the client wishes to replace the existing terminal boxes with replaceable terminal ceiling filter modules based on HEPA or ultra-low penetration air (ULPA) technology. Specify and select an appropriate fan from a manufacturer's catalog for this application. Details on the recommendation of filter modules should be provided for review by the client. Specify dampers, where required, to balance the airflow in the system. Most manufacturers may provide static pressure loss data for clean, new filters. Static pressure loss increases as the particulate matter accumulates on the filter over time. What impact will this have on operation of the fan? Will it stall? Comment.

2.11. A researcher at a local university has decided to install new equipment in a laboratory booth that will be used to fabricate fiber-reinforced polymer (FRP) composites for the construction industry on a pilot scale. The production process will produce gases (volatile organic compounds, VOCs) and non-flammable small-particle contaminants (carbon fiber particles) that will need to be exhausted. The researcher has engaged a mechanical engineer to design and layout a circular duct exhaust system, complete with a fan and other accessories such as dampers and filters. The researcher is also an engineer and has requested the use of HEPA filters to protect the fan from particle damage and to avoid their discharge to the open external ambient, a sidewall-mounted exhaust fan, and a damper at the inlet to the fan. The elevation plan, complete with the equipment, has been provided by the researcher. Four 10 in. diameter openings in the top of the booth were provided to allow installation of duct-work. The client will supply four exhaust hoods (dimensions: 5 ft long × 3 ft wide opening) for connection to the ductwork system, and as such, selection of the hoods is outside the scope of this problem. The tentative location of the exhaust fan has also been specified by the client in the drawing. Design the required system by referring to the client-supplied drawing of the elevation plan and through consultation with the International Building Code, the International Mechanical Code, and NFPA Standard 704. Most manufacturers may provide static pressure loss data for clean, new filters. Static pressure loss increases as the particulate matter accumulates on the filter over time. What impact will this have on operation of the fan? Will it stall? Comment.

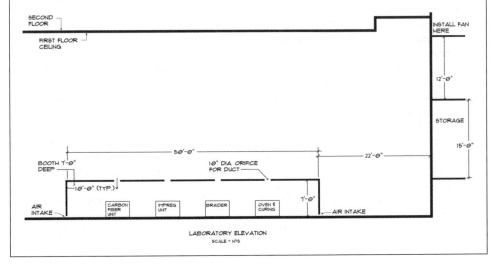

LABORATORY ELEVATION
SCALE = NTS

References and Further Reading

[1] Colebrook, C. (1939) Turbulent flow in pipes, with particular reference to the transition between the smooth and rough pipe laws. *Journal of the Institute of Civil Engineers London*, **11**, 133–156.

[2] Swamee, P. and Jain, A. (1976) Explicit equations for pipe-flow problems. *Proceedings of the ASCE, Journal of the Hydraulics Division*, **102** (HY5), 657–664.

[3] Miller, R. (1996) *Flow Measurement Engineering Handbook*, 3rd edn, McGraw-Hill, Inc., New York.

[4] Haaland, S. (1983) Simple and explicit formulas for the friction factor in turbulent pipe flow, *Journal of Fluids Engineering*, **105**, 89–90.

[5] Blasius, H. (1912) Das Aehnlichkeitsgesetz bei Reibungsvorgängen. *Zeitschrift des Vereines Deutscher Ingenieure*, **56**, 639–643.

[6] Churchill, S. (1977) Friction-factor equation spans all fluid-flow regimes. *Chemical Engineering*, **84**, 91–92.

[7] Petukhov, B. (1970) Heat transfer and friction in turbulent pipe flow with variable physical properties, in *Advances in Heat Transfer*, vol. 6 (eds T. Irvine and J. Hartnett), Academic Press, New York.

[8] Moody, L. (1944) Friction factors for pipe flows. *Transactions of the ASME*, **66**, 671–684.

[9] Howell, R., Sauer Jr. H., and Coad, W. (2005) *Principles of Heating, Ventilating, and Air-Conditioning*, American Society of Heating, Refrigerating, and Air-Conditioning Engineers, Inc., Atlanta, GA.

[10] International Code Council, Inc. (2006) *International Building Code*, Country Club Hills, IL.

[11] International Code Council, Inc. (2006) *International Mechanical Code*, Country Club Hills, IL.

[12] National Fire Protection Association (2012) *Standard for the Installation of Air-conditioning and Ventilating Systems*, Standard 90A, Quincy, MA.

[13] Sheet Metal and Air Conditioning Contractors National Association, Inc. (1985) *HVAC Duct Construction Standards: Metal and Flexible*, Vienna, VA.

[14] National Fire Protection Association (2011) *Standard for Ventilation Control and Fire Protection of Commercial Cooking Operations*, Standard 96, Quincy, MA.

[15] American Society of Heating, Refrigerating, and Air-Conditioning Engineers (2011) *Ventilation for Commercial Cooking Operations*, Standard 154-2011, Atlanta, GA.

[16] Carrier Corporation (1991) *Fan Application Manual*, Syracuse, NY.

[17] Pritchard, P. (2011) *Fox and McDonald's Introduction to Fluid Mechanics*, 8th edn, John Wiley & Sons, Hoboken, NJ.

[18] Çengel, Y. and Cimbala, J. (2010) *Fluid Mechanics: Fundamentals and Applications*, 2nd edn, McGraw Hill Companies, Inc., New York.

[19] Air-Conditioning, Heating, and Refrigeration Institute (formerly Air-Conditioning and Refrigeration Institute) (2005) *Standard for Performance Rating of Unitary Air-Conditioning and Air-Source Heat Pump Equipment*, Standard 210/240, Arlington, VA.

[20] American Society of Heating, Refrigerating, and Air-Conditioning Engineers, ASHRAE (2011) *Heating, Ventilating, and Air-Conditioning Applications*, Atlanta, GA.

[21] Kingsbury, H. (1995) Review and revision of room noise criteria. *Noise Control Engineering Journal*, **43**, 65–72.

[22] Carrier Corporation (1972) *System Design Manual: Part 2 – Air Distribution*, Chart 8, Syracuse, NY.

[23] Carrier Air Conditioning Co. (1974) *System Design Manual, Part 2: Air Distribution*, Syracuse, NY.

3

Liquid Piping Systems

3.1 Liquid Piping Systems

Piping systems are used to transport diverse liquids for a variety of different applications. These applications may range from water service for buildings to complex two-phase flow systems in industrial plants. The design of these systems requires consideration of several groups of specialties and accessories that will be needed for a functional system. This chapter will focus on fittings and accessories, pipe materials, fluid machines, and design considerations necessary for the successful design of practical piping systems for various applications.

3.2 Minor Losses: Fittings and Valves in Liquid Piping Systems

3.2.1 Fittings

Fittings are used to extend pipe lengths, expand the pipe network, or perform a selected function. Examples of fittings specific to liquid piping systems are plugs, unions, wyes, valves, tees, caps, ferrules, elbows, nipples, reducers, sleeves, couplings, adapters, fasteners, compression fittings, and bulkhead fittings, to name a few. All these add resistance to fluid flow. Tabulated K values (loss coefficients) are available for these and other fittings.

3.2.2 Valves

Valves are used to control the flow rate of fluid in piping systems. For valves, lower K values occur when they are fully open; thus, frictional losses will be low. The K values will increase as the valve is closed. A similar trend will apply to L_{equiv} values. There are many types of valves. Some of these valves and their drawing symbols are shown in Figure 3.1. Additional information and typical K values for valves are provided in

Introduction to Thermo-Fluids Systems Design, First Edition. André G. McDonald and Hugh L. Magande.
© 2012 André G. McDonald and Hugh L. Magande. Published 2012 by John Wiley & Sons, Ltd.

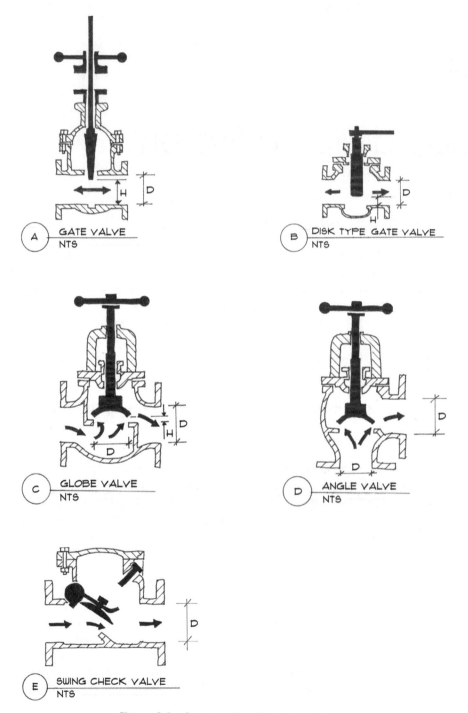

Figure 3.1   Some typical industrial valves

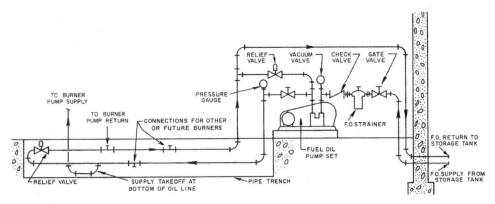

Figure 3.2 A typical fuel oil piping system complete with a pump set (*ASHRAE Handbook*, Fundamentals Volume, 2005; reprinted with permission)

Table A.14. Manufacturer's catalogs should be consulted to find other valves and/or pipe fittings and their K or L_{equiv} values.

3.2.3 A Typical Piping System—A Closed-Loop Fuel Oil Piping System

The schematic drawing shown in Figure 3.2 depicts a typical fuel oil piping system c/w a pump set. Note all the fittings and valves. Note also the symbols used for different equipment. Additional symbols are provided in Appendix B. The design engineer should be familiar with this type of drawing for piping systems.

Practical Note 3.1   Link Seals

Absent are link seals to protect the pipes that penetrate the concrete walls. The seals are used to fill voids between the pipe and the barrier through which it passes.

3.3 Sizing Liquid Piping Systems

3.3.1 General Design Considerations

Sizing liquid piping follows a similar procedure to that outlined in Example 2.1 of Chapter 2. Consider the following additional points when sizing and designing piping systems.

(A) *Pipe Materials*: Is the fluid corrosive? Are there particulates in the fluid (e.g., oil sand liquid slurries) that will erode the pipe? Is the fluid temperature high?
(B) *Pipe Thickness*: Is higher pipe strength required for high stress (high pressure) applications?

Table 3.1 Typical average velocities for selected pipe flows[a]

Fluid	Application	Velocity (fps)	Velocity (m/s)
Steam	Superheated process steam	148–328	45–100
	Auxiliary heat steam	98–246	30–75
	Saturated and low-pressure steam	98–164	30–50
Water	Centrifugal pump suction lines	3–4.9 (must be < 4.9 fps)[b]	0.9–1.5 (must be <1.5 m/s)[b]
	Power plant feedwater	7.9–15	2.4–4.6
	General building service	3.9–10.2	1.2–3.1
	Potable water	Up to 6.9 (must be < 9.8 fps)[b]	Up to 2.1 (must be <3.0 m/s)[b]

[a] Adapted from the US Department of the Army, TM 5-810-15, Central Boiler Plants, August 1995.
[b] Adapted from 2005 Fundamentals, American Society of Heating, Refrigerating, and Air-Conditioning Engineers, Atlanta, GA, 2005, pp. 36–11.

(C) *Plastic Piping*: Can plastic piping be used instead of metal pipes? Plastic pipes are lightweight, easy to join, and corrosion resistant.

Examples of Plastic Piping

(a) *Polyvinyl chloride (PVC)*: For applications in pipes for building cold water, drains, and condensate piping. **Not recommended** for hot water piping.
(b) *Chlorinated polyvinyl chloride (CPVC)*: Similar to PVC. This plastic material is able to withstand temperatures of up to 140°F.
(c) *Reinforced thermosetting resin plastic (RTRP)*: Recommended for hot water piping systems with temperatures on the order of 200°F.
(d) *Cross-linked polyethylene (PEX)*: For applications in hydronic-radiant heating systems, domestic water piping, natural gas and offshore oil applications, chemical transportation, and transportation of sewage and slurries. Recently, it has become a viable alternative to PVC, CPVC, and copper tubing for use as residential water pipes (particularly in Canada).
(e) *Acrylonitrile butadiene styrene (ABS)*: Used in building plumbing systems as drain, vent, and sewage piping. It may also be used to transport potable (drinking) water, chemicals, or chilled water.

(D) *Flow Velocities*: Different services and applications have different pipe velocity requirements and ranges. Table 3.1 gives typical average velocities for selected pipe flows.

Practical Note 3.2   Piping Systems Containing Air

Piping systems that contain dissolved air (such as cold water supply piping systems) require velocities that are lower and on the order of 4–8 fps or 1.2–2.4 m/s. Lower pump suction velocities are typically desired to minimize air entrapment in the fluid. Entrapped air will result in a drop in pump pressure, flow rate, power, and performance.

3.3.2 Pipe Data for Building Water Systems

As seen in Example 2.1 of Chapter 2, pipe and duct sizing can be laborious, requiring many iterations of complex correlation equations. In practice for building water systems, charts and tables are used to simplify the process of liquid pipe sizing. Of importance is pipe material selection, determination of pipe diameter, and installation specifications. While the focus is on building water systems, the procedures outlined here will apply directly to other types of piping systems such as wastewater/sludge systems, food processing systems, and chemical liquid systems.

(A) *Pipe Materials for Building Water Systems*: Type L copper tubing is widely used. Schedule 40 steel is used in steam heating systems. Schedule 80 steel, which is thicker than Schedule 40 steel, is used in high-pressure steam lines. PEX may also be used in cold or hot water systems in lieu of copper. It is **not recommended** for high-pressure steam lines. Table 3.2 shows some data for copper and steel pipes. Consult Tables A.6 through A.11 for more extensive information for a variety of materials, and refer to it often.

(B) *Pipe Sizing Considerations—Determination of Pipe Diameters*: Pipes in closed-loop building systems should be designed to limit friction losses to **3 ft of water per 100 ft of pipe**. This will reduce the final pump size and cost.

In addition, pipe velocities should be **less than 10 fps** or 3 m/s to reduce noise and pipe material erosion. Typically, volume flow rates in the pipes are known, which makes determination of the pipe diameter easier.

Practical Note 3.3   Higher Pipe Friction Losses and Velocities

The design engineer should bear in mind that the aforementioned points are guidelines. Flexibility exists in order to optimize the performance of a given system design. For example, higher friction losses may be acceptable for shorter run of pipes or for cases where smaller pipe sizes are mandatory by the client or application. Higher pipe velocities may be required for fluid systems that transport solid sediments. This ensures that the sediments do not clog the pipes by sticking to the pipe wall. However, for small diameter pipes, erosion becomes a concern when large velocities are used. For example, in stainless steel tubes, erosion becomes a concern when velocities are larger than 15 fps (consult Table A.13 for additional data). All deviations from the established guidelines should be justified by the design engineer.

After selection of the pipe friction loss and velocity, an appropriate chart may be used to determine the pipe diameter. One such chart for plastic piping systems is shown in Figure 3.3. Consult Appendix A for additional and more extensive charts.

(A) *Pipe Installation*: Pipes are typically hung between the slab and the dropped (finished) ceiling and are supported by **pipe hangers**. The hangers are fastened around the pipe and their support rods are anchored to the ceiling. Figure 3.4

Table 3.2   Pipe data for copper and steel

Material	Nominal	Inner	Outer	Weight per Linear Foot of Pipe and Water (lb)	Gallons of Water per Linear Foot
		Diameter (in.)			
Copper					
Type L	$3/8$	0.430	0.500	0.26	0.008
Type L	$1/2$	0.545	0.625	0.39	0.012
Type L	$3/4$	0.785	0.875	0.67	0.025
Type L	1	1.025	1.125	1.01	0.043
Type L	$1^{1}/4$	1.265	1.375	1.43	0.065
Type L	$1^{1}/2$	1.505	1.625	1.91	0.093
Type L	2	1.985	2.125	3.09	0.161
Type L	$2^{1}/2$	2.465	2.625	4.55	0.248
Type L	3	2.945	3.125	6.29	0.354
Type L	$3^{1}/2$	3.425	3.625	8.29	0.479
Type L	4	3.905	4.125	10.58	0.622
Steel					
Schedule 40	$1/4$	0.364	0.540	0.475	0.005
Schedule 40	$1/2$	0.622	0.840	0.992	0.016
Schedule 40	$3/4$	0.824	1.050	1.372	0.028
Schedule 40	1	1.049	1.315	2.055	0.045
Schedule 40	$1^{1}/4$	1.380	1.660	2.929	0.077
Schedule 40	$1^{1}/2$	1.610	1.900	3.602	0.106
Schedule 40	2	2.067	2.375	5.114	0.174
Schedule 40	$2^{1}/2$	2.469	2.875	7.873	0.248
Schedule 40	3	3.068	3.500	10.781	0.383
Schedule 40	$3^{1}/2$	3.548	4.000	13.397	0.513
Schedule 40	4	4.026	4.500	16.316	0.660
Schedule 80	$1/2$	0.546	0.840	1.189	0.012
Schedule 80	$3/4$	0.742	1.050	1.686	0.026
Schedule 80	1	0.957	1.315	2.483	0.037
Schedule 80	$1^{1}/4$	1.278	1.660	3.551	0.067
Schedule 80	$1^{1}/2$	1.500	1.900	4.396	0.092
Schedule 80	2	1.939	2.375	6.302	0.154
Schedule 80	$2^{1}/2$	2.323	2.875	9.491	0.220
Schedule 80	3	2.900	3.500	13.122	0.344
Schedule 80	$3^{1}/2$	3.364	4.000	16.225	0.458
Schedule 80	4	3.826	4.500	19.953	0.597

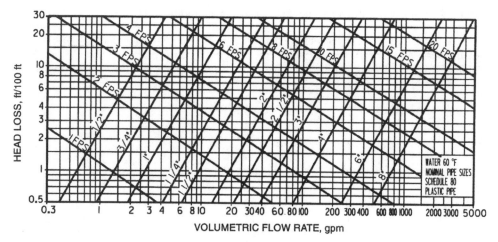

Figure 3.3 Plastic pipe (Schedule 80) friction loss chart (*ASHRAE Handbook*, Fundamentals Volume, 2005; reprinted with permission)

shows pipes supported on hangers. The weights per foot of piping filled with water are used to determine the spacing of the pipe hangers and the sizing of the support rods. Shown below (Tables 3.3 and 3.4) is data adapted from an Erico (Solon, OH, USA) catalog for steel and copper hangers for water or steam service.

Important Note: The hanger data presented in Tables 3.3 and 3.4 do not apply when span calculations are made or where concentrated loads from pipe accessories are present between the supports. Changes in pipe direction will require additional support. Section 308 of the International Plumbing Code provides further details on pipe hanger spacings.

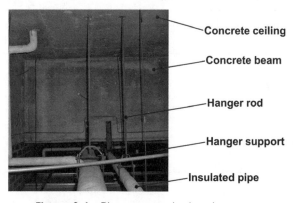

Figure 3.4 Pipes supported on hangers

Table 3.3 Hanger spacing for straight stationary pipes and tubes [1]

Nominal Pipe Diameter (in.)	Steel Pipe (ft)		Copper Pipe (ft)	
	Water Service	Steam Service	Water Service	Steam Service
$1/4$	7	8	5	5
$3/8$	7	8	5	6
$1/2$	7	8	5	6
$3/4$	7	9	5	7
1	7	9	6	8
$1^1/4$	7	9	7	9
$1^1/2$	9	12	8	10
2	10	13	8	11
$2^1/2$	11	14	9	13
3	12	15	10	14
$3^1/2$	13	16	11	15
4	14	17	12	16
5	16	19	13	18
6	17	21	14	20
8	19	24	16	23
10	20	26	18	25
12	23	30	19	28
14	25	32		
16	27	35		
18	28	37		
20	30	39		
24	32	42		
30	33	44		

Table 3.4 Minimum hanger rod size for straight stationary pipes and tubes [1]

Nominal Pipe Diameter (in.)	Minimum Rod Size (in.)
$1/4$–2	$3/8$
$2^1/2$–$3^1/2$	$1/2$
4–5	$5/8$
6	$3/4$
8–12	$7/8$
14–18	1
20–24	$1^1/4$

Figure 3.5   Pipes and an in-line pump mounted on brackets

Practical Note 3.4   Piping System Supported by Brackets

Not all pipes are installed by being hung on pipe hangers and supports. Pipes may be supported by mounting them on **brackets** that are attached to concrete walls. Shown in Figure 3.5 is a part of a piping system that is installed on stainless steel strut channels. A $1/4$-hp (horsepower) in-line pump is also supported by the pipes and the channels.

Example 3.1   Sizing and Installation of Pipes

Size and specify an appropriate pipe for a hot water heating system with flow rate of 10 gpm. The system will be used in a small building. Provide installation guidelines to the mechanical contractor.

Possible Solution

Definition

Size and specify a pipe for a small building hot water heating system.

Preliminary Specifications and Constraints

  (i) The pipe material should withstand high-temperature water.
 (ii) Flow rate limited to 10 gpm.
(iii) The system will be used in a small building.

Detailed Design

Objective

To size and specify an appropriate pipe.

Data Given or Known

(i) The water flow rate is 10 gpm.

Assumptions/Limitations/Constraints

(i) Choose type L copper tubing. This is most common in building piping systems, is used readily for hot water heating systems, and is widely available to contractors.
(ii) Limit the pipe friction losses to 3 ft of water per 100 ft of pipe, as per industry standard.
(iii) Limit the water velocity to 6 fps. The erosion limit of water in small copper tubes is 6 fps. Further, this velocity limit is within the range for general building service (4–10 fps).

Sketch

A drawing is not needed.

Analysis

Figure A.3 presents a copper tubing friction loss chart. From this chart, some possible pipe sizes, velocities, and friction losses are

Option *a*: 1 in. nominal, $1\frac{1}{8}$ in. OD, 4 fps, 7 ft of water per 100 ft of pipe
Option *b*: $1\frac{1}{4}$ in. nominal, $1\frac{3}{8}$ in. OD, 2.7 fps, 2.5 ft of water per 100 ft of pipe
Option *c*: $1\frac{1}{2}$ in. nominal, $1\frac{5}{8}$ in. OD, 1.7 fps, 1.2 ft of water per 100 ft of pipe

Option *a* is immediately discarded because the friction loss exceeds 3 ft of water per 100 ft of pipe.

Drawings

No drawings are required.

Conclusions

Options *b* and *c* meet the design requirements and satisfy the constraints. Since smaller pipes will result in lower material and installation costs, **option b is chosen**.
Therefore, choose

Copper type L, $1\frac{1}{4}$ in. nominal, $1\frac{3}{8}$ in. OD

Installation Guidelines and Specifications

Tables 3.3 and 3.4 should be consulted to prepare the installation specifications:

The contractor shall use copper hangers and rods. The hangers shall be spaced 7 ft apart. The rod size shall be $\frac{3}{8}$ in.

> Copper hangers were chosen to eliminate any thermal expansion coefficient differences between the pipe and the hanger. This could generate severe stresses on the pipes. Also, similar pipe and hanger material will eliminate the possibility of corrosion (galvanic/dissimilar metal corrosion).

3.4 Fluid Machines (Pumps) and Pump–Pipe Matching

3.4.1 Classifications and Terminology

Pumps serve to move liquids. The pump adds energy to the fluid to keep it moving, to overcome head losses, and/or to build pressure on the fluid to overcome elevation head in the line. As the fluid passes through the pump and energy is added, the **discharge pressure** of the exiting fluid becomes greater than the pressure of the inlet fluid. There are many types of pumps available on the market. The *Pump Handbook* [2] should be consulted for detailed information on pumps.

3.4.2 Types of Pumps

(a) *Gas pumps*: These pumps move gases. Examples of these types of pumps are fans and compressors.
(b) *Positive displacement pumps*: These pumps pressurize the fluid by contracting or changing their boundaries to force fluid to flow. Suction is achieved when the boundaries of the pump expand or the volume of the pump becomes larger. Examples of positive displacement pumps include the human heart, flexible-tube peristaltic pumps, and double-screw pumps. These pumps are capable of creating a significant vacuum pressure at their inlets, even when dry, and are able to lift fluids from long distances below the pump. These pumps are also called **self-priming pumps**.
(c) *Dynamic pumps*: Rotating blades are used to supply energy to the fluid. The blades impart momentum to the fluid. Fluid enters the eye of the impeller, is flung from the impeller blades into the scroll case (volute) of the pump where the fluid is pressurized. The fluid exits at high pressure.

Figure 3.6 shows schematic cross-sectional drawings of some typical pumps encountered in engineering practice. Figure 3.6a and 3.6b show examples of positive displacement pumps and Figure 3.6c shows an example of a dynamic pump (centrifugal pump).

3.4.3 Pump Fundamentals

There are several fundamental parameters that need to be mentioned before proceeding to pump sizing and selection.

(a) *Pump Capacity*: Volume flow rate of fluid through the pump:

$$\dot{V} = \frac{\dot{m}}{\rho}. \tag{3.1}$$

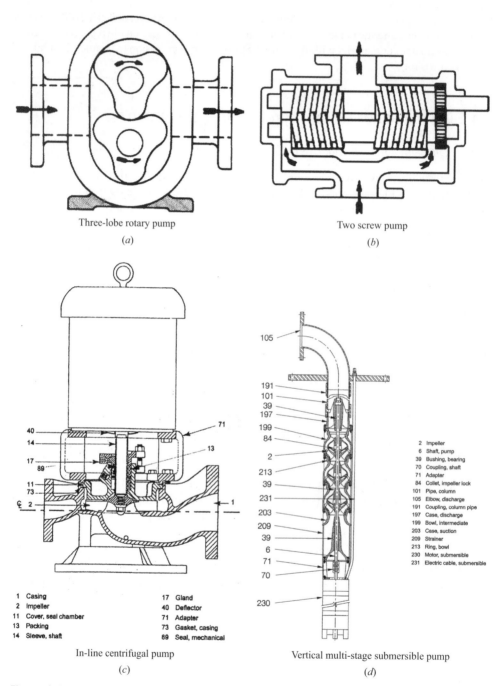

Three-lobe rotary pump

(*a*)

Two screw pump

(*b*)

1	Casing	17	Gland
2	Impeller	40	Deflector
11	Cover, seal chamber	71	Adapter
13	Packing	73	Gasket, casing
14	Sleeve, shaft	89	Seal, mechanical

In-line centrifugal pump

(*c*)

2	Impeller
6	Shaft, pump
39	Bushing, bearing
70	Coupling, shaft
71	Adapter
84	Collet, impeller lock
101	Pipe, column
105	Elbow, discharge
191	Coupling, column pipe
197	Case, discharge
199	Bowl, intermediate
203	Case, suction
209	Strainer
213	Ring, bowl
230	Motor, submersible
231	Electric cable, submersible

Vertical multi-stage submersible pump

(*d*)

Figure 3.6 Types of industrial pumps: (a) three-lobe rotary pump; (b) two-screw pump; (c) in-line centrifugal pump; (d) vertical mutistage submersible pump (Hydraulic Institute, Parsippany, NJ, www.pumps.org; reprinted with permission)

(b) *Pump Net Head*: Used to increase the fluid energy (usually in units of length)

An expression for the pump net head can be found by considering the energy equation.

Thus, the pump specific work is

$$w_{\text{pump}} = \left(\frac{p_2}{\rho} + \frac{V_2^2}{2} + gz_2\right) - \left(\frac{p_1}{\rho} + \frac{V_1^2}{2} + gz_1\right) + h_{\text{IT}}, \qquad (3.2)$$

where points 1 and 2 are points in the pipe system chosen by the engineer.

Dividing by g gives the pump net head (in units of length):

$$H_{\text{pump}} = \frac{w_{\text{pump}}}{g} = \left(\frac{p_2}{\rho g} + \frac{V_2^2}{2g} + z_2\right) - \left(\frac{p_1}{\rho g} + \frac{V_1^2}{2g} + z_1\right) + H_{\text{IT}}. \qquad (3.3)$$

If a **control volume** were drawn around the pump only, the following could be assumed:

$H_{\text{IT}} = 0$ (applies to pipe head loss, only); point 1 is the pump inlet; point 2 is the pump outlet; $z_2 = z_1$ (horizontally mounted pump); $D_2 = D_1$ (assuming suction and discharge pipe diameters are equal); $V_2 = V_1$ (no area changes yields equal velocities).

Therefore,

$$H_{\text{pump}} = \frac{p_2}{\rho g} - \frac{p_1}{\rho g} = \frac{p_2 - p_1}{\rho g} = \frac{p_{\text{outlet}} - p_{\text{inlet}}}{\rho g} = \frac{\Delta p_{\text{rise}}}{\rho g}. \qquad (3.4)$$

The energy required to generate a pressure rise across the pump is directly related to the pump net head.

(c) *Water Horsepower*: Power delivered directly to the fluid by the pump:

$$\dot{W}_{\text{water horsepower}} = \dot{m} g H_{\text{pump}} = \rho g \dot{V} H_{\text{pump}}. \qquad (3.5)$$

(d) *Brake Horsepower* (bhp): External power supplied to the pump by a mechanical shaft or an electrical motor.

For a rotating shaft that supplies the bhp:

$$\text{bhp} = \dot{W}_{\text{pump,shaft}} = \omega T_{\text{shaft}}, \qquad (3.6)$$

where ω is the rotational speed and T_{shaft} is the torque generated by the shaft.

(e) *Pump efficiency*: Ratio of useful power to supplied power:

$$\eta_{\text{pump}} = \frac{\dot{W}_{\text{water horsepower}}}{\text{bhp}} = \frac{\rho g \dot{V} H_{\text{pump}}}{\omega T_{\text{shaft}}}. \qquad (3.7)$$

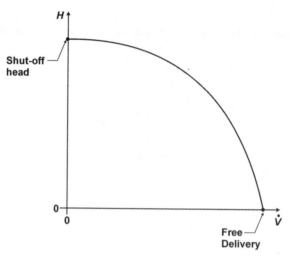

Figure 3.7 Schematic of a H_{pump} versus $\dot{V}$ curve for a centrifugal pump

3.4.4 Pump Performance and System Curves

Curves of H_{pump}, η_{pump}, bhp as functions of $\dot{V}$ are called **pump performance curves**. Consider the H_{pump} versus $\dot{V}$ curve shown in Figure 3.7 for a centrifugal pump. As the pump net head increases, total resistance to flow increases and the volume flow rate decreases. At the **shut-off head**, the pump net head is a maximum ($H_{\text{pump}} = H_{\text{pump,max}}$) and the volume flow rate or pump capacity is zero ($\dot{V} = 0$). There is no fluid flow at the pump shut-off head.

As the pump net head decreases, total resistance to flow decreases and the volume flow rate increases. At **free delivery**, the pump net head is zero ($H_{\text{pump}} = 0$) and the volume flow rate or pump capacity is a maximum ($\dot{V} = \dot{V}_{\text{max}}$). No energy is added to the fluid, and the fluid flows through the pump as though it were a pipe. Consider the η_{pump} versus $\dot{V}$ curve shown in Figure 3.8. At the pump shut-off head, $\dot{V} = 0$. Therefore,

$$\eta_{\text{pump}} = \frac{\rho g \dot{V} H_{\text{pump}}}{\omega T_{\text{shaft}}} = \frac{\rho g \, (0) \, H_{\text{pump}}}{\omega T_{\text{shaft}}} = 0. \tag{3.8}$$

The pump efficiency is zero.

At free delivery, $H_{\text{pump}} = 0$. Therefore,

$$\eta_{\text{pump}} = \frac{\rho g \dot{V} H_{\text{pump}}}{\omega T_{\text{shaft}}} = \frac{\rho g \dot{V} \, (0)}{\omega T_{\text{shaft}}} = 0. \tag{3.9}$$

The pump efficiency is zero.

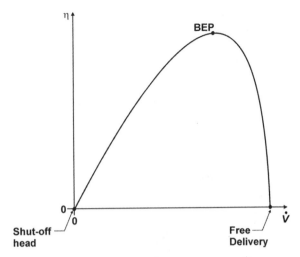

Figure 3.8 Schematic of a η_{pump} versus $\dot{V}$ curve

The **best efficiency point (BEP)** of the pump is the maximum pump efficiency ($\eta_{pump,max}$). Note that the performance curve will show the pump capacity that produces the BEP.

When selecting a pump, the design engineer needs to determine the total amount of head that is required to overcome all the losses in the piping system, build up the pressures required by the design, overcome elevation differences, and increase the fluid velocity (as required) across selected points in the system.

Thus, between two points, 1 and 2, the required head for the system is

$$H_{pump,required} = \left(\frac{p_2}{\rho g} + \frac{V_2^2}{2g} + z_2\right) - \left(\frac{p_1}{\rho g} + \frac{V_1^2}{2g} + z_1\right) + H_{lT}. \tag{3.10}$$

The pump performance curve shows the amount of head that is available ($H_{pump,available}$) and can be delivered by a specific pump. The pump head required must be calculated, and the performance curve must be checked to determine if the selected pump can provide the required head. Calculated values of the required pump head ($H_{pump,required}$) at different pump capacities (flow rates) can be plotted alongside the performance curve. These curves of $H_{pump,required}$ versus $\dot{V}$ are called **system curves**. The point of intersection between the pump performance curve and the system curve is the **operating point of the pipe system**.

Therefore, at the operating point,

$$H_{pump,available} = H_{pump,required}. \tag{3.11}$$

This indicates that the selected pump can provide the total head required by the system.

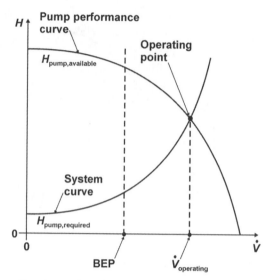

Figure 3.9 Schematic of a system curve intersecting a pump performance curve

Below is a schematic of a system curve intersecting a pump performance curve at the system operating point (Figure 3.9). This performance curve is for a pump of a fixed impeller diameter and rotational speed.

Practical Note 3.5   Manufacturers' Pump Performance Curves

The pump performance curves must be determined experimentally by the pump manufacturers. These are usually available in equipment catalogs online or in hard copy. The curves are usually (almost always) based on water as the working fluid. The system curve or the operating point must be determined by the design engineer by considering the pipe system flow rate and the total system pump head required ($H_{pump,required}$).

3.4.5 Pump Performance Curves for a Family of Pumps

Manufacturers will typically (almost always) provide a group of pump performance curves for a group of pumps, all on one plot. Below in Figure 3.10 is a schematic of a group of performance curves for a family of "**geometrically similar**" pumps.
   Note the following points:

(a) D_1, D_2, D_3 are impeller diameters of each pump.
(b) The casing enclosure is the same for each pump to satisfy the requirement of geometric similarity.
(c) Different pump efficiencies are shown. Note the shape of the efficiency curves. The BEP is as shown.

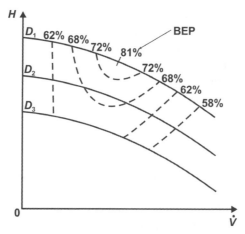

Figure 3.10 Performance curves for a family of geometrically similar pumps

(d) These curves will be determined experimentally by the manufacturer for a specific liquid. Typically, the liquid is water.

3.4.6 A Manufacturer's Performance Plot for a Family of Centrifugal Pumps

Figure 3.11 shows a real pump performance plot from a catalog provided by Taco Inc. Additional pump performance plots are provided in Appendix A for Bell and Gosset

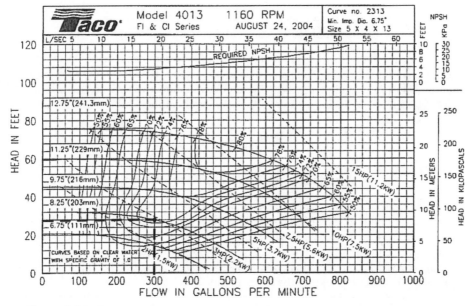

Figure 3.11 Pump performance plot (Taco, Inc.; reprinted with permission)

centrifugal pumps. The onus is on the design engineer to consult the websites of other pump manufacturers to select appropriate pumps that meet the requirements of the system design, should the Bell and Gosset pumps not meet the requirements of a particular system design.

Note the following points:

(a) *Model Information:* Model 4013, FI and CI Series centrifugal pumps.
(b) *Pump Speed:* 1160 rpm.
(c) *Casing Size:* 5 in. × 4 in. × 13 in. (5 in. suction diameter, 4 in. discharge diameter, 13 in. casing).
(d) *Identify the Axes:* Pump head available on the y-axis and pump capacity on the x-axis.
(e) Provided are performance curves for each impeller diameter.
(f) The BEP is approximately 80% for this family of pumps.
(g) Provided are curves for the bhp. The minimum pump power is 2 hp and the maximum pump power is 15 hp for this family of pumps.

Consider the following scenario:
Select an appropriate Taco pump to provide 300 gpm of fluid ($\dot{V}$) and overcome 28 ft of head ($H_{pump,required}$).

$$\text{Select a pump with:} \quad D_{impeller} = 8.25 \text{ in.}$$
$$\eta_{pump} = 74\%$$
$$bhp = 3 \text{ hp.}$$

Practical Note 3.6 "To-the-point" Design

Choosing a 3-hp motor for the pump above would be recommended for **"to the point"** design. However, what would happen if the flow rate spiked to 325 gpm (an 8% increase in flow rate), which is possible in practice? **The motor would overload.** Hence, for this design, we would select a 5-hp motor, which would be **nonoverloading** for the entire pump curve.

Practical Note 3.7 Oversizing Pumps

Do not oversize your pumps simply to be safe. After selecting the next larger motor size to ensure that the pump would be nonoverloading for the entire pump curve, do not proceed to oversize the pump further to be safe. Hence, do not choose a 10-hp motor, when a 5-hp motor is sufficient to provide nonoverloading. At 300 gpm, all that is needed is 3–5 hp. Choosing a 10-hp motor would simply waste energy, increase the installation cost, and require extra resources to support the larger pump.

Example 3.2   Pump Selection from Manufacturer's Performance Plots

You are a design engineer charged with the responsibility of selecting an appropriate in-line mounted centrifugal pump for an application. The flow rate will be constant at 80 gpm, but the required head could vary between 20 and 40 ft. Select an appropriate Bell & Gossett Series 60 pump for this application.

Solution. A study of each performance plot from the large Bell & Gossett Series 60 centrifugal pump catalog to find an appropriate pump could be time consuming. A **master pump selection chart** is used to identify a family (or families) of pumps that will meet the requirements of the design. The master pump selection chart for the Bell & Gossett Series 60 centrifugal pumps is shown below (from Appendix A).

SELECTION CHART

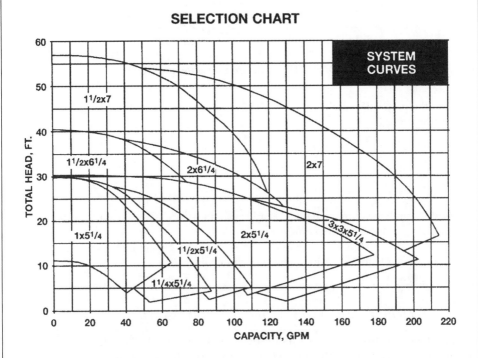

From the master pump selection chart, the family of pumps with a $1\frac{1}{2}$ in. $\times$ 7 in. casing is chosen. This family of pumps covers the range of interest for the head loss (20–40 ft).

The pump impeller diameter, motor power, and efficiency are found from the performance plot for the $1\frac{1}{2}$ in. $\times$ 7 in. family of pumps. The plot is shown below.

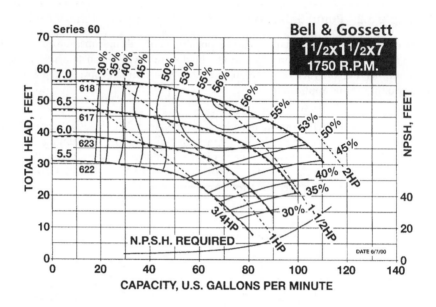

From the performance plot, and for $\dot{V} = 80$ gpm and $H_{\text{pump}} = 40$ ft (to cover the entire range of interest):

$D_{\text{impeller}} = 7.0$ in.
$\eta_{\text{pump}} \approx 55\%$
bhp ≈ 2 hp.

For this family of pumps, the BEP is approximately 56%. The efficiency of the selected pump is close to the maximum efficiency.

"To-the-point" design would require the selection of a $1\frac{1}{2}$-hp motor for this pump. To avoid motor overloading, a 2-hp motor was chosen instead.

3.4.7 Cavitation and Net Positive Suction Head

Suction into a pump may occur as reduced pressure and high velocity is created to promote the movement of fluid into the pump. The inlet pressures on the suction side of the pump may be significantly lower than the discharge pressures, and may be lower than atmospheric pressure. In some cases in liquid pumps, the inlet pressure to the pump may become lower than the **vapor pressure** of the liquid at the operating temperature. When this occurs, the liquid will vaporize to form bubbles. Pressure drop across the pump inlet passage or losses in the pump impeller may also decrease the pump inlet pressure to values lower than the vapor pressure of the fluid.

Hence, vapor bubbles will form in liquid pumps if

$$P_{\text{inlet}} < P_{\text{vapor,liquid}}. \tag{3.12}$$

In a liquid pump, these vapor-filled bubbles are called **cavitation bubbles**. In the high-pressure regions of the pump, these cavitation bubbles will collapse, resulting in damage to the pump and reduction in pump performance. Other negative consequences of cavitation include

(a) noise;
(b) vibration;
(c) pump efficiency reduction;
(d) pitting and erosion of pump impeller, casing, and blades due to high-pressure explosion of the bubbles.

The presence of vapor in the impeller can also result in a loss of pump pressure rise. A study of Equation (3.4) shows that, due to the low density of the vapor, the pressure rise across the pump will be small. In effect, for a vapor-filled impeller, a total loss of pump performance will occur.

It is necessary to ensure that $P_{inlet} > P_{vapor,liquid}$ to avoid cavitation and pump performance loss. **NPSH** can be used to verify if this requirement will be met.

Therefore, in units of length,

$$\text{NPSH} = \left(\frac{p}{\rho g} + \frac{V^2}{2g} \right)_{pump,inlet} - \frac{P_{vapor}}{\rho g}. \tag{3.13}$$

Pump manufacturers will provide the NPSH required (NPSHR) on the performance plots for a family of pumps (NPSHR vs. $\dot{V}$), and is determined experimentally by the manufacturer. This NPSHR value must be compared with the calculated NPSH.

Thus, to avoid cavitation,

$$\text{NPSHR} < \text{NPSH}. \tag{3.14}$$

The NPSH must be calculated by the design engineer for comparison with the NPSHR to ensure that sufficient NPSH is available.

Practical Note 3.8   NPSH

It is common to see NPSH referred to as net positive suction head available (NPSHA). The NPSHA is always calculated.

Example 3.3   Net Positive Suction Head

A centrifugal *Peerless Pump*® Type 4AE11 is tested at 1750 rpm using the flow system shown. The water level in the inlet reservoir is 3.5 ft above the pump centerline; the inlet line consists of 6 ft of 4-in. diameter Class 150 cast-iron pipe, a standard elbow, and a fully open gate valve. Calculate the NPSHA at the pump inlet. The volume flow rate is given as 1200 gpm of water at 60°F. Will cavitation occur at this temperature?

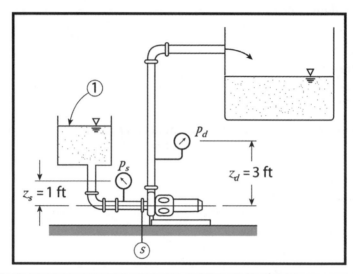

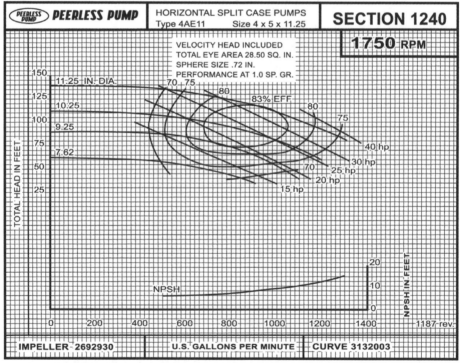

Source: Peerless Pump Company (reprinted with permission)

Solution. The NPSHA is

$$\mathrm{NPSHA} = \left(\frac{p_s}{\rho g} + \frac{V_s^2}{2g} \right)_{\text{pump,inlet}} - \frac{P_{\text{vapor}}}{\rho g}.$$

The energy equation will be used to find the pump inlet pressure, p_s:

$$\frac{P_1}{\rho g} + \alpha_1 \frac{V_1^2}{2g} + z_1 = \frac{p_s}{\rho g} + \alpha_s \frac{V_s^2}{2g} + z_s + H_{\text{lT}}.$$

Point 1 is the free surface of the water in the reservoir. Thus, $V_1 \approx 0$ and $P_1 = P_{\text{atm}}$. Therefore,

$$\frac{p_s}{\rho g} = \frac{P_{\text{atm}}}{\rho g} - \alpha_s \frac{V_s^2}{2g} + (z_1 - z_s) - H_{\text{lT}}.$$

Assume that the pipe flow is turbulent. That is, $\alpha_s \approx 1$. The Reynolds number will need to be verified. Hence,

$$\frac{p_s}{\rho g} = \frac{P_{\text{atm}}}{\rho g} - \frac{V_s^2}{2g} + (z_1 - z_s) - H_{\text{lT}}.$$

Then,

$$\mathrm{NPSHA} = \frac{P_{\text{atm}}}{\rho g} - \frac{V_s^2}{2g} + (z_1 - z_s) - H_{\text{lT}} + \frac{V_s^2}{2g} - \frac{P_{\text{vapor}}}{\rho g}$$

$$\mathrm{NPSHA} = \frac{P_{\text{atm}} - P_{\text{vapor}}}{\rho g} + (z_1 - z_s) - H_{\text{lT}}.$$

Find the total head loss in the pipe. Note that since the diameter of the pipe does not change, the law of conservation of mass requires that the pipe average velocity be constant. Therefore,

$$H_{\text{lT}} = H_l + H_{\text{lm}} = \left(f \frac{L}{D} + \sum K \right) \frac{V_s^2}{2g}.$$

For the 4-in. nominal diameter pipe, it will be assumed that the system is assembled with flanged fittings. The loss coefficients are shown in Table A.14 (or other sources) and are shown below. Note that the values for the 4-in. diameter pipe were used:

90° regular elbow: $K_{\text{elbow}} = 0.3$
Open gate valve: $K_{\text{gate}} = 0.16$
Sharp-edged inlet: $K_{\text{inlet}} = 0.5$.

The Reynolds number and the relative roughness of the pipe are needed to determine the friction factor, f:

$$\text{Re}_D = \frac{\rho V_s D}{\mu} = \frac{V_s D}{\nu} = \frac{4\dot{V}}{\nu \pi D}.$$

At 60°F, $\nu = 0.121 \times 10^{-4}$ ft²/s. The inner diameter of 4-in. nominal Class 150 cast-iron piping is 4.10 in. (see Table A.9):

$$\text{Re}_D = \frac{4\,(1200\ \text{gpm})}{\left(0.121 \times 10^{-4}\text{ft}^2/\text{s}\right) \pi\,(4.10\ \text{in.})} \times \frac{35.315\ \text{ft}^3/\text{s}}{15850\ \text{gpm}} \times \frac{12\ \text{in.}}{1\ \text{ft}} = 8.25 \times 10^5.$$

Since $\text{Re}_D > 4000$, the flow is turbulent.

The average roughness of cast-iron piping is 0.00085 ft (Table A.1). Thus, the relative roughness is

$$\frac{\varepsilon}{D} = \frac{0.00085\ \text{ft}}{4.10\ \text{in.}} \times \frac{12\ \text{in.}}{1\ \text{ft}} = 0.00249.$$

From the Moody chart, the friction factor is

$$f \approx 0.025.$$

Therefore,

$$H_{IT} = \left[(0.025)\frac{6\ \text{ft}}{4.10\ \text{in.}} \times \frac{12\ \text{in.}}{1\ \text{ft}} + (0.3 + 0.16 + 0.5) \right]$$

$$\left[\frac{4\,(1200\ \text{gpm})}{\pi\,(4.10\ \text{in.})^2} \times \frac{35.315\ \text{ft}^3/\text{s}}{15850\ \text{gpm}} \times \left(\frac{12\ \text{in.}}{1\ \text{ft}}\right)^2 \right]^2 \frac{1}{2\,\left(32.2\ \text{ft/s}^2\right)}$$

$H_{IT} = 18.5$ ft.

At 60°F, $P_{\text{vapor}} = 0.256$ psia. $P_{\text{atm}} = 14.7$ psia.

Therefore,

$$\text{NPSHA} = \frac{(14.7 - 0.256)\ \text{lbf/in.}^2}{\left(62.36\ \text{lb/ft}^3\right)\left(32.2\ \text{ft/s}^2\right)} \times \frac{32.2\ \text{lb-ft/s}^2}{1\ \text{lbf}} \times \left(\frac{12\ \text{in.}}{1\ \text{ft}}\right)^2 + (3.5\ \text{ft}) - 18.5\ \text{ft}$$

NPSHA = 18.4 ft.

Cavitation will not occur if NPSHA > NPSHR. From the performance plots for this pump operating at 1200 gpm,

NPSHR $\approx$ 12 ft.

Cavitation will not occur at this temperature.

Comment

The following expression for the NPSHA was derived specifically for this problem:

$$\text{NPSHA} = \frac{P_{\text{atm}} - P_{\text{vapor}}}{\rho g} + (z_1 - z_s) - H_{\text{IT}}.$$

From this equation, it is observed that shorter suction pipes will result in lower head loss (H_{IT}), which will increase NPSHA. This will reduce the possibility of the occurrence of cavitation.

The amount of fluid in the tank and the static pressure head will have an effect on the NPSHA and cavitation. The static pressure head will increase as (z_1-z_s) increases. As the static pressure head increases, the NPSHA will increase, and the possibility of cavitation will be reduced.

3.4.8 Pump Scaling Laws: Nondimensional Pump Parameters

Dimensional analysis, the Buckingham Pi (Π) theorem, and the method of repeating variables can be used to generate **nondimensional pump parameters** that include the pump head, pump capacity, and pump bhp.

For a typical pump, the following functional forms will apply:

$$g H_{\text{pump}} = f\left(\dot{V}, D, \varepsilon, \omega, \rho, \mu\right) \tag{3.15}$$
$$\text{bhp} = f\left(\dot{V}, D, \varepsilon, \omega, \rho, \mu\right), \tag{3.16}$$

where ε is the relative roughness and D is the diameter of the pump impeller.

The resulting dimensionless Π groups produce nondimensional pump parameters with special names. Following are the six nondimensional pump parameters of interest:

$$\text{(a)} \quad \frac{\rho \omega D^2}{\mu} = \text{Re} = \text{Reynold's number,} \tag{3.17}$$

where ωD is the characteristic velocity.

$$\text{(b)} \quad \frac{\varepsilon}{D} = \text{Relative roughness} \tag{3.18}$$

$$\text{(c)} \quad \frac{g H_{\text{pump}}}{\omega^2 D^2} = C_H = \text{Head coefficient} \tag{3.19}$$

$$\text{(d)} \quad \frac{\dot{V}}{\omega D^3} = C_Q = \text{Capacity coefficient} \tag{3.20}$$

$$\text{(e)} \quad \frac{\text{bhp}}{\rho \omega^3 D^5} = C_P = \text{Power coefficient} \tag{3.21}$$

$$\text{(f)} \quad \frac{g \times \text{NPSHR}}{\omega^2 D^2} = C_{\text{NPSH}} = \text{Suction head coefficient.} \tag{3.22}$$

3.4.9 Application of the Nondimensional Pump Parameters—Affinity Laws

Pumps that are geometrically similar may considered to be equivalent. **Pump equivalence** implies that the nondimensional pump parameters are equal for different pumps or for different states of the same pump.

Consider two pumps (pump A and pump B) or consider a pump at two different states of volume flow rate, speed, pump head, bhp, etc. The pump equivalence equations are:

$$\text{(a)} \qquad \frac{\dot{V}_A}{\omega_A D_A^3} = \frac{\dot{V}_B}{\omega_B D_B^3} = C_{Q,A} = C_{Q,B} \tag{3.23}$$

$$\text{(b)} \qquad \frac{\rho_A \omega_A D_A^2}{\mu_A} = \frac{\rho_B \omega_B D_B^2}{\mu_B} = Re_A = Re_B \tag{3.24}$$

$$\text{(c)} \qquad \frac{\varepsilon_A}{D_A} = \frac{\varepsilon_B}{D_B} \tag{3.25}$$

$$\text{(d)} \qquad \frac{g H_{\text{pump,A}}}{\omega_A^2 D_A^2} = \frac{g H_{\text{pump,B}}}{\omega_B^2 D_B^2} = C_{H,A} = C_{H,B} \tag{3.26}$$

$$\text{(e)} \qquad \frac{\text{bhp}_A}{\rho_A \omega_A^3 D_A^5} = \frac{\text{bhp}_B}{\rho_B \omega_B^3 D_B^5} = C_{P,A} = C_{P,B}. \tag{3.27}$$

The pump equivalence equations can be expressed as ratios in order to scale pumps of different sizes or determine the parameters at different states for the same pump. Thus, for different pumps,

$$\text{(a)} \qquad \frac{\dot{V}_B}{\dot{V}_A} = \frac{\omega_B}{\omega_A} \left(\frac{D_B}{D_A} \right)^3 \tag{3.28}$$

$$\text{(b)} \qquad \frac{H_B}{H_A} = \left(\frac{\omega_B}{\omega_A} \right)^2 \left(\frac{D_B}{D_A} \right)^2 \tag{3.29}$$

$$\text{(c)} \qquad \frac{\text{bhp}_B}{\text{bhp}_A} = \frac{\rho_B}{\rho_A} \left(\frac{\omega_B}{\omega_A} \right)^3 \left(\frac{D_B}{D_A} \right)^5. \tag{3.30}$$

The aforementioned equations are known as the **affinity laws**. The affinity laws also apply to a given pump moving the same liquid. For this case, $\rho_A = \rho_B$ and $D_A = D_B$. In general, $\omega = 2\pi N$, where N is the number of revolutions per minute of the pump impeller.

Hence, for the same pump that has experienced a speed change from N_A to N_B,

$$\text{(a)} \quad \frac{\dot{V}_B}{\dot{V}_A} = \frac{N_B}{N_A} \tag{3.31}$$

$$\text{(b)} \quad \frac{H_B}{H_A} = \left(\frac{N_B}{N_A}\right)^2 \tag{3.32}$$

$$\text{(c)} \quad \frac{\text{bhp}_B}{\text{bhp}_A} = \left(\frac{N_B}{N_A}\right)^3. \tag{3.33}$$

3.4.10 Nondimensional Form of the Pump Efficiency

Dimensional analysis can also be used to establish a nondimensional form of the pump efficiency. Consider the following analysis:

$$\eta_{\text{pump}} = \frac{\dot{W}_{\text{water horsepower}}}{\text{bhp}} = \frac{\rho g \dot{V} H_{\text{pump}}}{\text{bhp}} \tag{3.34}$$

Remember: $\quad \dfrac{g H_{\text{pump}}}{\omega^2 D^2} = C_H, \quad \dfrac{\dot{V}}{\omega D^3} = C_Q, \quad \dfrac{\text{bhp}}{\rho \omega^3 D^5} = C_P$

Therefore,

$$\eta_{\text{pump}} = \frac{\rho \left(\omega D^3 C_Q\right)\left(\omega^2 D^2 C_H\right)}{\rho \omega^3 D^5 C_P} = \frac{C_Q C_H}{C_P}. \tag{3.35}$$

Example 3.4   Manipulating the Affinity Laws

When operated at $N = 1170$ rpm, a centrifugal pump, with impeller diameter, $D = 8$ in., has shut-off head $H_o = 25.0$ ft of water. At the same operating speed, best efficiency occurs at $\dot{V} = 300$ gpm, where the head is $H = 21.9$ ft of water. Specify the discharge and head for the pump when it is operated at 1750 rpm at both the shut-off and BEPs.

Solution. The pump remains the same, so the two flow conditions are geometrically similar. If no cavitation occurs, the flows will also be kinematically similar.
   Let condition 1 be at 1170 rpm and condition 2 be at 1750 rpm.
   For the flow rates (discharge),

$$C_{Q1} = C_{Q2}$$

$$\frac{\dot{V}_1}{\omega_1 D_1^3} = \frac{\dot{V}_2}{\omega_2 D_2^3}$$

Note that $\omega = 2\pi N$. Thus,

$$\frac{\dot{V}_1}{2\pi N_1 D_1^3} = \frac{\dot{V}_2}{2\pi N_2 D_2^3}.$$

Since $D_1 = D_2$,

$$\frac{\dot{V}_1}{N_1} = \frac{\dot{V}_2}{N_2}$$

$$\dot{V}_2 = N_2 \frac{\dot{V}_1}{N_1}.$$

At the shut-off point,

$$\dot{V}_2 = 1750 \text{ rpm} \frac{0 \text{ gpm}}{1170 \text{ rpm}}$$

$$\dot{V}_2 = 0 \text{ gpm}.$$

At the BEP:

$$\dot{V}_2 = 1750 \text{ rpm} \frac{300 \text{ gpm}}{1170 \text{ rpm}}$$

$$\dot{V}_2 = 449 \text{ gpm}.$$

For the head,

$$C_{H1} = C_{H2}$$

$$\frac{g H_1}{\omega_1^2 D_1^2} = \frac{g H_2}{\omega_2^2 D_2^2}.$$

Note that $\omega = 2\pi N$. Thus,

$$\frac{g H_1}{(2\pi N)_1^2 D_1^2} = \frac{g H_2}{(2\pi N)_2^2 D_2^2}.$$

Since $D_1 = D_2$,

$$\frac{H_1}{N_1^2} = \frac{H_2}{N_2^2}$$

$$H_2 = H_1 \left(\frac{N_2}{N_1}\right)^2.$$

At the shut-off point,

$$H_2 = 25.0 \text{ ft} \left(\frac{1750 \text{ rpm}}{1170 \text{ rpm}}\right)^2$$

$$H_2 = 55.9 \text{ ft.}$$

At the BEP,

$$H_2 = 21.9 \text{ ft} \left(\frac{1750 \text{ rpm}}{1170 \text{ rpm}}\right)^2$$

$$H_2 = 49.0 \text{ ft.}$$

Example 3.5 Selecting a Pump Using Equivalent Performance Plots

Bell & Gossett manufacture 3 in. × 3 in. × $5\frac{1}{4}$ in. pumps that operate at 1750 rpm. Unfortunately, a design project requires a 1750 rpm pump that would provide 70 gpm of kerosene for a total head of 10 ft of kerosene, if it were to operate at 1150 rpm. Select an appropriate pump from the available Bell & Gossett performance curves. Prepare a pump schedule for the purposes of installation by the mechanical contractor.

Solution. Appendix A shows performance plots from the Bell & Gossett Series 60 centrifugal pump catalog. The performance plot for the 3 in. × 3 in. × $5\frac{1}{4}$ in. family of pumps is shown below.

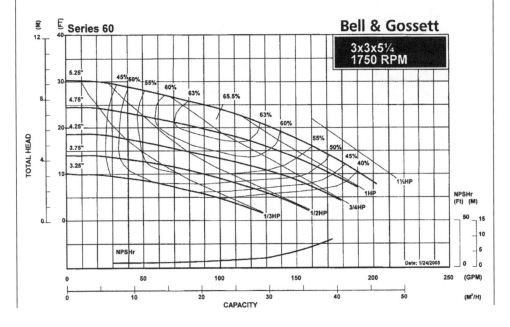

Only the performance curves for the 1750 rpm family of pumps are available. The plots are based on clean, pure water. The affinity laws and available data will be used to select a 1750 rpm pump that would give the same performance as an 1150 rpm pump.

The affinity law for the volume flow rate is

$$\frac{\dot{V}_B}{\dot{V}_A} = \frac{\omega_B}{\omega_A} \left(\frac{D_B}{D_A} \right)^3.$$

The impeller diameter is constant. Hence,

$$\frac{\dot{V}_B}{\dot{V}_A} = \frac{N_B}{N_A}.$$

Let state "A" correspond to the 1150 rpm speed. The flow rate required at this speed is 70 gpm.

$$\dot{V}_B = \dot{V}_A \frac{N_B}{N_A}$$

$$\dot{V}_B = (70 \text{ gpm}) \frac{1750 \text{ rpm}}{1150 \text{ rpm}} = 107 \text{ gpm}.$$

107 gpm would be produced in the 1750 rpm pump.

The pump head values in the Bell & Gossett pump performance plots are in terms of foot of water. The total head in terms of foot of kerosene must be converted to foot of water. Therefore,

$$H_A = 10 \text{ ft kerosene} \times SG_{\text{kerosene}} = 10 \text{ ft kerosene} \times \frac{51.2 \text{ lb/ft}^3}{62.4 \text{ lb/ft}^3} = 8.2 \text{ ft water.}$$

The total pump head required for the 1150 rpm speed is 8.2 ft of water. At the 1750 rpm speed, the total pump head is found from the affinity law for the pump head:

$$\frac{H_B}{H_A} = \left(\frac{\omega_B}{\omega_A} \right)^2 \left(\frac{D_B}{D_A} \right)^2.$$

For a constant impeller diameter,

$$H_B = H_A \left(\frac{N_B}{N_A} \right)^2$$

$$H_B = (8.2 \text{ ft water}) \left(\frac{1750 \text{ rpm}}{1150 \text{ rpm}} \right)^2 = 19 \text{ ft of water.}$$

This head corresponds to the total pump head required for the pump operating at 1750 rpm.

The pump performance plot can be used to select a 1750 rpm pump to deliver 107 gpm of kerosene at 19 ft of water. According to the plot, a *4.85-in. diameter impeller* would be satisfactory. Note that the manufacturer could trim a 5.25 in. impeller down to 4.85 in. on a lathe. Since that option may be costly, or cannot be accomplished by the manufacturer, choose a 5.25-in. diameter impeller.

The bhp required to drive the 1750 rpm pump is 1 hp.

Therefore, the final choice is a pump with the following operating parameters:

3 in. × 3 in. × 5¼ in. pump casing
5.25 in. impeller diameter
1-hp motor
1750 rpm

The pump schedule is shown below.

					Fluid			Electrical		
Pump Schedule										
Tag	Manufacturer and Model Number	Type	Construction	Flow Rate (gpm)	Working Fluid	Head Loss (ft)	Motor Size (hp)	Motor Speed (rpm)	V/Ph/ Hz	
P-1	Bell & Gossett Series 60, or Equal	Centrifugal, In-line Mounted	Iron 3 × 3 × 5¼ in. Casing, 5.25 in. φ	107	Kerosene	19	1	1750	208/3/ 60	

It should be noted that the electrical information (volt/phase/hertz) is usually provided in the manufacturer's catalog or on the pump submittal data sheet.

3.5 Design of Piping Systems Complete with In-Line or Base-Mounted Pumps

3.5.1 Open-Loop Piping System

In an **open-loop piping system**, some part of the circuit is open to the atmosphere. For example, a drain may be open to the atmosphere or the piping circuit may be open to the atmosphere as are typical of cooling towers. Figure 3.12 shows a typical open-loop condenser piping system for water.

Note the following regarding typical open-loop piping systems:

(a) A strainer (filter in liquid piping systems) is required to protect the pump from large sediments and potential blockage.
(b) Isolation valves may be installed around equipment in the piping system. These will allow for maintenance without the need for complete drainage of the piping system. These valves could be ball valves, globe valves, or gate valves.

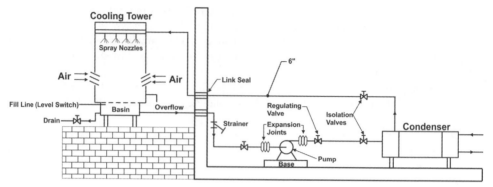

Figure 3.12   A typical open-loop condenser piping system for water

(c) Regulating valves are used to control the flow rate, especially on the discharge line from the pump. These valves are typically globe valves. This type of flow rate regulation is known as **discharge throttling**.

(d) Expansion (flexible connectors) joints are required to protect the pipes from expansion and contraction forces due to pumping thermal stress. They also isolate the piping from pump vibrations.

Practical Note 3.9   Bypass Lines

Not shown in the schematic are **bypass lines**. These are additional sections of piping that could be installed around equipment in the system to divert all or a part of the flow. The bypass line may include an auxiliary piece of equipment that is normally brought on-line during maintenance of the primary equipment. Isolation valves should be installed at the entrance and exit of the bypass line.

Practical Note 3.10   Regulation and Control of Flow Rate across a Pump

Control of the flow rate across a pump can be accomplished by varying the pump head, speed, or both simultaneously. While there are many different methods of regulating the fluid flow rate, only three of the methods will be considered.

   Discharge throttling involves the use of a partially closed valve installed on the discharge line of the pump. In this case, the system curve intersects the pump performance curve (operating point) at a higher pump head and lower flow rate. However, this will produce lower pump efficiencies. This is the cheapest and most common method of flow control.

Bypass regulation occurs with the use of a bypass line. Diversion of a portion of the flow will result in a decrease in the amount of power required. For this reason, it may be preferred to discharge throttling.

Speed regulation can be used to control the flow rate by varying the speed of the pump. Some options include variable-speed mechanical drives or variable frequency drives (VFD) on motors. By changing the pump speed, the power requirement varies, without adverse impact on the efficiency of the pump.

Practical Note 3.11   In-Line and Base-Mounted Pumps

Pumps in piping systems can either be mounted in-line with the piping system and supported by the pipes and brackets. Or, the pumps could be **base mounted** on concrete pads. In the case of base-mounted pumps, **vibration isolators** or **vibration isolation pads** should be specified. Isolators are typically installed between the pump and the concrete pad. This will reduce the transmission of vibration and noise to the main building structure. If vibration isolators are not desired, then the design engineer should ensure that the pump's concrete foundation is **$1^1/_2$–3 times** the total weight of the pump and motor assembly to provide sufficient vibration isolation and noise control.

Example 3.6   Designing an Open-Loop Piping System

Metal Mint, Inc. (located in Southern Mexico) has contracted the services of EBA Engineering Consultants, Ltd. to design a cooling system for their high-quality stainless steel bullion bars. The hot stainless steel bars will be transported slowly on a conveyor belt to allow for cooling. A design engineer at EBA Engineering Consultants, Ltd. has suggested the use of water in an evaporative cooling system. They wish to transport water from an existing 1200-gallon tank, which has a height of 50 in., to a large nozzle located 10 ft above grade (the ground). Once transported to the nozzle ($K_L = 14$), the water will be atomized to small droplets to produce shower-like streams of water to enhance evaporative cooling. Given that water is scarce in this area, the subsequent design of the conveyor cooling system will be determined by the piping system design. A sketch was submitted by the design engineer for consideration:

(a) Based on the sketch provided by the design engineer, design an appropriate piping system, complete with all necessary equipment, to transport the water to the atomization nozzle. Metal Mint, Inc. (the client) has an open business agreement with Bell & Gossett—ITT Industries. Equipment schedules, specifications, and installation guidelines are not required by the client.

(b) For the convenience of the client, it is recommended that a low-level switch be installed in the tank. Based on the piping system design and equipment selected, conduct an analysis to determine and/or justify the minimum height of water in the tank (i.e., minimum height of the low-level switch in the tank).

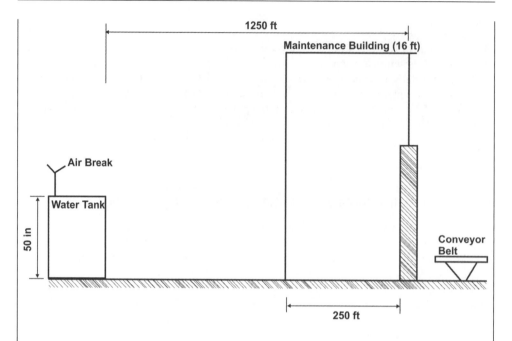

Possible Solution

Detailed Design

Objective

To determine the size of a pipe to move water. An appropriate pump will also be sized and selected, with selection from a Bell & Gossett pump performance curve.

Data Given or Known

(i) The water tank can hold 1200 gallons of liquid. The height of the tank is 50 in.
(ii) The distance from the tank to the wall adjacent to the conveyor belt is 1250 ft.
(iii) The maintenance building has a length of 250 ft and a height of 16 ft.
(iv) The water nozzle has to be located 10 ft above the ground.
(v) The K_L value of the nozzle is 14.
(vi) The tank is complete with an air break (vent) to the atmosphere.

Assumptions/Limitations/Constraints

(i) Let the pipe material be Schedule 40 steel. It appears as though a portion of the piping system will be outside the building. The thicker Schedule 40 steel will provide protection to the pipe, and will be more durable than copper piping. However, it should be noted that the steel may rust due to the presence of dissolved air in this open-loop piping. This could cause blockage of the nozzle.

(ii) Let the flow velocity in the pipe be lower than 10 fps, which is the erosion limit for small steel pipes. For the suction lines of centrifugal pumps (Bell & Gossett pumps), the velocity should not be more than 5 fps. Hence, any pipe velocity lower than 5 fps will be acceptable.

(iii) Limit the frictional losses in the straight run of pipe to 3 ft wg per 100 ft of pipe. This is based on industry standards/guidelines.

(iv) Let all pipe changes be gradual to reduce losses in the system.

(v) Install threaded fittings. This is common for smaller pipes.

(vi) Let the pipe be connected at the bottom of the water tank, approximately 6 in from the ground. The height of the fluid in the tank will provide some pressure head during pumping.

(vii) Isolation valves around the pump will be gate or globe valves. The globe valve should be installed on the pump discharge line to regulate flow, if required.

(viii) Assume that the pump is located in the maintenance building, 50 ft from the exterior wall closest to the water tank.

(ix) The water temperature is 70°F because the system is located in Southern Mexico.

Sketch

A sketch of the piping system, complete with valves and a pump, are shown.

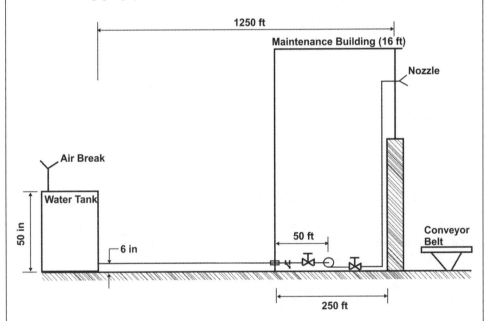

Analysis

Pipe Sizing

The friction loss is fixed at 3 ft wg per 100 ft of pipe. The pipe velocity cannot exceed 5 fps. From the friction loss chart for Schedule 40 steel in open piping systems (Figure A.4),

Pipe diameter: 2 in. (nominal)
Pipe velocity: 3 fps
Flow rate: 30 gpm.

$$\text{Reynolds number}: \frac{\rho V D_{inner}}{\mu} = \frac{(62.30\ \text{lb/ft}^3)\,(3.0\ \text{ft/s})\,(2.067\ \text{in.})}{6.556 \times 10^{-4}\ \text{lb/ft/s}} \times \frac{1\ \text{ft}}{12\ \text{in.}} = 49105.$$

The flow is fully turbulent.

Pump Sizing

The pump head is required to size the pump. The energy equation is

$$H_{pump} = \left(\frac{p_2}{\rho g} + \frac{V_2^2}{2g} + z_2 \right) - \left(\frac{p_1}{\rho g} + \frac{V_1^2}{2g} + z_1 \right) + H_{lT}.$$

This is an open-loop piping system. Let point 1 be at the entrance of the pipe attached to the water tank and point 2 be at the exit of the nozzle. To select the largest possible pump (most conservative design), assume that the water level in the water tank is 6 in. above grade. Thus, $z_1 = 6$ in. The air break on the tank ensures that the free surface of the water reservoir remains at normal atmospheric pressure. Since the free surface is level with the pipe entrance, $p_1 = p_{atm}$. At the exit of the nozzle, $p_2 = p_{atm}$. Changes in the velocity will be assumed to be negligible since the pipe diameter is constant. So, $V_2 \approx V_1$.
Therefore,

$$H_{pump} = (z_2 - z_1) + H_{lT}.$$

The total head loss in the pipe is

$$H_{lT} = H_l + H_{lm} = H_l + \sum K_L \frac{V^2}{2g}.$$

For the 2 in. pipe, the loss coefficient values are:

For the pipe entrance: $K_L = 0.5$ (sharp-edged entrance)
For the strainer: $K_L = 1.5$
For the gate valve: $K_L = 0.16$
For the globe valve: $K_L = 6.9$
For the 90° elbows: $K_L = 0.95$ (regular elbows)
For the nozzle: $K_L = 14$.

The total length of piping is approximately $(1250 + 10{-}0.5)\ \text{ft} = 1260\ \text{ft}$.
Thus,

$$H_{lT} = \frac{3.0\ \text{ft wg}}{100\ \text{ft}} \times 1260\ \text{ft} + [0.5 + 1.5 + 0.16 + 6.9 + 2\,(0.95) + 14] \frac{(3.0\ \text{ft/s})^2}{2(32.2\ \text{ft/s}^2)}$$

$$H_{lT} = 41\ \text{ft}.$$

Hence,

$$H_{pump} = (10 - 0.5)\,ft + 41\,ft$$
$$H_{pump} = 50.5\,ft \approx 51\,ft.$$

The flow rate is 30 gpm. From the performance plots for the Bell & Gossett Series 60 in-line mounted centrifugal pumps, select the following pump:

$1\frac{1}{2} \times 1\frac{1}{2} \times 7$ in. casing
7.0 in. impeller diameter
$1\frac{1}{2}$-hp motor
1750 rpm

Or, choose Bell & Gossett Series 60 pump **Model 618T**, which is in stock.

Drawings

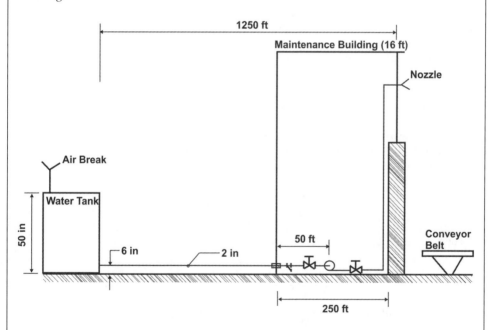

Conclusion

The piping system and pump has been sized and designed. The following points should be noted:

(i) A gate and globe valve are used as isolation valves. The globe valve on the discharge line of the pump may be used to control the flow rate, if needed. Another option could be to use a VFD to control the pump via the motor.

(ii) If the water tank is not available as a packaged unit, complete with a valve on the discharge line, a gate valve may be required at the tank discharge. In any case, this would add a small additional loss to system.

(iii) Though negligible, there will be changes in the velocity between points 1 and 2, since the diameter of the atomized fluid jets at the nozzle will be smaller than the pipe diameter.

(iv) The pump was slightly oversized to avoid overloading the motor.

(v) It should be noted that a 2 in. × 1½ in. reducing fitting will be required on the pump suction and discharge for installation of the pump.

(vi) The pump suction velocity should not be much larger than 5 fps. In this case, the velocity will be about 3 fps.

(vii) As requested by the client, no equipment schedules, specifications, or installation guidelines were provided.

A comparison of the NPSHR and the NPSHA will be used to determine the minimum height of water in the tank, below which cavitation will occur and pump performance will be adversely affected.

The NPSHA is

$$\text{NPSHA} = \left(\frac{p_s}{\rho g} + \frac{V_s^2}{2g} \right)_{\text{pump,inlet}} - \frac{P_{\text{vapor}}}{\rho g}.$$

The energy equation will be used to find the pump inlet pressure, p_s:

$$\frac{P_1}{\rho g} + \alpha_1 \frac{V_1^2}{2g} + z_1 = \frac{p_s}{\rho g} + \alpha_s \frac{V_s^2}{2g} + z_s + H_{\text{lT,suction}}.$$

Point 1 is the free surface of the water in the tank. Thus, $V_1 \approx 0$ and $P_1 = P_{\text{atm}}$. The pipe flow is turbulent. Hence, $\alpha_s \approx 1$.
Therefore,

$$\frac{p_s}{\rho g} = \frac{P_{\text{atm}}}{\rho g} - \frac{V_s^2}{2g} + (z_1 - z_s) - H_{\text{lT,suction}}.$$

Then,

$$\text{NPSHA} = \frac{P_{\text{atm}}}{\rho g} - \frac{V_s^2}{2g} + (z_1 - z_s) - H_{\text{lT,suction}} + \frac{V_s^2}{2g} - \frac{P_{\text{vapor}}}{\rho g}$$

$$\text{NPSHA} = \frac{P_{\text{atm}} - P_{\text{vapor}}}{\rho g} + (z_1 - z_s) - H_{\text{lT,suction}}.$$

Note that $H_{\text{lT,suction}}$ is the total head loss in the suction line of the pipe, only.
z_1 Is the height of the free surface of the water in the tank. The minimum height will be that which gives NPSHA = NPSHR.
So,

$$z_1 = \text{NPSHR} - \frac{P_{\text{atm}} - P_{\text{vapor}}}{\rho g} + z_s + H_{\text{lT}}.$$

From the performance plot of the $1\frac{1}{2} \times 1\frac{1}{2} \times 7$ in. casing pump, and at 30 gpm, the NPSHR is approximately 3 ft. At 70°F, $P_{vapor} = 0.363$ psia. $P_{atm} = 14.7$ psia. Based on the pump casing size, assume that the centerline of the pump is 1 ft above grade. Therefore, $z_s = 1$ ft.

The head loss in the suction line of the pipe is

$$H_{IT} = H_{l,\text{suction}} + \sum K_L \frac{V^2}{2g}.$$

The total length of suction piping is approximately 1050 ft.
Hence,

$$H_{IT} = \frac{3.0 \text{ ft wg}}{100 \text{ ft}} \times 1050 \text{ ft} + [0.5 + 1.5 + 0.16] \frac{(3.0 \text{ ft/s})^2}{2\left(32.2 \text{ ft/s}^2\right)}$$

$$H_{IT} = 32 \text{ ft}$$

and

$$z_1 = 3 \text{ ft} - \frac{(14.7 - 0.363)\,\text{lbf/in.}^2}{\left(62.3 \text{ lb/ft}^3\right)\left(32.2 \text{ ft/s}^2\right)} \times \frac{32.2 \text{ lb/ft/s}^2}{1 \text{ lbf}} \times \left(\frac{12 \text{ in.}}{1 \text{ ft}}\right)^2 + 1 \text{ ft} + 32 \text{ ft}$$

$$z_1 \approx 2.86 \text{ ft} = 2 \text{ ft } 10 \text{ in.}$$

According to this calculation, cavitation will only occur if the free surface of the fluid is less than **2 ft 10 in.** above grade. Therefore, the designer may choose any appropriate height above 2 ft 10 in. to mount the low-level switch. If this is not possible, and the low-level switch must be installed close to the base of the tank, the client may consider mounting the tank on a support that is at least 2 ft 10 in. high. This will generate the same static pressure head required to avoid cavitation. If mounting the tank is not an option, the client is advised to consider another tank with height greater than 50 in.

3.5.2 Closed-Loop Piping System

In a **closed-loop piping system**, there are no points open to the ambient. Figure 3.13 shows schematic drawings of two-pipe and four-pipe closed-loop piping systems.
Note the following regarding typical closed-loop piping systems:

(a) Filters are optional for this system. They may be included in the design if it is deemed necessary to protect equipment.
(b) An expansion tank is required for closed-loop piping systems. There should be only one expansion tank in each loop of a piping system.
Expansion tanks:
  (i) Protect the system from damage due to volume changes induced by temperature variations.
  (ii) Create a point of constant pressure in the system.

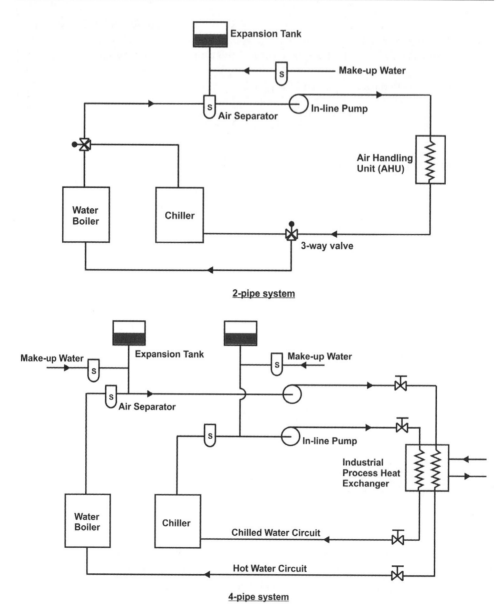

Figure 3.13 Diagrams of closed-loop piping systems

(iii) Provide a collecting space for entrapped air. Entrapped air is undesirable in the water systems since it makes the flow noisy and produces unwanted pipe vibrations—water hammer.

(c) Air separators for air elimination from the closed-loop piping system are required. Air will enter the closed-loop piping system when **makeup water** from external sources is introduced. Devices such as a vortex air separator create a high-speed fluid vortex in its center. This low-pressure region enhances the release of air bubbles from the liquid, which is released through an automatic vent or other device.

(d) Pressure regulators, isolation, control, and/or three-way valves, flow meters, and thermocouples may also be required.

Practical Note 3.12   Flanged or Screwed Pipe Fittings?

Flanged pipe is generally specified for aboveground service for air, water, sewage, oil, and other fluids where **rigid, restrained joints** are needed. It may be used for larger, heavier pipes and fluids. It is also widely used in industrial piping systems, water treatment plants, sewage treatment plants, and for other interior piping. American Water Works Association (AWWA) standards restrict the use of flanged joints underground due to the rigidity of the joints.

Example 3.7   Designing a Closed-Loop Piping System

A consulting engineering firm received the following sketch of a water piping system from the Ministry of Defense in Ottawa.

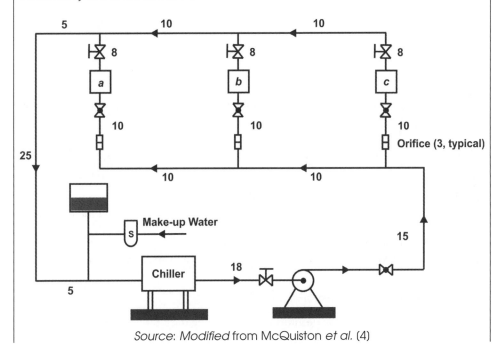

Source: Modified from McQuiston *et al.* (4)

Unit	Flow Rate (gpm)	Head Loss (ft)
a	30	15
b	40	12
c	50	10
Chiller	120	20

For security reasons, units a, b, and c are unspecified. However, a chart is provided (see above) that gives details on the flow rates and head losses across the units. The head loss across each orifice is 6 ft. All pipe lengths are in foot. Complete the design of this system.

Hint: A piping and pump schedule must be provided.

Possible Solution

Definition

Size the piping and specify the pumping requirements for a partially designed piping system provided by the Ministry of Defense.

Preliminary Specifications and Constraints

(i) The working fluid is cold water due to the presence of a chiller in the system.
(ii) Pipe velocity should be less than 10 fps for general building service. The choice of pipe material may limit the velocity further.
(iii) This is a two-pipe system with parallel piping, valves, orifices, undefined units, a chiller, and a pump.

Detailed Design

Objective

To size the pipes in the system and to size and select an appropriate pump. The piping material must be selected.

Data Given or Known

(i) Length of the pipe sections.
(ii) Total system flow rate is 120 gpm (from the chiller).
(iii) Head losses and flow rates for all the equipment are given (including the undefined units).
(iv) The orifice head loss is 6 ft.
(v) Preliminary layout of the piping system was provided by the Ministry of Defense.

Assumptions/Limitations/Constraints

(i) The Ministry did not provide information regarding the location of the system. Therefore, care will be taken to ensure that the system operates quietly.

(ii) Let the flow velocity be about 4 fps. This is acceptable for general building service or potable water. In addition, this velocity does not exceed the erosion limits of any general pipe material.

(iii) Limit pipe frictional losses to 3 ft of water per 100 ft of pipe.

(iv) Pipe changes should be gradual to reduce losses.

(v) All bends will be 90° flanged bends to facilitate maintenance and reduce losses. For smaller pipes, screwed/threaded bends will be used. This is more common in industry.

(vi) Branch and line flow tees are flanged. For smaller pipes, screwed/threaded tees will be used. This is more common in industry.

(vii) Negligible elevation head. Assume that all components are on the same level.

(viii) Assume that the piping material is **Schedule 40 steel**. This will provide durability and flexibility over type L copper pipes.

Sketch

The pipe sections are labeled and shown on the layout provided by the Ministry.

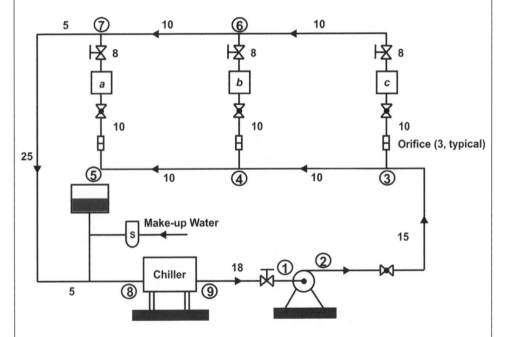

Analysis

Bear the following points in mind for this design problem:

(i) Pipe sizing for this system can be done quickly by using the appropriate friction loss charts for Schedule 40 pipe (Figure A.4). However, if the required sizes, flow rates, velocities, etc. are not included on the published charts or if the design is based on a specialized pipe material, then the designer should use Colebrook's equation (or other correlation equation) to find the friction factor (f) and iterate to find the pipe diameters.

(ii) Assign larger head losses per 100 ft of pipe to shorter pipe sections. The designer should attempt to have a constant velocity throughout the system (approximately).

Pipe Sections

The pipe sections in this design are:

Sections 2-3, 3-4, 4-5, 3-6, 4-6, 5-7, 6-7, 7-8, 9-1 (see sketch).

Consider Section 4-5. It has the shortest pipe length (10 ft) and the lowest flow rate (120–50–40 gpm = 30 gpm). Use a frictional loss of 3 ft of water per 100 ft of pipe as a guide to size the pipe.

With the friction loss and the flow rate, the friction loss chart for commercial steel pipe (Schedule 40) in a closed-loop piping system is consulted to find suitable pipe data (Figure A.4):

Nominal pipe size: 2 in.
Water velocity: 2.7 fps
Lost head: 1.6 ft per 100 ft of pipe
Major head loss: $\frac{1.6\ ft}{100\ ft} \times 10\ ft = 0.16\ ft$

The pipe system flow velocity should be close to 3 fps. This information will be used to size pipes in the other sections. See the Pipe Data table for the pipe sizes.

Now that the pipe diameters are known, the minor losses for each section can be estimated. See the Minor Losses table for information.

Minor Losses

The loss coefficients for the bends, fittings, and area changes for the piping in Section 4.5 (2 in. diameter) are given below from Table A.14:

Gate valve: $K_{gate} = 0.16$
Ball valve: $K_{ball} = 0.05$
Flanged 90° regular bends: $K_{90°\ bend} = 0.39$
Branch tee: $K_{branch\ tee} = 0.80$
Line tee: $K_{line\ tee} = 0.19$
Pipe contraction: $K_{contraction} = 0.07$ for a contraction angle of 60°
Pipe expansion: $K_{expansion} = 0.30$ for $d/D = 0.2$

$K_{expansion} = 0.25$ for $d/D = 0.4$
$K_{expansion} = 0.15$ for $d/D = 0.6$
$K_{expansion} = 0.10$ for $d/D = 0.8$

Minor Loss:

1 flanged 90° bend: $K_{90° \text{ bend}} = 0.39$
1 line tee: $K_{\text{line tee}} = 0.19$
1 pipe contraction from a 3 in. diameter in Section 3–4 to a 2 in. diameter in Section 4-5:
  $K_{\text{contraction}} = 0.07$

Therefore,

$$H_{\text{lm}} = \sum K_L \frac{V_{\text{ave}}^2}{2g} = (0.39 + 0.19 + 0.07) \left[\frac{(2.8 \text{ ft/s})^2}{2 \left(32.2 \text{ ft/s}^2 \right)} \right] = 0.079 \text{ ft.}$$

The total head loss in the Section 4-5 is

$$H_{\text{lT}} = H_l + H_{\text{lm}} = 0.16 \text{ ft} + 0.079 \text{ ft} = 0.24 \text{ ft.}$$

A similar procedure is followed for the other pipe sections. The results are shown in the tables below.

						Minor Losses			
Pipe Section No.	Gate Valves	Ball Valves	90° Regular Bends	Tees-Branch	Tees-Line	Gradual Expansion (K_L = vary)	Gradual Contrac-tion (K_L = 0.07)	Total Minor Loss (ft)
2-3		0.05 (one)	0.3 (three)	0.64 (one)				0.32
3-4					0.19 (one)		0.07 (one)	0.04
4-5			0.39 (one)		0.19 (one)		0.07 (one)	0.08
3-6	0.35 (one)	0.05 (one)	0.39 (one)	0.80 (one)		$K_L = 0.10$		0.27
4-6	0.35 (one)	0.05 (one)		0.80 (two)				0.23
5-7	0.35 (one)	0.05 (one)		0.80 (one)				0.15
6-7					0.19 (one)	$K_L = 0.10$		0.07
7-8			0.30 (two)		0.14 (two)			0.18
9-1	0.30 (one)		0.30 (one)					0.12

				Pipe Data				
Pipe Section No.	Pipe Length (ft)	Flow Rate (gpm)	Lost Head (ft/100 ft)	Fluid Velocity (ft/s)	Nominal Size (in.)	Minor Losses (ft)	Total Head Loss (ft)
2-3	15	120	1.4	3.6	3½	0.32	0.53
3-4	10	70	1.3	3	3	0.04	0.17
4-5	10	30	1.6	2.8	2	0.08	0.24
3-6	28	50	1.8	3.2	2½	0.27	0.77
4-6	18	40	1.3	2.7	2½	0.23	0.46
5-7	18	30	1.6	2.8	2	0.15	0.44
6-7	10	90	1.9	3.8	3	0.07	0.26
7-8	35	120	1.4	3.6	3½	0.18	0.67
9-1	18	120	1.4	3.6	3½	0.12	0.37

Next, consider the lost head for each of the three parallel circuits between points 3 and 7. Let: circuit a includes sections 3-4, 4-5, 5-7, the orifice, and unit a.

Circuit b includes sections 3-4, 4-6, 6-7, the orifice, and unit b.
Circuit c includes sections 3-6, 6-7, the orifice, and unit c.

The circuit head losses are

$$H_a = H_{34} + H_{45} + H_{57} + H_{orifice} + H_{unit\text{-}a} = 0.17 \text{ ft} + 0.24 \text{ ft} + 0.44 \text{ ft} + 6 \text{ ft} + 15 \text{ ft} = 21.9 \text{ ft}$$

$$H_b = H_{34} + H_{46} + H_{67} + H_{orifice} + H_{unit\text{-}b} = 0.17 \text{ ft} + 0.46 \text{ ft} + 0.26 \text{ ft} + 6 \text{ ft} + 12 \text{ ft} = 18.9 \text{ ft}$$

$$H_c = H_{36} + H_{67} + H_{orifice} + H_{unit\text{-}c} = 0.77 \text{ ft} + 0.26 \text{ ft} + 6 \text{ ft} + 10 \text{ ft} = 17.0 \text{ ft.}$$

The system will be balanced by installing balancing valves in circuits b and c to increase the head to 21.9 ft (for circuit a). The head loss for circuit a will be used to find the pump head (h_{pump}).

For this closed-loop system, let the starting and ending point be point 1. Hence, the pump head is

$$h_{pump} = \left(\frac{p_1}{\rho g} + \frac{V_1^2}{2g} + z_1 \right) - \left(\frac{p_1}{\rho g} + \frac{V_1^2}{2g} + z_1 \right) + H_{IT} = H_{IT}$$

$$h_{pump} = H_{23} + H_{34} + H_{45} + H_{57} + H_{78} + H_{91} + H_{orifice} + H_{unit\text{-}a} + H_{chiller}$$
$$h_{pump} = 0.53 \text{ ft} + 0.17 \text{ ft} + 0.24 \text{ ft} + 0.44 \text{ ft} + 0.67 \text{ ft} + 0.37 \text{ ft} + 6 \text{ ft} + 15 \text{ ft} + 20 \text{ ft}$$
$$h_{pump} = 43.4 \text{ ft} \sim 45 \text{ ft of water.}$$

For this system, a pump that is rated to produce 120 gpm at 45 ft of head is required. Use manufacturer's charts to select an appropriate pump. The drawing from the Ministry shows a base-mounted pump. In addition, the flow rate and pump head are large. A base-mounted pump will provide additional support. Choose a Taco FI/CI Series pump. A 4 in. × 3 in. × 7 in. casing pump is selected. From the performance plot for this family of pumps, the final choice is

4 in × 3 in × 7 in. casing
7 in. impeller diameter
3-hp motor
1760 rpm speed.

The pump performance curve is shown below.

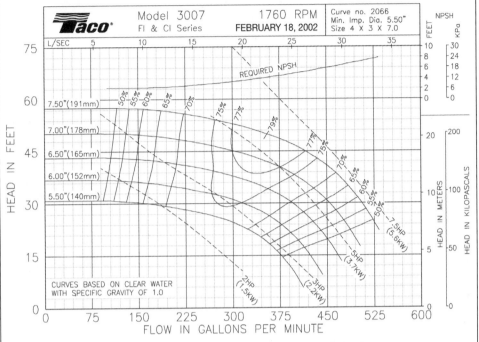

Source: Taco, Inc. (reprinted with permission)

Drawings

The drawing below shows the pipe sizes required.

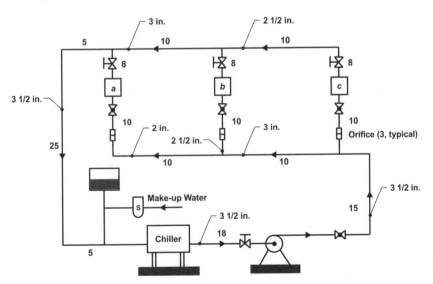

Conclusions

The piping system has been designed. All the pipe sizes are known, and the piping material has been selected. A pump has been specified and selected. It should be noted that for the pipe sizes in this design, threaded fittings could have been used. In that case, the losses would have increased. It should also be noted that 4 × 3½ in. and 3½ × 3 in. reducing fittings will be required on the pump suction and discharge, respectively, for installation of the pump.

The Pipe Data and Pump Schedule are shown below.

					Pipe Data		
Pipe Section No.	Pipe Length (ft)	Flow Rate (gpm)	Lost Head (ft/100 ft)	Fluid Velocity (ft/s)	Nominal Size (in.)	Minor Losses (ft)	Total Head Loss (ft)
2-3	15	120	1.4	3.6	3½	0.32	0.53
3-4	10	70	1.3	3	3	0.04	0.17
4-5	10	30	1.6	2.8	2	0.08	0.24
3-6	28	50	1.8	3.2	2½	0.27	0.77
4-6	18	40	1.3	2.7	2½	0.23	0.46
5-7	18	30	1.6	2.8	2	0.15	0.44
6-7	10	90	1.9	3.8	3	0.07	0.26
7-8	35	120	1.4	3.6	3½	0.18	0.67
9-1	18	120	1.4	3.6	3½	0.12	0.37

				Pump Schedule					
				Fluid			Electrical		
Tag	Manufacturer and Model Number	Type	Construction	Flow Rate (gpm)	Working Fluid	Head Loss (ft)	Motor Size (hp)	Motor Speed (rpm)	V/Ph/Hz
P-1	Taco, FI/CI Series, Model 3007, or equal	Centrifugal, base-mounted	Cast iron 4 × 3 × 7 in. Casing, 7 in.φ	120	Water	45	3	1760	208/ 3/60

Problems

3.1. A junior design engineer consults a senior engineer regarding a piping design problem. The junior design engineer has determined the size of a pump to be $3\frac{1}{2}$ hp. However, to avoid motor overloading due to small spikes in the design flow rate, they have chosen a 5-hp pump. The 5-hp pump is fully capable of providing higher flow rates, which may be detrimental to the process for which the piping system is designed. The senior engineer suggests using valves to control and reduce the flow. Which of the following valves would be *best suited* for this purpose, even when it is fully opened? Justify your response.
Choices

Gate valve; Globe valve; Ball valve; Butterfly valve; Swing-check valve; Angle valve

3.2. Cooling water is pumped from a reservoir for equipment at a construction job site by using the pipe system shown. The flow rate required is 600 gpm and water must leave the spray nozzle at 120 fps. Determine the minimum pressure needed at the pump outlet. Estimate the required motor power if the pump efficiency is 75%. Note that aluminum pipes have similar average roughness as drawn tubing. The diameter shown is the pipe inner diameter.

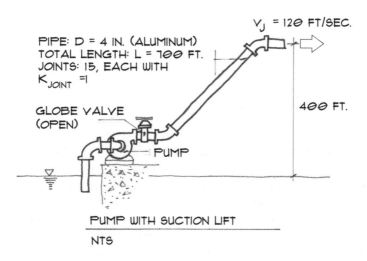

3.3. When operated at 1170 rpm, a centrifugal pump, with impeller diameter of 8 in., has a shut-off head of 25.0 ft of water. At the same operating speed, best efficiency occurs at 300 gpm, where the head is 21.9 ft of water. Specify the discharge and head for the pump when it is operated at 1750 rpm at both the shut-off and BEPs.

3.4. MDM Consulting Engineers, Inc. has prepared the following piping schematic to supply water to a chemical plant. The lengths shown are in foot. Piping is Schedule 40, commercial steel.

Complete the design of the primary circuit piping system.

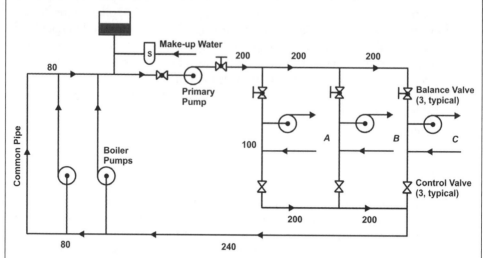

Source: Modified from McQuiston et al. (4)

Circuit	Flow Rate (gpm)	Control Valve Head Loss (ft)
A	60	40
B	70	50
C	70	50

3.5. The owner of a small office building has decided to design and install a piping network that includes five tankless water heaters to heat and transport water. The owner currently requires 18 gpm of hot water service. Cold water (feedwater) from the city will be supplied to each heater, and the piping connections were previously installed by the building contractor. The hot water, at 160°F, from the heaters will be transported through a main header for distribution in the rest of the building. Each heater has been specified to provide a maximum of 4.6 gpm of water. The heater unit has a dedicated pump that serves only to circulate the cold water through its internal finned-coil system. Unfortunately, the city supplies the feedwater at a low pressure of 40 psia, and the owner needs hot water at 80 psia for complete distribution though the building. Based on the sketch provided below, complete the design of the hot water supply lines in this system. Bear in mind that the owner would like to maintain flexibility to expand the system in the future to its maximum capabilities.

Further Information: In practice, the lengths of the hot water supply piping that connect directly to the heaters are small in comparison to that of the main supply header.

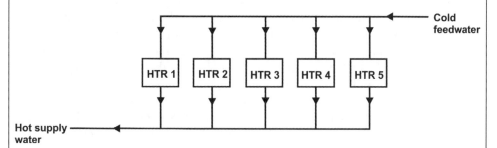

3.6. A *heating coil* is a *heat exchanger* that can be used to transfer energy from hot 50% ethylene glycol solution to air for the purposes of space heating. A proposal has been submitted that will focus on the development of a heating coil that contains one row with six loops of coils. Each loop will be 24 in. long and they will be connected to each other by way of 180° smooth return bends. The 50% ethylene glycol solution will be supplied at 200°F and 5 gpm to the heating coil component. A dedicated pump will be required for this component. Design the piping system in the heating coil and specify any major accessories that will be required.

3.7. A plant manager has hired a recent graduate of a mechanical engineering program to consider the design of a piping system to supply kerosene to three burners that will be located outdoors. Burner "A" will require 10 gpm of fuel, burner "B," 20 gpm, and burner "C," 25 gpm. For convenience, the manager will purchase a vented outdoor fuel oil storage tank to supply enough kerosene for 8 hours of operation. They wish to install the discharge pipe connection at the base of the tank to take advantage of the fuel static pressure head and to facilitate pipe installation. It is understood that codes will need to be verified. The fuel oil storage tank and piping should be supplied with sufficient heat tracing to maintain the kerosene temperature at about 68°F. Given that the fuel will be atomized and combusted completely in the burner, there will be no need for return piping from the burners to the storage tank. Design the burner kerosene supply piping system that will meet the requirements of the manager.

> *Further Information*: Consultation with NFPA 31 may provide guidance to justify some design decisions.

3.8. A pipe-pump assembly is being considered by a farmer for use in irrigating a field during summer. There is an underground water well that is 250 ft deep. The well is an open-hole type in which the free surface of water is about 8 ft below grade. To ensure that one of the pumps will operate for only a few hours per day, the farmer will fill a 2500-gallon tank in 4 hours, which will provide enough water for 2 days of farm operation. The tank cannot be more than 10 ft long and should come complete with valved piping connections whose locations will be specified by the farmer. The tank can be located anywhere in the line that the farmer deems fit. Due to location of the field and well, a pump must be located no more than 600 ft from the well. The pipe will be connected to an existing irrigation system that is 750 ft from the well and is 10 ft long. The farmer's land is flat, from the well to the field. Design the main piping system to meet the water requirements of the farmer. Specify the maximum depth of the pipe inlet in the well for the purposes of installation by the farmer.

> *Further Information*: The design engineer may consider the use of a *foot valve* (a type of check valve), complete with a strainer at the inlet to the piping system. This will ensure that the pumps are properly *primed*, that is, completely filled with water before the start of pumping. Other alternatives are available.

3.9. An engineer has presented a piping system layout that includes a pump, condenser, and cooling tower for a building cooling application. Size the piping and specify the pump required if the total volume flow rate of water is 450 gpm, the pressure drop in the condenser is 15 ft of water gage, the vertical distance from the pump centerline to the top of the cooling tower is 45 ft, the horizontal distance from the condenser to the tower is 70 ft, and the pump is 15 ft below the tower basin. All the necessary accessories are shown in the sketch provided by the engineer.

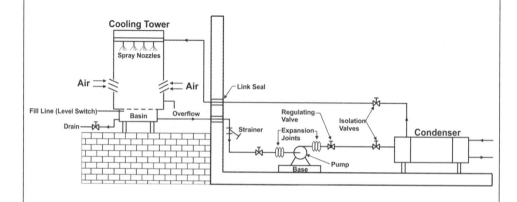

3.10. Independent Living Developments Inc. has recently awarded a project for the first phase of the detailed design of a system to deliver potable water from a bore-well to a 2-floor, 12-apartment residence to a mechanical engineering consulting firm. In this first phase of the project, the client has requested that the design be restricted to include elements associated with the extraction of water from a bore-well in sufficient quantity to meet the needs of their proposed residential complex and to deliver it at a pressure of 100 psia at the inlet to each floor riser. This phase of the project is not to include any aspects of the design of the bore-well or the layout of the internal plumbing systems of the building. Detailed drawings of items such as connections and penetrations are not required at this time. The proposed construction site is a rectangular plot of land with an area of $\frac{1}{2}$-acre that has road access on one side and land on the other three sides. The footprint of the 12-apartment complex is planned to be 60 ft × 120 ft. Any existing building structures on the plot of land have been slated for demolition. The complex will be located in an area where the average outdoor temperature in winter is −4°F, and in the summer, 80°F. The apartment complex will have a basement, which will serve as a storage area and mechanical room and will have a foundation that is 6 ft below grade and a clear height of 8 ft to meet the requirements of the International Building Code. The first floor has eight 2-bedroom apartments (floor area 800–1000 ft^2 each and height 10 ft) and the second floor has four 3-bedroom apartments (floor area 1000–1200 ft^2 each and height 10 ft). There is an existing bore-well located 50 ft from the proposed building. Standard draw down and recovery tests that were conducted at the well depth of 100 ft showed that the well could support continuous extraction of water at 10 gpm. Through consultation with the local utility company, it was found that each person uses approximately 60 gallons of water per day. In order to guarantee the availability of sufficient amounts of water, the client has requested the use of a water storage tank. With reference and adherence to the International Building Code and the International Plumbing Code (or Uniform Plumbing Code), design the piping system required to deliver water from the well to the floors of the complex (up to and including the floor risers). The designer should ensure that necessary components such as pumps, pipes, valves, piping accessories, the tank and its size, to name a few, are clearly specified.

> *Further Information*: The design engineer may consider the use of a *foot valve* (a type of check valve), complete with a strainer at the inlet to the piping system. This will ensure that the pumps are properly *primed*, that is, completely filled with water before the start of pumping. Other alternatives may include a submersible pump complete with a foot valve.

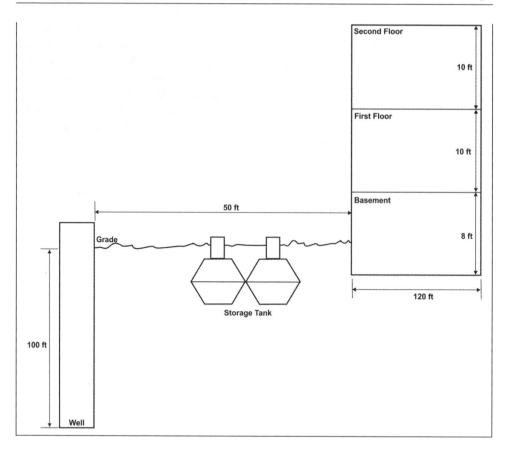

References and Further Reading

[1] Erico International Corp. (2010) *Pipe Hanger and Support Recommended Specifications*, Erico Corp., Solon, OH, pp. 107–118.
[2] Karassik, I., Messina, J., Cooper, P. *et al.* (eds) (2008) *Pump Handbook*, 4th edn, McGraw-Hill Companies, Inc., New York.
[3] Fox, R. and McDonald, A. (1998) *Introduction to Fluid Mechanics*, 5th edn, John Wiley & Sons, New York.
[4] McQuiston, F., Parker, J. and Spitler, J. (2000) *Heating, Ventilating, and Air Conditioning: Analysis and Design*, 5th edn, John Wiley & Sons, New York.

4

Fundamentals of Heat Exchanger Design

4.1 Definition and Requirements

Heat exchangers are devices that facilitate energy transfer between two fluids at different temperatures while keeping them from mixing with each other. Fundamental knowledge of heat conduction and heat convection are required for heat exchanger design and/or selection. Examples of heat exchangers include (i) car radiator; (ii) hydronic baseboard heaters; (iii) condensers; (iv) superheaters; (v) boilers; and (vi) regenerators/recurperators.

4.2 Types of Heat Exchangers

4.2.1 Double-Pipe Heat Exchangers

In this type of heat exchanger, one fluid flows through a pipe and the other fluid flows through an annular space that encloses the pipe. There are two types of flows in double-pipe heat exchangers: parallel flow and counter flow. Figure 4.1 shows schematics of double-pipe heat exchangers:

Parallel flow: Both fluids enter and exit at the same point (ie. fluids flow in the same direction). For very long systems, the temperatures of the two fluids will eventually become equal at the exit.
Counter flow: The fluids enter and leave at opposite ends (ie. fluids flow in opposite directions). The flows are in opposite directions. A larger temperature difference exists over the length of this heat exchanger compared to the parallel flow heat exchanger.

Introduction to Thermo-Fluids Systems Design, First Edition. André G. McDonald and Hugh L. Magande.
© 2012 André G. McDonald and Hugh L. Magande. Published 2012 by John Wiley & Sons, Ltd.

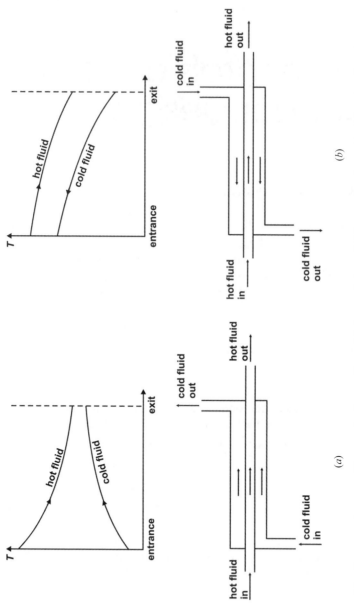

Figure 4.1 Temperature profiles and schematics of (a) parallel and (b) counter flow double-pipe heat exchangers

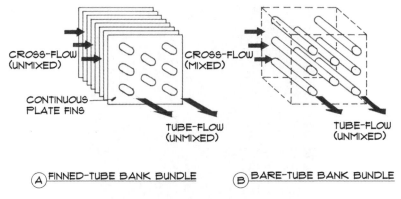

Figure 4.2   Cross-flow heat exchangers

4.2.2 Compact Heat Exchangers

Compact heat exchangers were developed to provide large surface areas for heat transfer (per unit volume). A heat exchanger is considered compact if the **area density** (β) is large. Area density is

$$\beta = \frac{A_s}{V}, \tag{4.1}$$

where A_s is the surface area and V is the volume.
A heat exchanger is considered compact if: $\beta > 200 \ \text{ft}^2/\text{ft}^3$.

   Cross-flow heat exchangers are excellent examples of compact heat exchangers. The addition of fins to extend the heat transfer surface area will make the heat exchanger more compact. Figure 4.2 shows some cross-flow heat exchangers with fins (finned-tube) and without fins (bare). The fins and tubes separate the working fluids into distinct sections, producing **unmixed flow**. **Mixed flow** occurs when the working fluid is not separated into smaller subsections (see Figure 4.2b). Figure 4.3 shows a picture of a continuous plate-fin-tube type cross-flow heat exchanger.

4.2.3 Shell-and-Tube Heat Exchangers

Shell-and-tube heat exchangers have a large number of tubes (**tube bank**) packed in a shell casing. Heat transfer occurs as the fluid (at temperature A) in the shell flows across the tubes. Fluid flows through the tubes (at temperature B). **Baffles** are typically used to force the fluid to flow across the tubes in the shell in a serpentine fashion, thus enhancing the heat transfer. The fluid in the tubes collects in **headers** before it enters or leaves the system. Figure 4.4 shows schematics of several shell-and-tube heat exchangers. While outside the scope of this textbook, shell-and-tube heat exchanger design has been treated extensively by Lee [1] and Kays and London [2].

Figure 4.3   Picture of a continuous plate-fin-tube type cross-flow heat exchanger

4.3 The Overall Heat Transfer Coefficient

The rate of heat transfer across the surfaces in a heat exchanger will be governed, in part, by the **thermal resistance** to heat transfer across those surfaces. Determination of the thermal resistance will require simplified heat conduction and heat convection analyses.

In heat exchangers, heat transfer occurs primarily by conduction and convection. The schematic drawing below shows these modes of heat transfer through an ideal plane wall:

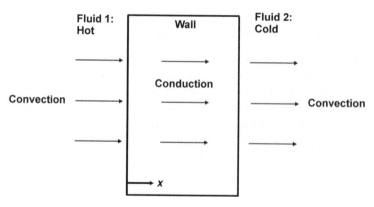

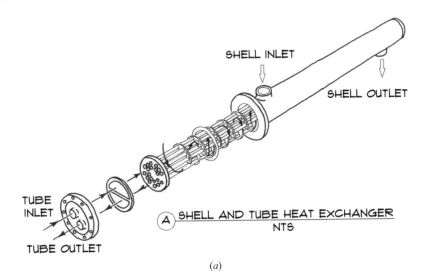

(a)

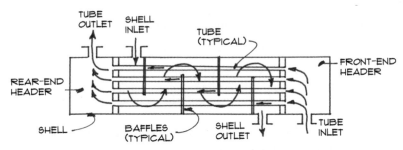

(b)

Figure 4.4   Schematics of shell-and-tube heat exchangers

Note the following assumptions:

(a) Heat transfer is one-dimensional.
(b) Steady state exists.
(c) Radiation effects are negligible or are included in the convection terms through the heat transfer coefficients.
(d) Isotropic material.

4.3.1 The Thermal Resistance Network for Plane Walls—Brief Review

Figure 4.5 shows a schematic drawing that presents the temperature distribution around and through a one-dimensional plane wall that experiences convection on both sides.

The schematic diagram shows that the temperature decreases through the wall from the hotter to the colder fluid. Note that $h_{fluid,1}$ and $h_{fluid,2}$ are the convective heat transfer coefficients of the two fluids. Of interest will be the heat transferred across the wall between the two fluids.

The heat flux (heat transfer rate per unit area) through the wall is given by Fourier's law:

$$q_x'' = -k\frac{dT}{dx}, \tag{4.2}$$

where q_x'' is the heat flux, k is the constant thermal conductivity of the wall, and T is the temperature distribution of the wall.

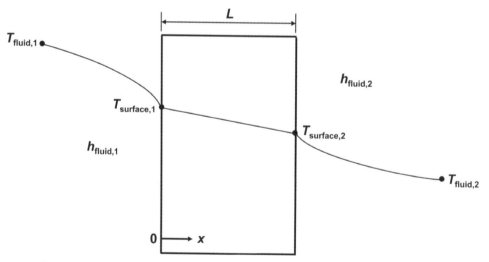

Figure 4.5   Temperature distribution around and through a 1D plane wall

The heat transfer rate is

$$q_x'' = \frac{q_x}{A} = -k\frac{dT}{dx} \tag{4.3}$$

$$q_x = -kA\frac{dT}{dx}, \tag{4.4}$$

where q_x is the heat transfer rate.

To determine the heat transfer rate through the wall, the temperature distribution in the wall (T) must be known. For one-dimensional, steady heat transfer through an isotropic wall, the governing equation for the temperature distribution is

$$\frac{d^2T}{dx^2} = 0. \tag{4.5}$$

Solution of this ordinary, one-dimensional differential equation gives

$$T(x) = Cx + D, \tag{4.6}$$

where C and D are constants of integration.

Two boundary conditions are needed to determine the constants of integration:

(i) $\quad -k\dfrac{dT(0)}{dx} = h_{\text{fluid},1}\left[T_{\text{fluid},1} - T_{\text{surface},1}\right]$ \hfill (4.7)

(ii) $\quad -k\dfrac{dT(L)}{dx} = h_{\text{fluid},2}\left[T_{\text{surface},2} - T_{\text{fluid},2}\right].$ \hfill (4.8)

At this point, make a note of the following points:

$$T_{\text{surface},1} = T(0) \text{ at } x = 0 \tag{4.9}$$

$$T_{\text{surface},2} = T(L) \text{ at } x = 0. \tag{4.10}$$

Therefore, the boundary conditions become

(i) $\quad -k\dfrac{dT(0)}{dx} = h_{\text{fluid},1}\left[T_{\text{fluid},1} - T(0)\right]$ \hfill (4.11)

(ii) $\quad -k\dfrac{dT(L)}{dx} = h_{\text{fluid},2}\left[T(L) - T_{\text{fluid},2}\right].$ \hfill (4.12)

Use the boundary conditions and the temperature distribution equation to find the constants of integration:

$$-kC = h_{\text{fluid},1}\left[T_{\text{fluid},1} - D\right] \tag{4.13}$$

$$-kC = h_{\text{fluid},2}\left[CL + D - T_{\text{fluid},2}\right]. \tag{4.14}$$

Solve the equations for the C constant of integration:

$$C = \frac{T_{\text{fluid},2} - T_{\text{fluid},1}}{k\left[\dfrac{1}{h_{\text{fluid},1}} + \dfrac{L}{k} + \dfrac{1}{h_{\text{fluid},2}}\right]}. \tag{4.15}$$

Thus,

$$T(x) = \frac{T_{\text{fluid},2} - T_{\text{fluid},1}}{k\left[\dfrac{1}{h_{\text{fluid},1}} + \dfrac{L}{k} + \dfrac{1}{h_{\text{fluid},2}}\right]}x + D, \tag{4.16}$$

and

$$\frac{dT}{dx} = \frac{T_{\text{fluid},2} - T_{\text{fluid},1}}{k\left[\dfrac{1}{h_{\text{fluid},1}} + \dfrac{L}{k} + \dfrac{1}{h_{\text{fluid},2}}\right]}. \tag{4.17}$$

The heat transfer rate $\left(q_x = -kA\frac{dT}{dx}\right)$ becomes

$$q_x = \frac{T_{\text{fluid},1} - T_{\text{fluid},2}}{\dfrac{1}{Ah_{\text{fluid},1}} + \dfrac{L}{Ak} + \dfrac{1}{Ah_{\text{fluid},2}}}. \tag{4.18}$$

The terms in the denominator of the expression for the heat transfer rate represent the resistances to heat transfer. Figure 4.6 shows this **resistance network** around the wall.

Therefore,

$$\text{Convective Resistance 1:} \quad R_{\text{cv},1} = \frac{1}{Ah_{\text{fluid},1}} \tag{4.19}$$

$$\text{Conductive Resistance:} \quad R_{\text{cd}} = \frac{L}{Ak} = R_{\text{wall}} \tag{4.20}$$

$$\text{Convective Resistance 2:} \quad R_{\text{cv},2} = \frac{1}{Ah_{\text{fluid},2}}. \tag{4.21}$$

Thus,

$$q_x = \frac{T_{\text{fluid},1} - T_{\text{fluid},2}}{\frac{1}{Ah_{\text{fluid},1}} + R_{\text{wall}} + \frac{1}{Ah_{\text{fluid},2}}}. \tag{4.22}$$

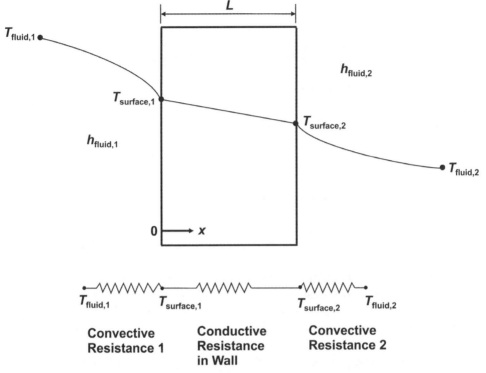

Figure 4.6 Thermal resistance network around a plane wall

The heat transfer rate may be rewritten as

$$q_x = \frac{\Delta T_{\text{fluid}}}{R_T},$$ (4.23)

where $\Delta T_{\text{fluid}} = T_{\text{fluid},1} - T_{\text{fluid},2}$ and $R_T = \frac{1}{Ah_{\text{fluid},1}} + R_{\text{wall}} + \frac{1}{Ah_{\text{fluid},2}} = $ total resistance to heat transfer.

It is typical to see the heat transfer rate through the wall written in a form similar to that of Newton's law of cooling:

$$q_x = \frac{\Delta T_{\text{fluid}}}{R_T} = UA\Delta T_{\text{fluid}},$$ (4.24)

where U **is the overall heat transfer coefficient.**

Therefore,

$$R_T = \frac{1}{UA} = \frac{1}{Ah_{\text{fluid},1}} + R_{\text{wall}} + \frac{1}{Ah_{\text{fluid},2}}$$ (4.25)

and

$$\frac{1}{U} = \frac{1}{h_{\text{fluid},1}} + R_{\text{wall}} A + \frac{1}{h_{\text{fluid},2}}. \tag{4.26}$$

It should be noted at this point that the area, A is the **surface area** of the wall. It should also be noted that the expression for the wall resistance (R_{wall}) will be different for plane walls and cylindrical walls (pipes, tubes):

$$\text{For plane walls:} \quad R_{\text{wall}} = \frac{L}{Ak}. \tag{4.27}$$

$$\text{For cylindrical walls:} \quad R_{\text{wall}} = \frac{1}{2\pi kL} \ln\left(\frac{r_{\text{o}}}{r_{\text{i}}}\right). \tag{4.28}$$

For cylindrical walls, r_{o} is outer wall thickness, r_{i} is the inner wall thickness, and $A = 2\pi rL$.

Later in this chapter, it will be shown that determination of the overall heat transfer coefficient is essential to the design of heat exchangers. Tables 4.1 and 4.2 provide some representative values for some types of heat exchangers. Note that **heat exchangers are defined based on the working fluids**.

4.3.2 Thermal Resistance from Fouling—The Fouling Factor

An additional resistance may be encountered on heat exchanger tubes or plates that have been in operation for prolonged periods of time. Materials or other contaminants may deposit on the surface, restricting heat transfer. This deposition of material

Table 4.1   Values of the overall heat transfer coefficient (US)

Type of Heat Exchanger	U, Btu/(h ft² °F)
Water-to-water	150–300
Water-to-oil	18–60
Water-to-gasoline or kerosene	55–180
Feedwater heaters	180–1500
Steam-to-light fuel oil	35–70
Steam-to-heavy fuel oil	10–35
Steam condenser	180–1060
Freon condenser (water cooled)	55–180
Ammonia condenser (water cooled)	140–250
Alcohol condensers (water cooled)	45–125
Gas-to-gas	2–7
Water-to-air in finned tubes (water in tubes)	5–10 (air); 70–150 (water)
Steam-to-air in finned tubes (steam in tubes)	5–50 (air); 70–705 (water)

Table 4.2 Values of the overall heat transfer coefficient (SI)

Type of Heat Exchanger	U (W/(m^2 °C))
Water-to-water	850–1700
Water-to-oil	100–350
Water-to-gasoline or kerosene	300–1000
Feedwater heaters	1000–8500
Steam-to-light fuel oil	200–400
Steam-to-heavy fuel oil	50–200
Steam condenser	1000–6000
Freon condenser (water cooled)	300–1000
Ammonia condenser (water cooled)	800–1400
Alcohol condensers (water cooled)	250–700
Gas-to-gas	10–40
Water-to-air in finned tubes (water in tubes)	30–60 (air); 400–850 (water)
Steam-to-air in finned tubes (steam in tubes)	30–300 (air); 400–4000 (water)

Source: Çengel [3].

or contaminants on the heat exchanger surfaces is known as **fouling**. Fouling will deteriorate the performance of the heat exchanger by adding an additional resistance.
Some sources of fouling include

(i) precipitation of salts from hard water;
(ii) chemical reactions which form solid deposits;
(iii) biological species growing on surfaces;
(iv) sedimentation of solid deposits;
(v) corrosion, which forms low thermal conductivity metal oxides on the heat exchanger surfaces.

The design engineer must make allowances to account for fouling, where applicable. In this case, the designer must consider the fouling factor, R_f, or the additional resistance due to fouling in the calculation of the overall heat transfer coefficient.
Therefore, with fouling, the overall heat transfer coefficient becomes

$$\frac{1}{U} = \frac{1}{h_{fluid,1}} + R_{f,1} + R_{wall} A + R_{f,2} + \frac{1}{h_{fluid,2}}, \tag{4.29}$$

where $R_{f,1}$ is the fouling resistance on the side of the wall in contact with fluid 1 and $R_{f,2}$ is the fouling resistance on the side of the wall in contact with fluid 2.
Table 4.3 provides some representative values of fouling factors for some working fluids.

Table 4.3 Representative fouling factors in heat exchangers

Fluid	R_f (ft^2 h °F)/Btu
Gas oil	0.00051
Transformer oil	0.00102
Lubrication oil	0.00102
Heat transfer oil	0.00102
Hydraulic oil	0.00102
Fuel oil	0.0051
Hydrogen	0.00999
Engine exhaust	0.00999
Steam (oil-free)	0.00051
Steam with oil traces	0.0010
Cooling fluid vapors with oil traces	0.00199
Organic solvent vapors	0.0010
Alcohol vapors	0.00057
Refrigerants (vapor)	0.0023
Compressed air	0.00199
Natural gas	0.0010
Distilled water, seawater, river water, boiler feedwater: below 122°F	0.00057
Distilled water, seawater, river water, boiler feedwater: above 122°F	0.0011
Refrigerants (liquid)	0.0011
Cooling fluid	0.0010
Organic heat transfer fluids	0.0010
Salts	0.00051
Liquefied petroleum gas (LPG), liquefied natural gas (LNG)	0.0010
MEA and DEA (Amines) solutions	0.00199
DEG and TEG (glycols) solutions	0.00199
Vegetable oils	0.0030

4.4 The Convection Heat Transfer Coefficients—Forced Convection

Determination of the convective heat transfer coefficients, $h_{fluid,1}$ and $h_{fluid,2}$, is needed to calculate values of the overall heat transfer coefficient. The heat transfer coefficient is a parameter (not a property) that depends on the surface geometry, fluid velocity, fluid properties, and temperatures of the solid surface and fluid.

Heat transfer coefficients are typically nondimensionalized as the **Nusselt number**. The Nusselt number is defined as

$$Nu = \frac{h\delta}{k_{fluid}},$$

(4.30)

where δ is a characteristic length.

For heat exchangers with circular tubes, $\delta = D$. For noncircular tubes, the hydraulic diameter is used. The hydraulic diameter is $D_h = \frac{4A}{P}$.

Therefore,

$$Nu = \frac{hD}{k_{fluid}}.$$ (4.31)

Fundamental Note: The thermal conductivity (k) in the Nusselt number is that of the fluid at a **mean temperature (T_m)**. This mean temperature is typically the average of the inlet and outlet temperatures of the fluid in the tube.

4.4.1 Nusselt Number—Fully Developed Internal Laminar Flows

For fully-developed (hydrodynamic and thermal boundary layers have merged) laminar flows in channels (circular tubes, square ducts, and pipes), the Nusselt numbers will be different for constant surface temperature (T_s = constant) and constant surface heat flux (q_s'' = constant) conditions. Table 4.4 shows values of the Nusselt number for a variety of channel geometries.

4.4.2 Nusselt Number—Developing Internal Laminar Flows—Correlation Equation

A correlation equation exists for the case of internal laminar flows with fully developed velocity profiles (hydrodynamic boundary layers have merged) and developing temperature profiles (thermal boundary layers have not merged). Note that a correlation equation is an equation that was partially developed analytically, and experimental data was used to develop the final form of the equation.

For developing temperature profiles,

$$L_{tube} \leq L_{t,laminar}$$

where $L_{t,laminar} \approx 0.05 \, Re \, Pr D$ = thermal entry length and $Pr = \frac{c_p \mu}{k_{fluid}}$ = Prandtl number. Therefore, the correlation equation is [4]

$$Nu = 3.66 + \frac{0.065 \left(\frac{D}{L}\right) Re \, Pr}{1 + 0.04 \left[\left(\frac{D}{L}\right) Re \, Pr\right]^{2/3}}.$$ (4.32)

This correlation equation is applied to the following:

(i) Entrance region of the tube.
(ii) Uniform surface temperature.
(iii) Fully developed velocity profile in laminar flows. This correlation equation will give good approximations for hydrodynamically developing laminar flows.
(iv) Developing temperature profiles in laminar flow.
(v) Fluid properties that are at the mean temperature of the bulk fluid.

Table 4.4 Nusselt numbers and friction factors for fully developed laminar flow in tubes of various cross sections: constant surface temperature and surface heat flux [3]

Pipe Geometry	L/H ratio or θ	Nusselt Number, Nu_{D_h}		Friction Factor, f
		$T_s = $ constant	$q_s'' = $ constant	
Circle	—	3.66	4.36	64.00/Re
Square/Rectangle	L/H ratio			
	1	2.98	3.61	56.92/Re
	2	3.39	4.12	62.20/Re
	3	3.96	4.79	68.36/Re
	4	4.44	5.33	72.92/Re
	6	5.14	6.05	78.80/Re
	8	5.60	6.49	82.32/Re
	∞	7.54	8.24	96.00/Re
Ellipse	L/H ratio			
	1	3.66	4.36	64.00/Re
	2	3.74	4.56	67.28/Re
	4	3.79	4.88	72.96/Re
	8	3.72	5.09	76.60/Re
	16	3.65	5.18	78.16/Re
Isosceles Triangle	Θ			
	10°	1.61	2.45	50.80/Re
	30°	2.26	2.91	52.28/Re
	60°	2.47	3.11	53.32/Re
	90°	2.34	2.98	52.60/Re
	120°	2.00	2.68	50.96/Re

In some cases, the temperature of the fluid in contact with the surface of the tube may differ greatly from that of the bulk fluid in the rest of the tube. In that case, viscosity differences in the fluid will affect the Nusselt number. The appropriate correlation equation to use is [5]:

$$Nu = 1.86 \left(\frac{Re\, Pr D}{L} \right)^{1/3} \left(\frac{\mu_{\text{bulk fluid}}}{\mu_{\text{surface fluid}}} \right)^{0.14}, \tag{4.33}$$

where $\mu_{\text{surface fluid}}$ is the viscosity of the fluid at the surface temperature.

This correlation equation is applied to

(i) $0.48 < \text{Pr} < 16700$;
(ii) $0.0044 < \frac{\mu_{\text{bulk fluid}}}{\mu_{\text{surface fluid}}} < 9.75$.

4.4.3 Nusselt Number—Turbulent Flows in Smooth Tubes: Dittus–Boelter Equation

For fully developed turbulent flows in smooth tubes, the Dittus–Boelter correlation equation [6] may be used to determine the Nusselt number.
Therefore,

$$\text{Nu} = 0.023\text{Re}^{0.8}\text{Pr}^n. \tag{4.34}$$

Note: $n = 0.4$ for heating the fluid flowing in the tube.
$n = 0.3$ for cooling the fluid flowing in the tube.

This correlation equation is applied to

(i) fully developed turbulent flow;
(ii) smooth tubes;
(iii) $\text{Re}_D > 10000$;
(iv) $0.7 < \text{Pr} < 160$;
(v) $\frac{L}{D} > 60$.

In some cases, the temperature of the fluid in contact with the surface of the tube may differ greatly from that of the bulk fluid in the rest of the tube. In that case, viscosity differences in the fluid will affect the Nusselt number. The appropriate correlation equation to use is [5]:

$$\text{Nu} = 0.027\text{Re}^{0.8}\text{Pr}^{1/3} \left(\frac{\mu_{\text{bulk fluid}}}{\mu_{\text{surface fluid}}} \right)^{0.14}, \tag{4.35}$$

where $\mu_{\text{surface fluid}}$ is the viscosity of the fluid at the surface temperature.
This correlation equation is applied to

(i) $0.7 < \text{Pr} < 17600$;
(ii) $\text{Re}_D > 10000$.

4.4.4 Nusselt Number—Turbulent Flows in Smooth Tubes: Gnielinski's Equation

For fully developed turbulent flows in smooth tubes, Gnielinski's correlation equation [7] may be used as an alternative to the Dittus–Boelter correlation equation to determine the Nusselt number. Though more complex, Gnielinski's correlation equation will give more accurate values of the Nusselt number, especially at lower Reynolds number flows ($\text{Re}_D > 2300$).

Therefore,

$$\text{Nu} = \frac{\left(\dfrac{f}{8}\right)(\text{Re}_D - 1000)\,\text{Pr}}{1.0 + 12.7\left(\dfrac{f}{8}\right)^{0.5}\left(\text{Pr}^{2/3} - 1\right)}\left[1 + \left(\frac{D}{L}\right)^{2/3}\right], \qquad (4.36)$$

where f is the Darcy friction factor. Consult Chapter 2 for correlation equations for the friction factor.

Note: For fully developed flow, $\frac{D}{L} \approx 0$.

This correlation equation is applied to

(i) developing or fully developed turbulent flow;
(ii) smooth tubes;
(iii) $2300 < \text{Re}_D < 5 \times 10^6$;
(iv) $0.5 < \text{Pr} < 2000$;
(v) $0 < \frac{D}{L} < 1$.

Practical Note 4.1   Industrial Flows

For most industrial applications involving flow through tubes in heat exchangers, the flow is turbulent.

Practical Note 4.2   Flow in Rough Pipes

In practice, the Dittus–Boelter and Gnielinski's correlation equations are also used with rough surfaces. This is due to a lack of appropriate, simple equations for rough tubes.

4.5 Heat Exchanger Analysis

4.5.1 Preliminary Considerations

Heat exchanger analysis involves the determination of the size (area, volume, length of tubes, where applicable) of the heat exchanger required to transfer a specified amount of heat and the quantification of the performance of the heat exchanger.

Some fundamental assumptions are as follows:

(i) Heat exchangers are steady-flow devices. That is, the system is in steady state.
(ii) All fluid and thermal properties are constant.

(iii) The overall heat transfer coefficient (U) is constant.

(iv) There is no heat exchange between the heat exchanger and the surroundings.

 (v) All heat exchange occurs between the fluids through the solid surfaces.

On the basis of these assumptions, the first law of thermodynamics may be written as

$$\dot{Q}_{hot} = \dot{Q}_{cold}, \tag{4.37}$$

where $\dot{Q}_{hot}$ is the heat transfer rate from the hot fluid and $\dot{Q}_{cold}$ is the heat transfer rate to the cold fluid.

For flowing fluids,

$$\dot{Q}_{hot} = \dot{m}_h c_{ph}(T_{h,in} - T_{h,out}) \tag{4.38}$$

$$\dot{Q}_{cold} = \dot{m}_c c_{pc}(T_{c,out} - T_{c,in}), \tag{4.39}$$

where the subscripts "in" and "out" represent the fluid entering and leaving the heat exchanger system, respectively.

Therefore,

$$\dot{Q}_{hot} = \dot{Q}_{cold} = \dot{m}_h c_{ph}(T_{h,in} - T_{h,out}) = \dot{m}_c c_{pc}(T_{c,out} - T_{c,in}). \tag{4.40}$$

In heat exchanger analysis, the heat capacity rates are used.

$$C_h = \dot{m}_h c_{ph} \text{ and } C_c = \dot{m}_c c_{pc}. \tag{4.41}$$

Thus,

$$C_h(T_{h,in} - T_{h,out}) = C_c(T_{c,out} - T_{c,in}). \tag{4.42}$$

4.5.2 Axial Temperature Variation in the Working Fluids—Single Phase Flow

As heat is transferred along the length of the heat exchanger, the temperature of the working fluids will change along the length of heat exchanger (axial direction). So, the temperature of the hot fluid may decrease from the inlet to the outlet of the heat exchanger while the temperature of the cold fluid may increase from the inlet to the outlet as heat is transferred.

Single-pass parallel flow heat exchangerc

In this case, the working fluids enter and exit at the same end. At the exit point, the temperatures of the hot and cold fluids approach equality, since the temperature difference (ΔT) becomes smaller (see Figure 4.7).

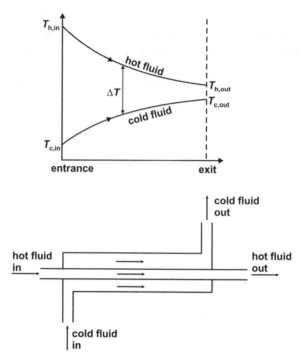

Figure 4.7 Axial temperature variation in parallel flow heat exchanger

Single-pass counter flow heat exchanger

In this case, the working fluids enter and leave at opposite ends. It requires longer heat exchangers to achieve equality of the temperatures at the exit points of any of the fluids. This improves the ability of counter flow heat exchangers to transfer more heat compared to parallel flow heat exchangers since ΔT remains larger (see Figure 4.8).

Single-pass counter flow heat exchanger with ΔT = constant

In this single-pass counter flow heat exchanger case, the temperature difference between the hot and cold fluids are maintained constant along the length of the heat exchanger (see Figure 4.9). For this to occur, the specific heat rates must be equal. Thus,

$$C_h = C_c = \dot{m}_h c_{ph} = \dot{m}_c c_{pc} \text{ (``balanced'' heat exchanger).} \quad (4.43)$$

Then,

$$(T_{h,in} - T_{h,out}) = (T_{c,out} - T_{c,in}) \quad (4.44)$$
$$(T_{h,in} - T_{c,out}) = (T_{h,out} - T_{c,in}) = \Delta T = \text{constant.} \quad (4.45)$$

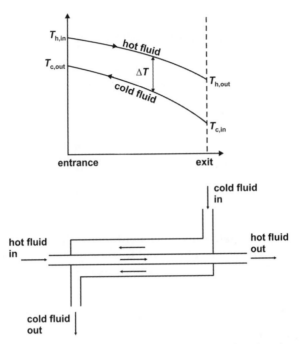

Figure 4.8   Axial temperature variation in counter flow heat exchanger

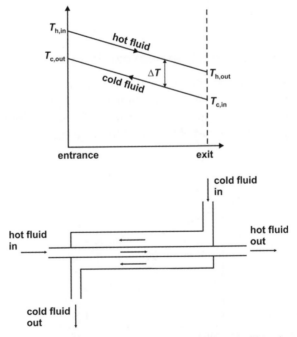

Figure 4.9   Axial temperature variation in a balanced heat exchanger

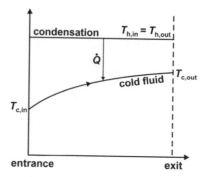

Figure 4.10 Axial temperature variation in a heat exchanger with condensation

Fundamental Note: The temperatures of the hot fluid (including $T_{h,in}$ and $T_{h,out}$) must be greater than the temperatures of the cold fluid (including $T_{c,in}$ and $T_{c,out}$) for heat transfer in accordance with the second law of thermodynamics.

Condensation in counter flow or parallel heat exchangers

In this case, a **phase change** occurs in one of the working fluids. The hot fluid will condense by transferring heat to the cold fluid. Since this is a phase change situation, the temperature of the hot fluid will remain constant. As the cold fluid becomes hotter, its temperature will approach that of the hot fluid (see Figure 4.10).

Boiling in counter flow or parallel heat exchangers

In this case, a phase change occurs in one of the working fluids. The cold fluid will boil due to the transfer of heat from the hot fluid. Since this is a phase change situation, the temperature of the cold fluid will remain constant. As the hot fluid becomes colder, its temperature will approach that of the cold fluid (see Figure 4.11).

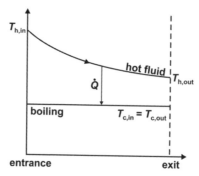

Figure 4.11 Axial temperature variation in a heat exchanger with boiling

> **Practical Note 4.3   Condensers and Boilers**
>
> **Condensers** and **boilers** are heat exchangers in which condensation and boiling phase changes occur, respectively. **Pinch point analysis** and other Nusselt number correlation equations for phase changes should be used for heat exchangers with phase changes.

4.6 Heat Exchanger Design and Performance Analysis: Part 1

Heat exchanger **design or sizing** refers to the specification of the construction and flow arrangement of the heat exchanger.

This may involve determination of the following:

(i)   Heat transfer surface area (A)
(ii) Tube diameters (D)
(iii) Tube lengths (L)
(iv) Number of tubes (n)

The objective of the design will be to provide a heat exchanger that will provide a specified outlet temperature of the working fluids or transfer a specified amount of heat.

Heat exchanger **performance** analysis serves to determine the outlet temperatures (T_{out}) of the working fluids and/or the amount of heat ($\dot{Q}$) exchanged by the heat exchanger. As a design engineer, it may be necessary to analyze the performance of a newly designed or existing heat exchanger within a system.

4.6.1 The Log-Mean Temperature Difference Method

This method may be used to determine the size (heat transfer surface area), the outlet temperatures of the working fluid, or the heat transfer rate of/within the heat exchanger. This method can be used to find the amount of heat exchanged in the heat exchanger by using Newton's law of cooling:

$$\dot{Q} = UA\Delta T_{lm},\qquad(4.46)$$

where U is the overall heat transfer coefficient, A is the heat exchanger surface area, and ΔT_{lm} is the log-mean temperature difference (LMTD).

This method works best in heat exchanger design and sizing problems when **all** the inlet and outlet temperatures of the working fluids are known.

In heat exchanger performance analysis where all the temperatures of the working fluids are not known and the outlet temperatures must be determined, the LMTD method becomes iterative, tedious, and long. In this course, emphasis will be placed on the more powerful **effectiveness-number of transfer units (ε-NTU) method** for both heat exchanger design and performance analysis.

4.6.2 The Effectiveness-Number of Transfer Units Method: Introduction

This method was developed by Kays and London [2] to allow for easier design and performance analysis of heat exchangers without exhaustive iterations. The method is more straightforward than the LMTD method.

The dimensionless heat transfer effectiveness (ε) is defined as

$$\varepsilon = \frac{\dot{Q}}{\dot{Q}_{max}} = \frac{\text{actual heat transfer rate}}{\text{maximum possible heat transfer rate}}. \tag{4.47}$$

This expression may be considered as the efficiency of the heat exchanger to exchange energy between the fluid streams.

The maximum heat transfer rate ($\dot{Q}_{max}$) will occur over the maximum possible temperature difference. The maximum possible temperature difference is

$$\Delta T_{max} = T_{h,in} - T_{c,in}. \tag{4.48}$$

In real heat exchangers, the maximum heat transfer and maximum temperature difference will occur in the working fluid with smaller heat capacity. The fluid with the smaller heat capacity will experience faster temperature changes, will store less energy, and will transfer heat faster.

Therefore,

$$\dot{Q}_{max} = \dot{m}c_{p,min} \left(T_{h,in} - T_{c,in} \right) = C_{min} \left(T_{h,in} - T_{c,in} \right). \tag{4.49}$$

The working fluid with the lower heat capacity ($c_{p,min}$) or lower heat capacity rate (C_{min}) could be either hot or cold.

Hence,

$$C_{min} = C_c \text{ or } C_{min} = C_h. \tag{4.50}$$

The effectiveness becomes

$$\varepsilon = \frac{C_c}{C_{min}} \frac{\left(T_{c,out} - T_{c,in} \right)}{\left(T_{h,in} - T_{c,in} \right)}$$

or

$$\varepsilon = \frac{C_h}{C_{min}} \frac{\left(T_{h,in} - T_{h,out} \right)}{\left(T_{h,in} - T_{c,in} \right)}. \tag{4.51}$$

The actual heat transfer rate is

$$\dot{Q} = \varepsilon \dot{Q}_{max} = \varepsilon C_{min} \left(T_{h,in} - T_{c,in} \right). \tag{4.52}$$

4.6.3 The Effectiveness-Number of Transfer Units Method: ε-NTU Relations

A variety of ε-NTU relations have been developed to facilitate heat exchanger design and performance analysis. The derivation of the ε-NTU relation for a simple double-pipe parallel flow heat exchanger is shown below. For other types of heat exchangers, the derivations become complicated. As a result, designers rely on referenced relation equations or charts.

Derivation of the ε-NTU relation for a simple double-pipe parallel flow heat exchanger

Consider a single-pass parallel-flow heat exchanger with a schematic drawing shown below:

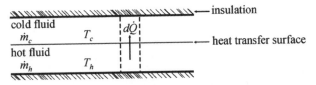

Take an infinitesimal section of the system for analysis

$$d\dot{Q} = -\dot{m}_h c_{ph} dT_h \tag{4.53}$$
$$d\dot{Q} = +\dot{m}_c c_{pc} dT_c \tag{4.54}$$

Rearrangement gives

$$dT_h = -\frac{d\dot{Q}}{\dot{m}_h c_{ph}}; \quad dT_c = +\frac{d\dot{Q}}{\dot{m}_c c_{pc}} \tag{4.55}$$

Substract the differential temperatures:

$$dT_h - dT_c = -d\dot{Q}\left[\frac{1}{\dot{m}_h c_{ph}} + \frac{1}{\dot{m}_c c_{pc}}\right]. \tag{4.56}$$

The differential heat transfer rate is governed by Newton's Law of Cooling:

$$d\dot{Q} = U(T_h - T_c)dA \tag{4.57}$$

Substitution:

$$dT_h - dT_c = -U(T_h - T_c)\,dA\left[\frac{1}{\dot{m}_h c_{ph}} + \frac{1}{\dot{m}_c c_{pc}}\right] \qquad (4.58)$$

Or

$$\frac{d(T_h - T_c)}{T_h - T_c} = -U\,dA\left[\frac{1}{\dot{m}_h c_{ph}} + \frac{1}{\dot{m}_c c_{pc}}\right] \qquad (4.59)$$

Integrate between the inlet and outlet of the heat exchanger:

$$\ln(T_{h,out} - T_{c,out}) - \ln(T_{h,in} - T_{c,in}) = -UA\left[\frac{1}{\dot{m}_h c_{ph}} + \frac{1}{\dot{m}_c c_{pc}}\right] \qquad (4.60)$$

$$\ln\left[\frac{T_{h,out} - T_{c,out}}{T_{h,in} - T_{c,in}}\right] = -UA\left[\frac{1}{\dot{m}_h c_{ph}} + \frac{1}{\dot{m}_c c_{pc}}\right]. \qquad (4.61)$$

Remember: $C_h = \dot{m}_h c_{ph}$ and $C_c = \dot{m}_c c_{pc}$.
Therefore,

$$\ln\left[\frac{T_{h,out} - T_{c,out}}{T_{h,in} - T_{c,in}}\right] = -\frac{UA}{C_c}\left[1 + \frac{C_c}{C_h}\right]. \qquad (4.62)$$

From the expressions for heat transfer rate,

$$C_h(T_{h,in} - T_{h,out}) = C_c(T_{c,out} - T_{c,in})$$

$$T_{h,out} = T_{h,in} - \frac{C_c}{C_h}(T_{c,out} - T_{c,in}). \qquad (4.63)$$

Substitution:

$$\ln\left[\frac{T_{h,in} - \dfrac{C_c}{C_h}(T_{c,out} - T_{c,in}) - T_{c,out}}{T_{h,in} - T_{c,in}}\right] = -\frac{UA}{C_c}\left[1 + \frac{C_c}{C_h}\right] \qquad (4.64)$$

$$\ln\left[\frac{T_{h,in} - T_{c,in} - \dfrac{C_c}{C_h}(T_{c,out} - T_{c,in}) + T_{c,in} - T_{c,out}}{T_{h,in} - T_{c,in}}\right] = -\frac{UA}{C_c}\left[1 + \frac{C_c}{C_h}\right] \qquad (4.65)$$

$$\ln\left[1 - \left(1 + \frac{C_c}{C_h}\right)\frac{T_{c,out} - T_{c,in}}{T_{h,in} - T_{c,in}}\right] = -\frac{UA}{C_c}\left[1 + \frac{C_c}{C_h}\right]. \qquad (4.66)$$

Remember that the effectiveness is defined as

$$\varepsilon = \frac{C_c}{C_{min}}\frac{T_{c,out} - T_{c,in}}{T_{h,in} - T_{c,in}}.$$

Therefore,

$$\frac{T_{c,out} - T_{c,in}}{T_{h,in} - T_{c,in}} = \varepsilon \frac{C_{min}}{C_c}. \tag{4.67}$$

Then,

$$\ln\left[1 - \left(1 + \frac{C_c}{C_h}\right)\varepsilon \frac{C_{min}}{C_c}\right] = -\frac{UA}{C_c}\left[1 + \frac{C_c}{C_h}\right]. \tag{4.68}$$

Solve for ε:

$$\varepsilon\frac{C_{min}}{C_c} = \frac{1 - \exp\left[-\dfrac{UA}{C_c}\left(1 + \dfrac{C_c}{C_h}\right)\right]}{\left(1 + \dfrac{C_c}{C_h}\right)}. \tag{4.69}$$

C_{min} can be either C_c or C_h. Let C_{min} be C_c and C_{max} be C_h.
Thus, effectiveness is

$$\varepsilon = \frac{1 - \exp\left[-\dfrac{UA}{C_{min}}\left(1 + \dfrac{C_{min}}{C_{max}}\right)\right]}{\left(1 + \dfrac{C_{min}}{C_{max}}\right)}. \tag{4.70}$$

Let the **number of transfer units** and the **capacity ratio** be

$$NTU = \frac{UA}{C_{min}} \text{ and } c = \frac{C_{min}}{C_{max}}. \tag{4.71}$$

Therefore,

$$\varepsilon = \frac{1 - \exp\left[-NTU\left(1 + c\right)\right]}{(1 + c)} \text{ for double-pipe, parallel flow heat exchangers.} \tag{4.72}$$

Table 4.5 and the charts of Figure 4.12 provide additional ε-NTU relations for other types of heat exchangers.

4.6.4 Comments on the Number of Transfer Units and the Capacity Ratio (c)

(i) The following are some general points regarding the number of transfer units and the capacity ratio (c).

Table 4.5 Effectiveness relations for heat exchangers

Heat Exchanger Type	Effectiveness Relation
Double pipe: parallel flow	$\varepsilon = \dfrac{1 - \exp\left[-NTU\left(1 + c\right)\right]}{1 + c}$
Double pipe: counter flow	$\varepsilon = \dfrac{1 - \exp\left[-NTU\left(1 - c\right)\right]}{1 - c\,\exp\left[-NTU\left(1 - c\right)\right]}$ for $c < 1$
	$\varepsilon = \dfrac{NTU}{1 + NTU}$ for $c = 1$
Shell-and-tube: one-shell pass and an even number of tube passes	$\varepsilon_1 = 2\left[1 + c + \sqrt{1 + c^2}\,\dfrac{1 + \exp\left[-NTU_1\sqrt{1 + c^2}\right]}{1 - \exp\left[-NTU_1\sqrt{1 + c^2}\right]}\right]^{-1}$
Shell-and-tube: N shell passes and $2N, 4N \ldots$ tube passes	$\varepsilon = \dfrac{\left(\dfrac{1 - \varepsilon_1 c}{1 - \varepsilon_1}\right)^N - 1}{\left(\dfrac{1 - \varepsilon_1 c}{1 - \varepsilon_1}\right)^N - c}$
	ε_1 and NTU_1 are for one-shell pass
Cross-flow (single-pass): both fluids unmixed	$\varepsilon = 1 - \exp\left[\dfrac{NTU^{0.22}}{c}\left[\exp\left(-cNTU^{0.78}\right) - 1\right]\right]$
Cross-flow (single-pass): both mixed	$\varepsilon = \left[\dfrac{1}{1 - \exp\left(-NTU\right)} + \dfrac{c}{1 - \exp\left(-cNTU\right)} - \dfrac{1}{NTU}\right]^{-1}$
Cross-flow (single-pass): C_{max} mixed; C_{min} unmixed	$\varepsilon = \dfrac{1}{c}\left(1 - \exp\left[1 - c\left[1 - \exp\left(-NTU\right)\right]\right]\right)$
Cross-flow (single-pass): C_{min} mixed; C_{max} unmixed	$\varepsilon = 1 - \exp\left[-\dfrac{1}{c}\left[1 - \exp\left(-cNTU\right)\right]\right]$
All heat exchangers with $c = 0$ (boiling or condensation)	$\varepsilon = 1 - \exp(-NTU)$

(ii) The $NTU = \frac{UA}{C_{min}}$ is a measure of the heat transfer surface area of the heat exchanger. It is directly proportional to the surface area. Larger NTU values indicate larger heat exchanger surface areas. With larger surface areas, fabrication costs will increase. For NTU values larger than 10, economic factors should drive the fabrication of such heat exchangers.

(iii) The ε-NTU relation charts shown in Section 4.6.3 show that the effectiveness increases rapidly for smaller values of NTU. For NTU greater than 1.5, increases in ε are small. The design engineer may be hard pressed to justify larger heat exchanger and larger NTU values. Therefore, **NTU values are typically lower than 3 to 5** (depending on the heat exchanger configuration).

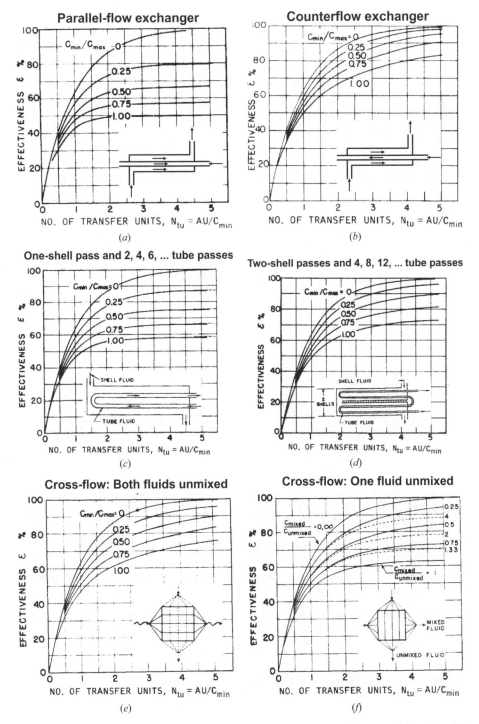

Figure 4.12 Effectiveness charts for some heat exchangers (Kays and London (2))

(iv) NTU is also a measure of the ability of the heat exchanger to transfer energy. Higher NTU values indicate higher heat transfer rates.

(v) For the special case of $c = \frac{C_{min}}{C_{max}} = 0$, $C_{max} \rightarrow \infty$. This means that one of the working fluids absorbs or rejects heat without a temperature change. This is most typical of a phase change situation (condensation or boiling). This case produces the maximum possible ε for a given NTU value and heat exchanger configuration. Boilers, condensers, and refrigeration systems are examples of equipment/systems in which this will occur.

Example 4.1   Heating Water in a Counter Flow Heat Exchanger

A counter flow double-pipe heat exchanger is used to heat water from 70°F to 180°F for use in a specialized self-service laundromat. The water will flow at a rate of 20 gpm. The heating is to be accomplished by hot, compressed geothermal brine available at 320°F at a flow rate of 35 gpm. The inner tube is thin-walled and has a diameter of ⅝ in. If the overall heat transfer coefficient of the heat exchanger is 110 Btu/(h ft² °F), determine the length of the heat exchanger required to achieve the desired heating.

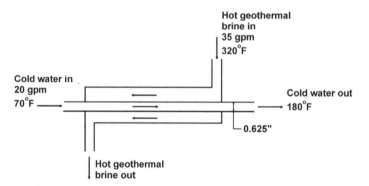

Solution. Several assumptions are relevant and will serve to guide the analysis:

  (i) Steady state operation of the heat exchanger.
 (ii) Constant fluid and thermal properties.
(iii) There is no fouling. So, the fouling resistance is zero.
(iv) The heat exchanger is well insulated. There is no heat loss from the device.
 (v) The brine solution is sufficiently dilute to assume that its properties are equal to those of pure water.

The total length of the heat exchanger is found from the surface area:

$$A = npD_{tube}L.$$

For this single-pipe heat exchanger, $n = 1$.
Therefore,

$$L = \frac{A}{\pi D_{tube}}.$$

The area is needed. The ε-NTU method can be used in the analysis of the heat exchanger. The NTU is

$$NTU = \frac{UA}{C_{min}}.$$

The area is

$$A = \frac{C_{min}NTU}{U}.$$

Consider each term in the area equation. To find C_{min}, the average temperatures will be used. In this case, the average cold fluid temperature is

$$T_c = \frac{T_{c,in} + T_{c,out}}{2} = \frac{70°F + 180°F}{2} = 125°F.$$

Find the cold-water properties at 125°F and the hot water properties at 320°F. The heat capacity rates are:

$$C_c = \dot{m}_c c_{pc} = \rho_c \dot{V}_c c_{pc} = \left(61.63 \text{ lb/ft}^3\right) (20 \text{ gpm}) (0.999 \text{ Btu/(lb R)}) \left(\frac{35.315 \text{ ft}^3/\text{s}}{15850 \text{ gpm}}\right)$$

$$= 2.74 \text{ Btu/(s R)} = C_{min}$$

$$C_h = \dot{m}_h c_{ph} = \rho_h \dot{V}_h c_{ph} = \left(56.65 \text{ lb/ft}^3\right) (35 \text{ gpm}) (1.036 \text{ Btu/(lb R)}) \left(\frac{35.315 \text{ ft}^3/\text{s}}{15850 \text{ gpm}}\right)$$

$$= 4.58 \text{ Btu/(s R)} = C_{max}$$

Find the NTU. The capacity ratio is

$$c = \frac{C_{min}}{C_{max}} = \frac{C_c}{C_h} = \frac{2.74 \text{ Btu/(s R)}}{4.58 \text{ Btu/(s R)}} = 0.599.$$

The effectiveness is

$$\varepsilon = \frac{C_c}{C_{min}} \frac{T_{c,out} - T_{c,in}}{T_{h,in} - T_{c,in}} = \frac{T_{c,out} - T_{c,in}}{T_{h,in} - T_{c,in}} = \frac{(180 - 70) °F}{(320 - 70) °F} = 0.44 = 44\%.$$

This is a counter flow heat exchanger problem. So, with c and ε known, the charts of Figure 4.12 or the equations in Table 4.5 can be used to find the NTU. From the charts,

$$NTU \approx 0.70.$$

The effectiveness equation for a double-pipe counter flow heat exchanger with $c < 1$ gives

$$\varepsilon = \frac{1 - \exp\left[-NTU\left(1 - c\right)\right]}{1 - c\exp\left[-NTU\left(1 - c\right)\right]} = \frac{1 - \exp\left[-0.401\,NTU\right]}{1 - 0.599\exp\left[-0.401\,NTU\right]} = 0.44.$$

Therefore,

$$\text{NTU} = 0.68.$$

With the NTU known, the area can be determined.

$$A = \frac{(2.74 \text{ Btu}/(\text{s R}))(0.68)}{110 \text{ Btu}/(\text{h ft}^2 \text{ R})} \times \frac{3600 \text{ s}}{1 \text{ h}} = 60.98 \text{ ft}^2$$

Thus,

$$L = \frac{60.98 \text{ ft}^2}{\pi (0.625 \text{ in.})} \times \frac{12 \text{ in.}}{1 \text{ ft}}$$

$$L = 373 \text{ ft.}$$

4.6.5 Procedures for the ε-NTU Method

The ε-NTU method can be used in both heat exchanger design and performance analysis problems in a direct, straightforward manner, without the need for iterations/trial-and-error.

(A) Consider the design of a heat exchanger to find the heat transfer surface area, A. In a design problem, the following may/will be known:

$$T_{h,in}; T_{h,out}; T_{c,in}; \dot{m}_c; \dot{m}_h.$$

The ε-NTU solution method approach is as follows:
   (i) Compute ε and c from available data.
   (ii) Determine the NTU from an appropriate relation or chart.
   (iii) Compute A from NTU $= \frac{UA}{C_{min}}$.
   (iv) For circular tubes, the total tube length is found from $A = n\pi D_{tube}L$.
(B) Consider a performance analysis of a heat exchanger to find $T_{c,out}; T_{h,out}; \dot{Q}$. In a performance analysis, the following may/will be known:

$$A; U; \dot{m}_c; \dot{m}_h; T_{h,in}; T_{c,in}.$$

The ε-NTU solution method approach is as follows:
   (i) Calculate the NTU from given data.
   (ii) Use relations or charts to find ε using the NTU and c.
   (iii) Calculate one of the outlet temperatures with $\varepsilon = \frac{C_c}{C_{min}}\frac{(T_{c,out}-T_{c,in})}{(T_{h,in}-T_{c,in})}$ or $\varepsilon = \frac{C_h}{C_{min}}\frac{(T_{h,in}-T_{h,out})}{(T_{h,in}-T_{c,in})}$.
   (iv) Calculate $\dot{Q}$ and the other outlet temperature.

4.6.6 Heat Exchanger Design Considerations

During the design of heat exchangers, consider the following:

(i) Properties of the working fluids must be determined at their mean temperatures. That is,

$$\overline{T}_c = \frac{T_{c,in} + T_{c,out}}{2} \tag{4.73}$$

$$\overline{T}_h = \frac{T_{h,in} + T_{h,out}}{2}. \tag{4.74}$$

In some cases (performance analysis), it may be necessary to assume some temperature values.

(ii) Correlation equations may be needed for the heat transfer coefficients to find the overall heat transfer coefficient. Be prepared to use correlation equations for the Nusselt numbers for internal and external flows.

(iii) The design engineer is responsible for the selection of an appropriate heat exchanger construction, tube material, and flow arrangement. That is, one-shell-pass, two-tube-passes-copper tube, etc.

(iv) It may be necessary to use concepts from fundamental fluid mechanics or charts to size the tubes, if the diameter is not given.

(v) Sizing and selection of an appropriate pump may be necessary. Calculation of the pressure drops in the system would be needed.

4.7 Heat Exchanger Design and Performance Analysis: Part 2

4.7.1 External Flow over Bare Tubes in Cross Flow—Equations and Charts

External flow over tubes is of great importance in heat exchanger design and performance analysis. For example, in shell-and-tube heat exchangers, fluid in the shell flows over the tubes in a similar fashion to **external flow over a tube bank**. In some cases, the tubes in the tube bank will be devoid of **fins—extended heat transfer surfaces**—and they will be **bare**. Figure 4.13 shows finned-tube and bare tube bank bundles.

Heat transfer coefficients (h_o) for external flow over bare tube banks are found through the *j*-**factor** developed by Chilton and Colburn:

$$j = \frac{\overline{h}_o}{G_m c_p} Pr^{2/3}, \tag{4.75}$$

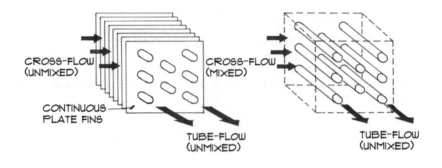

Figure 4.13   (a) Finned tube and (b) bare tube bank bundles

where

$\dfrac{\bar{h}_o}{G_m c_p}$ = St = Stanton number, which is a dimension-less heat transfer coefficient;

$\bar{h}_o$ = average heat transfer coefficient for external flow over the tube bank;

G_m = mass flow rate per unit minimum flow area between the tubes in the bank;

j = j-factor. This is found from charts showing j versus Re_{G_m}.

The mass flow rate per unit minimum flow area through the tube bank passages (G_m) is

$$G_m = \frac{\rho_f V_f A_T}{A_m} = \frac{\dot{m}}{A_m}, \tag{4.76}$$

where "f" refers to the free-stream working fluid flowing towards the front face of the tube bank, A_T is the total front face area of the coil normal to the direction of flow, and A_m is the minimum flow area between the tubes.

Typically, G_m is written as

$$G_m = \frac{\rho_f V_f}{\dfrac{A_m}{A_T}} = \frac{\rho_f V_f}{\sigma}, \tag{4.77}$$

where σ is the ratio of the minimum flow area between the tubes to the total area of the heat exchanger. σ will be presented with the j versus Re_{G_m} charts for different arrangements of tubes in the bank.

The Reynolds number (Re_{G_m}) is based on G_m, and is

$$\mathrm{Re}_{G_m} = \frac{G_m D_h}{\mu}, \tag{4.78}$$

**Flow
direction**

Figure 4.14   Flow pattern for an in-line tube bank (Çengel (3), reprinted with permission)

where D_h is the hydraulic diameter of the flow passage space between the tubes in the bank, and is found from j versus $\mathbf{Re}_{G_m}$ charts.

Charts of j versus $\mathbf{Re}_{G_m}$ for In-line Tube Bank Arrangements

Figure 4.14 shows the pattern of fluid flow through a bank of tubes arranged in line.

Charts of j versus Re_{G_m} must be prepared experimentally. Figure 4.15 shows such a chart for flow through an in-line tube bank. The chart and those presented in Appendix C were developed with experimental data from Kays and London [2], and are approximate. The design engineer is encouraged to consult Reference [2] for more detailed information.

The following observations are made regarding Figure 4.15:

(i) This chart was developed from transient tests. So, the number of rows of tubes is variable, and must be **determined through the design process**. In addition, curves of this type have wider application.
(ii) A curve of the friction factor (f) versus Re_{G_m} is provided at the top of the chart.
(iii) A curve of the j-factor versus Re_{G_m} is provided at the bottom of the chart.
(iv) D is the tube outer diameter, and is given as 0.375 in.
(v) X_t is the traverse tube-pitch ratio, $\frac{x_t}{D}$.

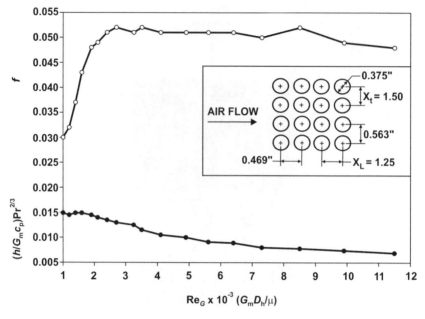

DATA
Tube outside diameter: 0.375 in.

Hydraulic diameter, D_h: 0.0248 ft

Free-flow area/Frontal area, σ: 0.333

Heat transfer area/Total volume, α: 53.6 ft²/ft³

Figure 4.15   Data for flow normal to an in-line tube bank (Kays and London (2))

(vi) X_L is the longitudinal tube-pitch ratio, $\frac{x_L}{D}$.
(vii) σ is given.
(viii) α is the ratio of the total heat transfer area on the external side of the tubes (outer surface area) to the total volume of the exchanger.
(ix) D_h is given.

Additional charts are provided in Appendix C. Refer to the appendix for a more extensive listing of charts.

Charts of *j* versus Re$_{G_m}$ for Staggered Tube Bank Arrangements

Figure 4.16 shows the pattern of fluid flow through a bank of tubes arranged in a staggered fashion.

Figure 4.17 shows a chart for flow through staggered tube banks.

**Flow
direction**

Figure 4.16 Flow pattern for a staggered tube bank (Çengel (3), reprinted with permission)

The following observations are made regarding Figure 4.17:

(i) This chart was developed from transient tests. So, the number of rows of tubes is variable, and must be **determined through the design process**. In addition, curves of this type have wider application.

Additional charts are provided in Appendix C. Refer to the appendix for a more extensive listing of charts for staggered tube banks.

Practical Note 4.5   Heat Transfer from Staggered Tube Banks

Observation of the staggered tube banks show that the heat transfer coefficient is larger for a given value of Re_{G_m} compared to the in-line tube banks, especially for lower values of Re_{G_m}. Therefore, the rate of heat transfer will be higher with staggered tube banks.

Fundamental Note: Churchill and Bernstein [8] have presented a correlation equation for external flow over **single cylinders**. The equation applies when RePr greater than 0.2 and the film temperature must be used to find the fluid properties.
    Therefore,

$$\text{Nu}_{\text{cyl}} = 0.3 + \frac{0.62\,\text{Re}^{1/2}\text{Pr}^{1/3}}{\left[1+(0.4/\text{Pr})^{2/3}\right]^{1/4}}\left[1+\left(\frac{\text{Re}}{282000}\right)^{5/8}\right]^{4/5}. \qquad (4.79)$$

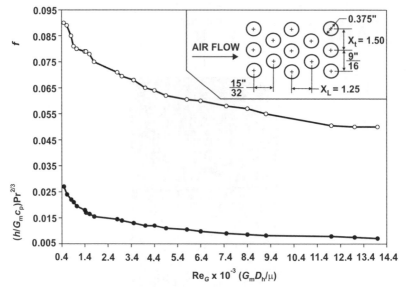

Figure 4.17 Data for flow normal to a staggered tube bank (Kays and London (2))

This correlation equation cannot be used for a bank of tubes since the fluid flows over the tubes will have an influence on each other. However, this equation may apply to the case of a single row of circular tubes oriented such that the flow over them will not be affected by the presence of others.

4.7.2 External Flow over Tube Banks—Pressure Drop

Pressure drop for **external flow over bare and finned-tubes in tube banks** is given empirically as [2]

$$\Delta P_{\text{bank}} = \frac{G_m^2}{2\rho_i}\left[\left(K_i + 1 - \sigma^2\right) + 2\left(\frac{\rho_i}{\rho_e} - 1\right) + f\frac{A_T}{A_c}\frac{\rho_i}{\rho_{\text{mean}}} - \left(1 - \sigma^2 - K_e\right)\frac{\rho_i}{\rho_e}\right],$$

(4.80)

where

"i" = inlet to the heat exchanger tube bank;
"e" = outlet of the heat exchanger tube bank;
K_i = loss coefficient at the inlet to the tube bank;
K_e = loss coefficient at the outlet of the tube bank;
f = friction factor found from the charts presented in Figures 4.15, 4.17, and Appendix C;
ρ_{mean} = average fluid density, $\dfrac{\rho_i + \rho_e}{2}$;
$\dfrac{A_T}{A_c}$ = $\dfrac{\text{total heat transfer area}}{\text{flow cross – sectional area}}$.

Note that

$$\frac{A_T}{A_c} = \frac{4L_{total}}{D_h}, \tag{4.81}$$

where L_{total} is the total length of the heat exchanger in the direction of flow. Figure 4.18 shows a schematic drawing of this total length.

For most industrial applications, the inlet and exit losses are small compared to the other losses in the core of the heat exchanger. So, $K_i = K_e \approx 0$.

Therefore,

$$\Delta P_{bank} = \frac{G_m^2}{2\rho_i}\left[(1+\sigma^2)\left(\frac{\rho_i}{\rho_e}-1\right) + f\frac{A_T}{A_c}\frac{\rho_i}{\rho_{mean}}\right]. \tag{4.82}$$

In terms of the head loss in units of length ($l_{h,bank}$), remember: $\Delta P_{bank} = \rho_{mean}g l_{h,bank}$. Thus,

$$l_{h,bank} = \frac{G_m^2}{2\rho_{mean}g\rho_i}\left[(1+\sigma^2)\left(\frac{\rho_i}{\rho_e}-1\right) + f\frac{A_T}{A_c}\frac{\rho_i}{\rho_{mean}}\right]. \tag{4.83}$$

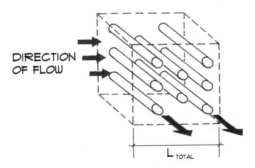

Figure 4.18   Schematic drawing of tube bank showing the total length, L_{total}

Practical Note 4.6   Coil Arrangement in Air-to-Water Heat Exchangers

For most small applications that involve air-to-water heat exchangers, where airflows over a tube bank and water flows in the tubes, the **number of rows of tubes is usually 4 to 6**. While oversizing the heat exchanger may be necessary due to equipment availability or design requirements, the design engineer should be aware that increased number of tubes in a tube bank will require greater powers from a fan (for air) or a pump (for liquids) to move the fluid through the bank. If the number of rows become larger than 6, the design engineer may wish to increase the center-to-center spacing (longitudinal tube pitch) between the tube rows to reduce pressure losses. An even number of rows is usually preferred to ensure that the tube entrance and exit points are on the same side and level of the heat exchanger. This will approach a counter flow heat exchanger system.

Practical Note 4.7   Pressure Drop Over Tube Banks

The equations presented earlier for the determination of the pressure drop and head loss for external flow over the tube bank apply to both bare tube and finned-tube heat exchangers. **Finned-tube heat exchangers are more widely used in industry.**

Example 4.2   Design of a Heating Coil Heat Exchanger

Mr. Chin has hired a junior design engineer to design a small heating coil for his fresh air intake duct line. Mr. Chin is a frugal entrepreneur, and has requested a bare-tube type heating coil arrangement. The HVAC engineer has presented the following **duty** and specifications for the coil:

 (i) Heat fresh air from 50°F to about 100°F.
 (ii) Duct airflow rate is 2000 cfm.
 (iii) Entering hot water temperature is about 150°F.
 (iv) This is a low-velocity duct section.
 (v) Water connections must be on the same end of the coil system to facilitate easy installation and to approach a counter flow heat exchanger system.
 (vi) One large inlet and exit header is required. The tubes will be attached to the headers.

Further Information: This design problem should address and present the following points:

 (i) Exit water temperature
 (ii) Heating coil configuration and number of coils
 (iii) Final dimensions of the heat exchanger (length, width, and depth)
 (iv) Pressure loss for the airflow through the coil bank
 (v) Head loss in the waterside of the tube

Possible Solution

Definition

Design a heating coil to heat fresh air in a duct line.

Preliminary Specifications and Constraints

(i) Heat fresh in-take outdoor air.
(ii) Heat air to 100°F.
(iii) Bare-tube heating coils are required by the client.
(iv) The airflow rate is constrained to 2000 cfm.
(v) This is a low-velocity duct section.
(vi) The water connections on the coiled heat exchanger must be on the same end of the system.

Detailed Design

Objective

To determine the coil configuration, number of coils, size, and performance of a heat exchanger.

Data Given or Known

(i) Heat fresh air from 50 to 100°F.
(ii) The airflow rate is 2000 cfm.
(iii) The entering hot water temperature is 150°F.

Assumptions/Limitations/Constraints

(i) Let the air velocity over the tubes be 1000 fpm. This is lower than 1200 fpm, which meets the requirement for a low-velocity duct section.
(ii) Let the flow velocity of water in the tubes be about 4 fps. This is acceptable for general building service or potable water. In addition, this velocity does not exceed the erosion limits of any general pipe material.
(iii) The tubes are arranged in a staggered fashion. This will enhance heat transfer.
(iv) Let the pipe material be Type L copper. Copper has high heat transfer properties and availability.
(v) The 180° return bends (regular) will be soldered to make the tube (hairpin) connections. Soldering or brazing is typically done for heating/cooling coils.
(vi) The tube (hairpin) wall thickness is small compared to its outer diameter.
(vii) Let the tube (hairpin) outer diameter be $3/8$ in. This is a reasonable tube size for heating coils used for this application.
(viii) Negligible elevation head. Assume that all components are on the same level.
(ix) This will be a counter flow arrangement.
(x) Entrance/exit losses of the air over the coils will be neglected. Other losses will be much larger.

Sketch

A complete drawing will be provided after the design to show the tube flow circuitry in the heating coil.

Analysis

Determine the overall heat transfer coefficient (U)

The ε-NTU method will be used in this design. To determine the effectiveness and NTU, the overall heat transfer coefficient (U) must be determined first. The overall heat transfer coefficient is

$$\frac{1}{U} = \frac{1}{h_i} + R_{fi} + R_{wall}A + R_{fo} + \frac{1}{h_o}.$$

The wall resistance is

$$R_{wall} = \frac{1}{2\pi k L} \ln\left(\frac{r_o}{r_i}\right).$$

It was assumed that the wall thickness is small. Therefore, $r_o \approx r_i$. So, $R_{wall} \approx 0$. Thus,

$$\frac{1}{U} = \frac{1}{h_i} + R_{fi} + R_{fo} + \frac{1}{h_o}.$$

The heat transfer coefficient for flow in the tube (hairpin) is found from the Dittus–Boelter correlation equation. Note that, though more complex, the Gnielinski correlation could have been used:

$$\text{Nu} = \frac{\overline{h}_i D_i}{k} = 0.023\text{Re}_{D_i}^{0.8}\text{Pr}^{0.3}.$$

The inlet and outlet temperatures of the working fluids will needed to find appropriate properties for the determination of Re_D and Pr.

Given: $T_{a,i} = 50°\text{F}, T_{w,i} = 150°\text{F}$
$\quad\quad\quad T_{a,o} = 100°\text{F}, T_{w,o} = $ unknown.

An educated guess of $T_{w,o}$ will be needed. Water has a very high specific heat capacity compared to air. Therefore, the water temperature changes slowly as heat is transferred. In general,

$$\frac{c_{p,w}}{c_{p,a}} \approx 4.$$

Therefore, if $\Delta T_{air} = (100 - 50)°\text{F} = 50°\text{F}$, then $\Delta T_{water} \approx 10°\text{F}$. Let $T_{w,o} = (150 - 10)°\text{F} = 140°\text{F}$.

The average temperatures are:

$$\overline{T}_a = \frac{(50 + 100)°F}{2} = 75°F$$

$$\overline{T}_w = \frac{(150 + 140)°F}{2} = 145°F.$$

For water at 145°F, the Reynolds number is

$$Re_{D_i} = \frac{\rho \overline{V} D_i}{\mu} = \frac{\left(61 \text{ lb/ft}^3\right)(4 \text{ ft/s})(0.315 \text{ in.})}{2.9 \times 10^{-4} \text{ lb/(fts)}} \times \frac{1 \text{ ft}}{12 \text{ in.}} = 22086.$$

Note that for $\frac{3}{8}$ in. outer diameter Type L copper, the inside diameter is 0.315 in. (Table A.6).

The Prandtl number is $Pr = 2.73$.

The flow in the tubes is turbulent and the Pr is between 0.7 and 160. Therefore, the Dittus–Boelter equation is valid for this analysis.

Thus,

$$\overline{h}_i = \frac{0.023k}{D_i} Re_{D_i}^{0.8} Pr^{0.3}$$

$$\overline{h}_i = \frac{0.023(0.38 \text{ Btu/(h ft R)})}{0.315 \text{ in.}} \times \frac{12 \text{ in.}}{1 \text{ ft}} \times (22086)^{0.8} (2.73)^{0.3}$$

$$\overline{h}_i = 1344 \text{ Btu/(h ft}^2 \text{ R)} = 1344 \text{ Btu/(h ft}^2 \text{ °F)}.$$

The fouling resistances can be found from Table C.3.

Assume temperature of distilled water to be above 122°F: $R_{fi} = 0.00114 \text{ (h ft}^2 \text{ °F)/Btu}$. Assume compressed air: $R_{fo} = 0.00199 \text{ (h ft}^2 \text{ °F)/Btu}$.

The average heat transfer coefficient for airflow over the bare tube bank is found from charts and appropriate correlation equations:

$$\overline{h}_o = \frac{jG_m c_p}{Pr^{2/3}},$$

and the Chilton–Colburn j-factor is

$$j = \frac{\overline{h}_o}{G_m c_p} Pr^{2/3}.$$

The following j-factor versus Re_{G_m} chart (Figure C.2b) will be used since the tube outer diameter is $\frac{3}{8}$ in. In addition, the ratio of the free-flow area to the frontal area (σ) is small. This implies that the mass flow rate per unit minimum flow area between the tubes (G_m)

will be large, resulting in a higher $\bar{h}_o$ and improved heat transfer. Consult Figure C.2 for additional charts.

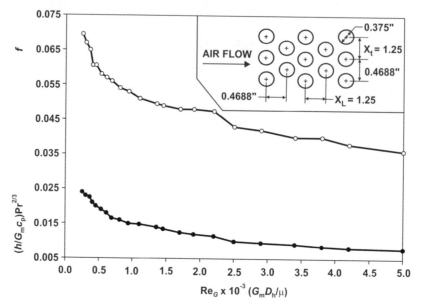

DATA

Tube outside diameter: 0.375 in.

Hydraulic diameter, D_h: 0.0125 ft

Free-flow area/Frontal area, σ: 0.200

Heat transfer area/Total volume, α: 64.4 ft²/ft³

Note: Minimum free-flow area is in the spaces
transverse to the flow

Remember that air properties are found at $\overline{T}_a = 75°F$. So, G_m is

$$G_m = \frac{\rho_a \overline{V}_a}{\sigma} = \frac{\left(0.074 \text{ lb/ft}^3\right)(1000 \text{ ft/min})}{0.200} \times \frac{1 \text{ min}}{60 \text{ s}} = 6.17 \text{ lb/(s ft}^2).$$

The Reynolds number based on G_m is

$$Re_{G_m} = \frac{G_m D_h}{\mu_a} = \frac{\left(6.17 \text{ lb/(s ft}^2)\right)(0.0125 \text{ ft})}{1.25 \times 10^{-5} \text{ lb/(ft s)}} = 6167 \approx 6.0 \times 10^3.$$

From the chart, $j \approx 0.007$. Note that the j-factor changes only slightly beyond $Re_{G_m} = 4.0 \times 10^3$.

Therefore,

$$\bar{h}_o = \frac{(0.007)\left(6.17 \text{ lb}/(\text{s ft}^2)\right)(0.24 \text{ Btu}/(\text{lb }^\circ\text{F}))}{(0.73)^{2/3}} \times \frac{3600 \text{ s}}{1 \text{ h}}$$

$$\bar{h}_o = 46.0 \text{ Btu}/(\text{h ft}^2 \,^\circ\text{F}).$$

As expected, the heat transfer coefficient of the liquid water is higher than that of gaseous air.

The overall heat transfer coefficient is

$$\frac{1}{U} = \frac{1}{1344 \text{ Btu}/(\text{h ft}^2 \,^\circ\text{F})} + 0.00114 \text{ (h ft}^2 \,^\circ\text{F})/\text{Btu} + 0.00199 \text{ (h ft}^2 \,^\circ\text{F})/\text{Btu} + \frac{1}{46.0 \text{ (h ft}^2 \,^\circ\text{F})/\text{Btu}}$$

$$\frac{1}{U} = 0.0256 \text{ (h ft}^2 \,^\circ\text{F})/\text{Btu}$$

$$U = 39 \text{ Btu}/(\text{h ft}^2 \,^\circ\text{F}).$$

Determine the configuration, number of coils, and dimensions of the heat exchanger. Find the total surface area of the tubes by using the ε-NTU method.

$$\text{NTU} = \frac{UA}{C_{min}}$$

$$A = \frac{C_{min}\text{NTU}}{U}$$

The minimum heat capacity is needed. For air,

$$C_a = \dot{m}_a c_{pa} = \rho_a \dot{V}_a c_{pa} = (0.074 \text{ lb}/\text{ft}^3)(2000 \text{ ft}^3/\text{min})(0.24 \text{ Btu}/(\text{lb }^\circ\text{F})) = 35.5 \text{ Btu}/(\text{min }^\circ\text{F}) = C_c.$$

For water, the law of conservation of energy can be used:

$$C_w = C_h = C_c \frac{T_{a,o} - T_{a,i}}{T_{w,i} - T_{w,o}} = (35.5 \text{ Btu}/(\text{min }^\circ\text{F}))\left[\frac{100^\circ\text{F} - 50^\circ\text{F}}{150^\circ\text{F} - 140^\circ\text{F}}\right] = 177.5 \text{ Btu}/(\text{min }^\circ\text{F}).$$

Therefore, $C_a = C_c = C_{min}$ and $C_w = C_h = C_{max}$.

The capacity ratio is

$$c = \frac{C_{min}}{C_{max}} = \frac{35.5 \text{ Btu}/(\text{min }^\circ\text{F})}{177.5 \text{ Btu}/(\text{min }^\circ\text{F})}$$

$$c = 0.20.$$

The effectiveness is

$$\varepsilon = \frac{C_c}{C_{min}} \frac{(T_{a,o} - T_{a,i})}{(T_{w,i} - T_{a,i})} = \frac{(T_{a,o} - T_{a,i})}{(T_{w,i} - T_{a,i})} = \frac{100^\circ\text{F} - 50^\circ\text{F}}{150^\circ\text{F} - 50^\circ\text{F}}$$

$$\varepsilon = 0.50.$$

At this point the thermal capacity or heat exchanged in the heat exchanger can be determined:

$$\dot{Q} = \varepsilon C_{min}\left(T_{w,i} - T_{a,i}\right) = 0.50\,(35.5\ \text{Btu}/(\text{min}\ ^\circ\text{F}))\,(150^\circ\text{F} - 50^\circ\text{F})$$

$$\dot{Q} = \textbf{1775 Btu/min} = \textbf{106500 Btu/h}.$$

The system will be designed to approach a counter flow heat exchanger. With c and ε known, the NTU value can be read directly from Figure 4.12. Hence,

$$\textbf{NTU} \approx \textbf{0.75}.$$

The total heat transfer surface area of the tubes is

$$A = \frac{(35.5\ \text{Btu}/(\text{min}\ ^\circ\text{F}))\,(0.75)}{39\ \text{Btu}/(\text{h}\ \text{ft}^2\ ^\circ\text{F})} \times \frac{60\ \text{min}}{1\ \text{h}}$$

$$A = 41\ \text{ft}^2.$$

The total heat transfer volume is found from the ratio of the heat transfer area to the total volume:

$$\alpha = \frac{A}{\Omega}.$$

Therefore,

$$\Omega = \frac{A}{\alpha} = \frac{41\ \text{ft}^2}{64.4\ \text{ft}^2/\text{ft}^3} = 0.64\ \text{ft}^3.$$

The depth of the heat exchanger system (dimension in the direction of airflow) is

$$W = \frac{\Omega}{A_f}.$$

A_f is the face area of the heat exchanger box normal to the airflow direction.

$$\text{Let: } A_f = \frac{\dot{V}_a}{V_a} = \frac{2000\ \text{ft}^3/\text{min}}{1000\ \text{ft}/\text{min}} = 2\ \text{ft}^2.$$

Thus,

$$W = \frac{0.64\ \text{ft}^3}{2\ \text{ft}^2} \times \frac{12\ \text{in.}}{1\ \text{ft}}$$

$$W = \textbf{3.8 in.} \approx \textbf{4 in.}$$

From the j-factor versus Re_{Gm} chart, the longitudinal tube pitch is 0.4688 in. The number of tube rows is

$$N_r = \frac{W}{x_L} = \frac{3.8 \text{ in.}}{0.4688 \text{ in.}}$$

$$N_r = 8.1 \approx 8 \text{ tube rows.}$$

Determine the number of tubes per row. The cross-sectional area for the tube can be found from the mass flow rate equation (for water):

$$\dot{m}_w = \rho_w \overline{V}_w A_{tube}.$$

The tube area is $A_{tube} = N_{tube} \frac{\pi D_i^2}{4}$. The mass flow rate can be determined from the definition of the heat capacity rate, $C_h = \dot{m}_w c_{p,w}$.
Therefore,

$$\frac{C_h}{c_{p,w}} = \rho_w \overline{V}_w N_{tube} \frac{\pi D_i^2}{4}$$

$$N_{tube} = \frac{4 C_h}{c_{p,w} \rho_w \overline{V}_w \pi D_i^2}$$

$$N_{tube} = \frac{4 \, (177.5 \text{ Btu/(min } ^\circ\text{F))}}{(1.0 \text{ Btu/(lb } ^\circ\text{F))} \left(61 \text{ lb/ft}^3\right) (4 \text{ ft/s}) \, \pi \, (0.315 \text{ in.})^2} \times \left(\frac{12 \text{ in.}}{1 \text{ ft}}\right)^2 \times \frac{1 \text{ min}}{60 \text{ s}}$$

$$N_{tube} = 22.4 \text{ tubes per row} \approx 22 \text{ tubes per row.}$$

The total number of tubes along the length of the heat exchanger will be 176 tubes (8 rows × 22 tubes per row). If each row has 22 tubes, the height of the heating coil is

$$H = N_{tube} x_t = (22)(0.4688 \text{ in.})$$

$$H = 10.3 \text{ in.} \approx 11 \text{ in.}$$

The length of the heat exchanger is found by considering an alternate definition of the face area normal to the direction of airflow:

$$A_f = LH.$$

The heat exchanger length is

$$L = \frac{A_f}{H} = \frac{2 \text{ ft}^2}{10.3 \text{ in.}} \times \left(\frac{12 \text{ in.}}{1 \text{ ft}}\right)^2$$

$$L = 27.96 \text{ in.} \approx 28 \text{ in.}$$

The total length of the tubes in the heat exchanger is 410 ft (28 in. per tube × 176 tubes).

Pressure loss of air across the tube coils

The pressure drop of the air across the tube coil bank is given by

$$\Delta P_{\text{bank}} = \frac{G_m^2}{2\rho_{a,i}} \left[(1 + \sigma^2) \left(\frac{\rho_{a,i}}{\rho_{a,o}} - 1 \right) + f \frac{A_T}{A_c} \frac{\rho_{a,i}}{\rho_{\text{mean}}} \right].$$

Remember: $G_m = 6.17$ lb/(s ft^2) and $\text{Re}_{G_m} = 6167 \approx 6.0 \times 10^3$. From the j-factor versus Re_{G_m} chart, the friction factor (f) is

$$f \approx 0.035.$$

The area ratio is

$$\frac{A_T}{A_c} = \frac{4W}{D_h} = \frac{4\,(4 \text{ in.})}{0.0125 \text{ ft}} \times \frac{1 \text{ ft}}{12 \text{ in.}} = 106.7.$$

The mean density (ρ_{mean}) is

$$\rho_{\text{mean}} = \frac{\rho_{a,i} + \rho_{a,o}}{2} = \frac{(0.078 + 0.071) \text{ lb/ft}^3}{2} = 0.075 \text{ lb/ft}^3.$$

Therefore,

$$\Delta P_{\text{bank}} = \frac{\left(6.17 \text{ lb/(s ft}^2) \right)^2}{2 \left(0.078 \text{ lb/ft}^3 \right)} \left[(1 + 0.20^2) \left(\frac{0.078 \text{ lb/ft}^3}{0.071 \text{ lb/ft}^3} - 1 \right) + (0.035)\,(106.7) \frac{0.078 \text{ lb/ft}^3}{0.075 \text{ lb/ft}^3} \right]$$

$$\Delta P_{\text{bank}} = 972.8 \text{ lb/(ft s}^2) \times \frac{1 \text{ lbf}}{32.2 \text{ (lb ft)/s}^2} = 30.2 \text{ lbf/ft}^2 = 30.2 \text{ psf.}$$

In practice, the pressure drop is reported in inches of water. Thus,

$$\Delta P_{\text{bank}} = 972.8 \text{ lb/(ft s}^2) \times \frac{1}{0.075 \text{ lb/ft}^3} \times \frac{1}{32.2 \text{ ft/s}^2} = 402.8 \text{ ft of air}$$

$$\Delta P_{\text{bank}} = 402.8 \text{ ft of air} \times SG_{\text{air,75°F}} = 402.8 \text{ ft of air} \times \frac{0.075 \text{ lb/ft}^3}{62 \text{ lb/ft}^3} \times \frac{12 \text{ in.}}{1 \text{ ft}}$$

$$\Delta P_{\text{bank}} = \textbf{5.9 in. of water} = \textbf{5.9 in. wg.}$$

Note: This is a very large pressure drop across the tube coils. This is due to the large airflow rate and velocity, large value of G_m, and small values of x_L and σ.

Pressure loss of water in the tube coils

Determination of the pressure loss of the water in the tubes is needed to find the pump power required to move the fluid through the heat exchanger system. The total head loss is

$$H_{IT} = \left(f \frac{L_{\text{tube}}}{D_i} + K \right) \frac{\overline{V}_w^2}{2g}.$$

From the Moody chart, for $Re_{D_i} = 22086$ and for copper tubes with relative roughness, $\frac{\varepsilon}{D_i} = 0.00019$, the friction factor is $f \approx 0.0285$.

For the primary piping circuitry through the heat exchanger, there are 8 tube rows, including one supply run and one return run, each 28 in. long.

Therefore,

$$L_{tube} = (8 \text{ rows})(28 \text{ in. per row}) = 224 \text{ in.} = 18.7 \text{ ft.}$$

For the minor losses, $K = 2.0$ for the soldered/brazed 180° regular return bends. Note that the K value for soldered/brazed 180° regular return bends is probably lower than 2.0, and more on the order of the value for flanged 180° regular return bends.

Thus,

$$H_{IT} = \left((0.0285)\frac{224 \text{ in.}}{0.315 \text{ in.}} + (7)(2.0) \right) \frac{(4 \text{ ft/s})^2}{2\left(32.2 \text{ ft/s}^2\right)}$$

$$H_{IT} = 8.5 \text{ ft wg.}$$

It should be noted that only the longest run of piping is needed to determine the total head loss that will be used to find the total pump power required. In this case, the use of inlet and outlet headers in which each row of tubes has its inlet and outlet attached to a header will give the longest run of pipe with a length of 224 in. If this were not the case, and only one tube inlet and one tube outlet were available for the entire heating coil unit, the longest run of pipe would be the total length of all the tubes. In that case, the length of piping would be 4928 in., the total number of 180° return bends would be 154, and the total head loss would be 187 ft wg. The total head loss would be more than 20 times larger than that of the design with headers.

Drawings

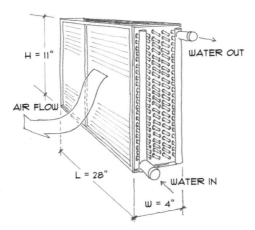

Conclusions

A heating coil heat exchanger with a bare-tube bank has been designed. The following points should be noted:

(i) The Re_{G_m} value that was used with the j-factor versus Re_{G_m} chart was slightly off the curve. The error incurred was small.
(ii) The pressure drop of the air across the tube bank and the number of rows of tubes were large. The tubes were too close, since $\sigma = 0.200$. $\sigma = 0.333$ may have been a better choice for improvement of the design.
(iii) A pressure drop of 5.9 in. wg on the air side is too large. This pressure drop should be limited to an order of 1 in. wg across the tube bank, if possible. Increasing x_L and σ would reduce the pressure drop.

A heat exchanger design data sheet is shown below.

Heat Exchanger Design Data Sheet	
Type	Counter Flow
Section: Tube Bank	
Working fluid	Air
Volume flow rate	2000 cfm
Inlet temperature	50°F
Outlet temperature	100°F
Pressure drop	5.9 in. wg
Section: Tube	
Tube material	Copper
Working fluid	Water
Velocity	4 ft/s
Tube inner diameter	0.315 in.
Tube outer diameter	0.375 in.
Number of tube rows	8
Number of tubes per row	22
Tube spacing ($x_t \times x_L$)	0.4688 in. × 0.4688 in.
Total tube length	410 ft
Inlet temperature	150°F
Outlet temperature	140°F
Head loss	8.5 ft wg
Heat Exchanger Parameters	
Thermal capacity	106500 Btu/h
Effectiveness	0.50
Capacity ratio	0.20
Overall heat transfer coefficient	39 Btu/(h ft² °F)
Number of transfer units (NTU)	0.75
Heat exchanger dimensions ($L \times H \times W$)	28 in. × 11 in. × 4 in.

4.7.3 External Flow over Finned-Tubes in Cross Flow—Equations and Charts

External flow over finned-tube tube banks is more prevalent in industrial applications. The addition of **fins** extends the heat transfer surface area of the bare tubes in the bank.

A staggered arrangement of the tubes, coupled with continuous fins across the length of the tube bank and heat exchanger will increase the heat transfer capability of the exchanger. Figure 4.19 shows some examples of finned heat exchangers that may be encountered by the design engineer in industry.

Of importance in the design and performance analysis of heat exchangers is the determination of the overall heat transfer coefficient (U). To determine the overall heat transfer coefficient for finned-tubes, consider the following assumptions and observations:

(i) Constant area, straight fins (e.g., continuous fins, plate-fin-tube geometry).
(ii) The base of the fin transfers heat also. Figure 4.20 shows a schematic of general constant area, straight fins attached to a surface (the base).
(iii) So, the combined fin/base surface effectiveness, η_s, is used in heat exchanger calculations, instead of the fin efficiency, η.
(iv) The fins are very thin. So, $l \gg t$. This is typical of fins used in industrial heat exchangers.
(v) The fins are rigidly and perfectly attached to the base. So, there is no contact resistance to heat transfer between the fin and the base.
(vi) The heat transfer coefficient, fluid properties, and thermal properties are constant.
(vii) The temperature distribution is uniform and steady (steady state) through the fin. So, the Biot number is small compared to unity $\left(\text{Bi} = \frac{h\delta}{k} \ll 1\right)$. This assumption will hold for very thin fins ($\delta = t$ or $\delta = t/2$) and/or high fin material thermal conductivities (k), which can be easily incorporated into any design. Low heat transfer coefficients (h) reduce Bi also. This is less desirable from a fin design/performance perspective.

Without derivation, the overall heat transfer coefficient for finned-tube heat exchangers is

$$\frac{1}{U_{FT}} = \frac{1}{h_o \eta_{so}} + R_{fo} + \frac{R_{fi}}{\left(A_i/A_o\right)} + \frac{1}{h_i \eta_{si} \left(A_i/A_o\right)}, \tag{4.84}$$

where "o" refers to the working fluid outside the tube in the tube bank, and "i" refers to the working fluid inside the tubes.

For most industrial applications, the tubes are completely filled with fluid. So, $\eta_{si} = 1$.

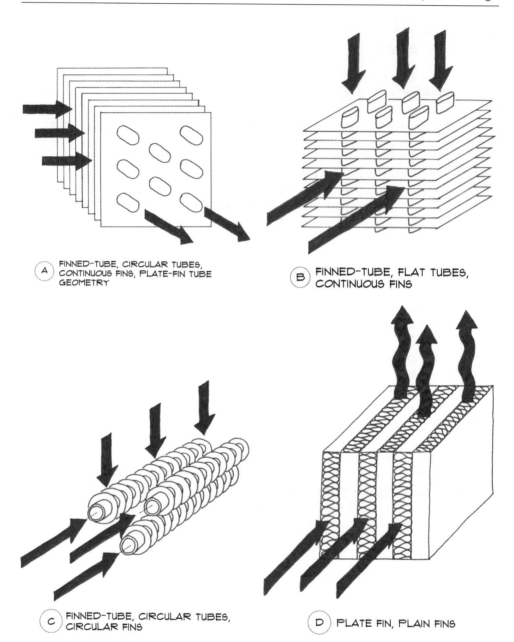

Figure 4.19 Examples of finned heat exchangers

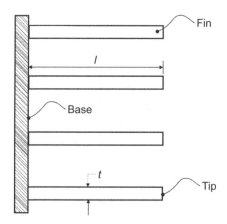

Figure 4.20   General constant area, straight fins attached to a surface

Therefore,

$$\frac{1}{U_{FT}} = \frac{1}{h_o \eta_{so}} + R_{fo} + \frac{R_{fi}}{\left(A_i / A_o \right)} + \frac{1}{h_i \left(A_i / A_o \right)}.$$ (4.85)

Fin Surface Effectiveness, η_s

The fin surface effectiveness is defined as

$$\eta_s = \frac{\text{Actual heat transfer from fin and base}}{\text{Heat transfer from fin and base when fin is at the base temperature}}$$ (4.86)

$$= 1 - \frac{A_f}{A}(1 - \eta),$$

where

$A_f =$ area of the fin, only;
$A =$ the total area of the fin and base;
$\frac{A_f}{A} =$ is determined from available and appropriate charts;
$\eta =$ fin efficiency.

For continuous fins that connect tubes in a staggered tube bank, it may be assumed that the tubes are arranged in a hexangular tube array, with a tube centered in each array. It is difficult to determine the fin efficiency (η) for a hexangular tube array. So, **empirical expressions** for η for circular fins are used to find an equivalent η for continuous plate hexangular finned-tube arrays. The empirical relations were developed by Schmidt [9].

Figure 4.21 shows a schematic drawing of a staggered tube bank with a hexangular finned-tube array. Note that the traverse tube pitch (x_t) and the longitudinal tube pitch (x_L) are shown.

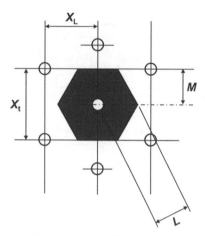

Figure 4.21   Staggered tube bank with a hexangular finned-tube array

The fin efficiency for this situation is defined as

$$\eta = \frac{\tanh{(mr\phi)}}{mr\phi},\tag{4.87}$$

where

$$r = \text{tube outer radius,}$$

$$m = \left(\frac{2h_\text{o}}{k_\text{fin}t}\right)^{1/2},\tag{4.88}$$

$$k_\text{fin} = \text{thermal conductivity of the fin material,}$$

$$\phi = \left(\frac{R_\text{EQ}}{r} - 1\right)\left[1 + 0.35\ln\left(\frac{R_\text{EQ}}{r}\right)\right],\tag{4.89}$$

R_EQ = the equivalent fin radius that will give an equivalent η for continuous plate hexangular fins.
$\frac{R_\text{EQ}}{r}$ is given empirically as

$$\frac{R_\text{EQ}}{r} = 1.27\psi\,(\beta - 0.3)^{1/2},\tag{4.90}$$

where

$$\psi = \frac{M}{r} = \frac{x_\text{t}}{2}\times\frac{1}{r} = \frac{x_\text{t}}{D} = X_\text{t},\tag{4.91}$$

$$\beta = \frac{L}{M}.\tag{4.92}$$

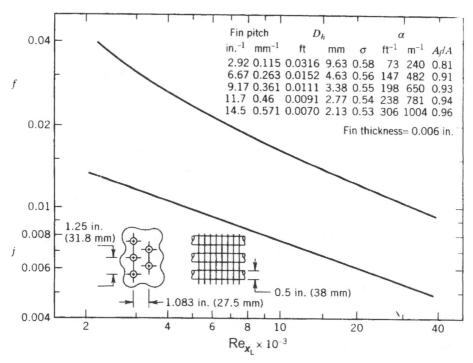

Figure 4.22   Data for flow normal to a finned staggered tube bank (*ASHRAE Transactions*, Vol. 79, Part II, 1973; reprinted with permission)

Analysis of the geometry shown in Figure 4.22 will give

$$L = \frac{\left[\left(\frac{x_t}{2}\right)^2 + x_L^2\right]^{1/2}}{2}. \tag{4.93}$$

The efficiency and the surface effectiveness of the fin can be determined.

Practical Note 4.8   L and M values

In practice, it is typical that $L \geq M$ and $\beta \geq 1$. The traverse tube pitch (x_t) and the longitudinal tube pitch (x_L) that are used to calculate L and M will be available from charts. The design engineer should verify that $L \geq M$ and $\beta \geq 1$.

Heat Transfer Coefficient, h_o

Heat transfer coefficients for flow through finned tube banks are also found by using the j-factor and appropriate charts. Remember: $j = \frac{\bar{h}_o}{G_m c_p} \mathrm{Pr}^{2/3}$.

Two types of charts may be presented to find the j-factor: j versus Re_{G_m} and j versus Re_{x_L}, where $\mathrm{Re}_{G_m} = \frac{G_m D_h}{\mu}$ and $\mathrm{Re}_{x_L} = \frac{G_m x_L}{\mu}$.

Charts of j versus Re_{x_L} for Staggered Tube Bank Arrangements (Finned Tubes)

Figure 4.22 shows a j versus Re_{x_L} chart for a staggered tube bank of finned tubes (plate-fin coils).

The following observations are made regarding Figure 4.22:

(i) This chart applies to a finned-tube bank with five rows.
(ii) The fin pitch is the number of fins per in. The inverse of the fin pitch will give the space between each fin.
(iii) The hydraulic diameter, D_h, varies with the fin pitch.
(iv) σ, α and $\frac{A_f}{A}$ (fin area-to-total area ratio) are all provided.
(v) The chart is based on Re_{x_L}.
(vi) The friction factor (f) curve is given.

More general charts may be used, which show varying numbers of tube rows or fin pitch. Figure 4.23 shows a group of j versus Re_{x_L} curves on one chart. Each curve represents a different number of rows of tubes.

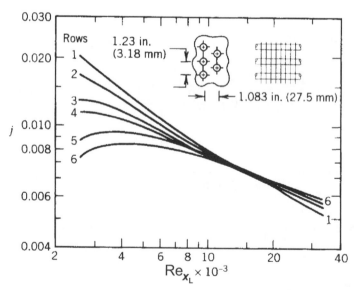

Figure 4.23   Data for flow normal to staggered tube banks: multiple tube rows (*ASHRAE Transactions*, Vol. 81, Part I, 1975; reprinted with permission)

The following observations are made regarding Figure 4.23. Additional charts are available in Appendix C:

(i) This chart applies to various numbers of tube rows.
(ii) There are large deviations in the j-factor for low values of Re_{x_L}.
(iii) At about $\text{Re}_{x_L} \approx 8000$, the j-factor becomes approximately independent of the number of rows.
(iv) Hence, for $\text{Re}_{x_L} > 8000$, any chart or curve may be used independently of the number of rows for other specified data such as tube diameter, fin pitch, σ, α, and $\frac{A_f}{A}$.

Example 4.3   Design of a Heating Coil Heat Exchanger c/w Finned Tubes

Mr. Chin is concerned about the large pressure drops across the bare tube bank, and has decided that he would prefer a small heating coil c/w finned tubes for his fresh air in-take duct line. A HVAC engineer had presented the following **duty** for the coil:

(i) Heat fresh air from 50 to about 100°F.
(ii) Duct airflow rate is 2000 cfm.
(iii) Entering hot water temperature is about 150°F.
(iv) This is a low-velocity duct section.
(v) Water connections must be on the same end of the coil system to facilitate easy installation and to approach a counter flow heat exchanger system.

Further Information: This design problem should address and present the following points:

(i) Exit water temperature
(ii) Heating coil configuration, number of coils, fin pitch
(iii) Final dimensions of the heat exchanger (length, width, and depth)
(iv) Pressure loss for the airflow through the coil bank
(v) Head loss in the waterside of the tube

Possible Solution

Definition

Design a heating coil complete with finned tubes to heat fresh air in a duct line.

Preliminary Specifications and Constraints

Same as Example 4.2, in addition to

(i) finned-tube heating coils required by the client.

Detailed Design

Objective

To determine the coil configuration, number of coils, size, and performance of a heat exchanger that has finned tubes.

Data Given or Known

(i) Same as Example 4.2.

Assumptions/Limitations/Constraints

(i) Same as Example 4.2, in addition to/except.
(ii) Let the tube outside diameter be 0.676 in. After studying the data in Table A.6, assume that the tube thickness is 0.04 in.
(iii) The tubes longitudinal pitch is 1.75 in. A large pitch is chosen (close to 1 in.) to reduce possible pressure losses on the air side.
(iv) Let the fin material be aluminum. This material is typically used in industry for the fabrication of fins for heating or cooling coils. Copper could also be used.
(v) Assume 7.75 fins per inch.

Sketch

A sketch has been provided with the problem preamble. The sketch will be modified to show the tube flow circuitry after the design.

Analysis

Determine the overall heat transfer coefficient (U_{FT})

The ε-NTU method will be used in this design. To determine the effectiveness and NTU, the overall heat transfer coefficient (U_{FT}) must be determined first. The overall heat transfer coefficient for finned tubes is

$$\frac{1}{U_{FT}} = \frac{1}{h_i \, (A_i/A_o)} + \frac{R_{fi}}{(A_i/A_o)} + R_{fo} + \frac{1}{h_o \eta_{so}}.$$

The heat transfer coefficient for flow in the tube is found from the Dittus–Boelter correlation equation:

$$\mathrm{Nu} = \frac{\overline{h}_i D_i}{k} = 0.023 \mathrm{Re}_{D_i}^{0.8} \mathrm{Pr}^{0.3}.$$

Appropriate properties for the determination of Re_D and Pr will be found at 145°F (see Example 4.2). The inner diameter of the copper tube is approximately 0.596 in.

Therefore, the Reynolds number is

$$\mathrm{Re}_{D_i} = \frac{\rho \overline{V} D_i}{\mu} = \frac{\left(61 \text{ lb/ft}^3\right)(4 \text{ ft/s})(0.596 \text{ in.})}{2.9 \times 10^{-4} \text{ lb/(ft s)}} \times \frac{1 \text{ ft}}{12 \text{ in.}} = 41789.$$

$$\mathrm{Pr} = 2.73.$$

The flow in the tubes is turbulent and the Pr is between 0.7 and 160. Thus, the Dittus–Boelter equation is valid for this analysis.
Therefore,

$$\overline{h}_i = \frac{0.023k}{D_i} \mathrm{Re}_{D_i}^{0.8} \mathrm{Pr}^{0.3}$$

$$\overline{h}_i = \frac{0.023 \, (0.38 \text{ Btu/(h ft R)})}{0.596 \text{ in.}} \times \frac{12 \text{ in.}}{1 \text{ ft}} \times (41789)^{0.8} \, (2.73)^{0.3}$$

$$\overline{h}_i = 1183 \text{ Btu/(h ft}^2 \text{ °F)}.$$

The fouling resistances can be found from Table C.3.

Assume distilled water above 122°F: $R_{fi} = 0.00114 \text{ (h ft}^2 \text{ °F)/Btu}$.
Assume compressed air: $R_{fo} = 0.00199 \text{ (h ft}^2 \text{ °F)/Btu}$.

The average heat transfer coefficient for airflow over the finned-tube bank is found from charts and appropriate correlation equations:

$$\overline{h}_o = \frac{j G_m c_p}{\mathrm{Pr}^{2/3}} \text{ for air at 75°F,}$$

and the Chilton–Colburn j-factor is

$$j = \frac{\overline{h}_o}{G_m c_p} \mathrm{Pr}^{2/3}.$$

The following j-factor versus Re_{G_m} chart (Figure C.4b) will be used. Note that the longitudinal pitch is 1.75 in. and the traverse pitch is 1.50 in. Consult Figure C.4 for additional charts.

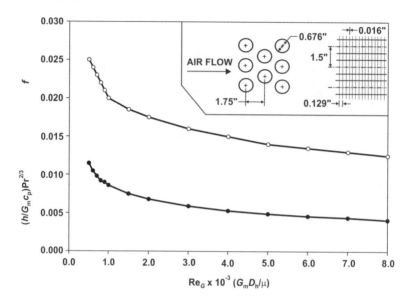

DATA

Tube outside diameter: 0.676 in.

Fin pitch: 7.75 fins per in.

Fin thickness: 0.016 in.

Hydraulic diameter, D_h: 0.0114 ft

Free-flow area/Frontal area, σ: 0.481

Heat transfer area/Total volume, α: 169 ft²/ft³

Fin area/Total area: 0.950

Note: Minimum free-flow area is in the spaces
transverse to the flow

Remember that air properties are found at $\overline{T}_a = 75°$F. So, G_m is

$$G_m = \frac{\rho_a \overline{V}_a}{\sigma} = \frac{\left(0.074 \text{ lb/ft}^3\right)(1000 \text{ ft/min})}{0.481} \times \frac{1 \text{ min}}{60 \text{ s}} = 2.56 \text{ lb/s ft}^2.$$

The Reynolds number based on G_m is

$$\text{Re}_{G_m} = \frac{G_m D_h}{\mu_a} = \frac{\left(2.56 \text{ lb/(s ft}^2)\right)(0.0114 \text{ ft})}{1.25 \times 10^{-5} \text{ lb/(ft s)}} = 2335 \approx 2.3 \times 10^3.$$

From the chart, $j \approx 0.0065$.

Therefore,

$$\bar{h}_o = \frac{(0.0065)\left(2.56 \text{ lb}/(\text{s ft}^2)\right)(0.24 \text{ Btu}/(\text{lb} \,^\circ\text{F}))}{(0.73)^{2/3}} \times \frac{3600 \text{ s}}{1 \text{ h}}$$

$$\bar{h}_o = 17.7 \text{ Btu}/(\text{h ft}^2 \,^\circ\text{F}).$$

As expected, the heat transfer coefficient of the liquid water is higher than that of gaseous air.

The fin surface effectiveness is

$$\eta_{so} = 1 - \frac{A_{\text{fin}}}{A}(1 - \eta_o).$$

The fin efficiency is

$$\eta_o = \frac{\tan h\,(mr\phi)}{mr\phi}.$$

For aluminum, $k_{\text{fin}} = 100$ Btu/(h ft $^\circ$F). For the fins, the thickness is $t = 0.016$ in., as obtained from the j-factor versus Re_{Gm} chart. Thus,

$$m = \frac{(2\bar{h}_o)^{1/2}}{(k_{\text{fin}}t)^{1/2}} = \left[\frac{2\left(17.7 \text{ Btu}/(\text{h ft}^2 \,^\circ\text{F})\right)}{(100 \text{ Btu}/(\text{h ft}\,^\circ\text{F}))(0.016 \text{ in.})\left(\dfrac{1 \text{ ft}}{12 \text{ in.}}\right)}\right]^{1/2} = 16.3 \text{ ft}^{-1}$$

The tube outer radius is

$$r = \frac{0.676 \text{ in.}}{2} \times \frac{1 \text{ ft}}{12 \text{ in.}} = 0.0282 \text{ ft}.$$

Find ϕ.

$$\phi = \left(\frac{R_{EQ}}{r} - 1\right)\left[1 + 0.35 \ln\left(\frac{R_{EQ}}{r}\right)\right]$$

$$\frac{R_{EQ}}{r} = 1.27\psi\,(\beta - 0.3)^{1/2}$$

$$\psi = \frac{M}{r} = \frac{x_t}{2} \times \frac{1}{r} = \frac{x_t}{D_o} = X_t = \frac{1.50 \text{ in.}}{0.676 \text{ in.}} = 2.22$$

$$\beta = \frac{L}{M}$$

$$M = \frac{x_t}{2} = \frac{1.50 \text{ in.}}{2} = 0.75 \text{ in.}$$

$$L = \frac{\left[\left(\frac{x_t}{2}\right)^2 + x_L^2\right]^{1/2}}{2} = \frac{\left[\left(\frac{1.50 \text{ in.}}{2}\right)^2 + (1.75 \text{ in.})^2\right]^{1/2}}{2} = 0.952 \text{ in.}$$

Thus,

$$\beta = \frac{0.952 \text{ in.}}{0.750 \text{ in.}} = 1.27.$$

Note that $L \geq M$ and $\beta \geq 1$, as observed in practice. Therefore,

$$\frac{R_{EQ}}{r} = 1.27 \, (2.22) \, (1.27 - 0.3)^{1/2} = 2.78.$$

$$\phi = (2.78 - 1) \, [1 + 0.35 \ln (2.78)] = 2.42.$$

The fin efficiency is

$$\eta_o = \frac{\tanh \left(16.3 \text{ ft}^{-1} \times 0.0282 \text{ ft} \times 2.42 \right)}{16.3 \text{ ft}^{-1} \times 0.0282 \text{ ft} \times 2.42} = 0.724.$$

Thus, the fin effectiveness is

$$\eta_{so} = 1 - 0.950 \, (1 - 0.724) = 0.737.$$

Find the $\frac{A_i}{A_o}$ ratio. Remember that $\alpha = \frac{A_o}{\Omega}$. From the j-factor versus Re_{G_m} chart, $\alpha = 169 \text{ ft}^2/\text{ft}^3$. Assume that the tubes fill the volume bounded by $x_t x_L L_{tube}$. Hence,

$$\frac{A_i}{\Omega} \approx \frac{\pi D_i L_{tube}}{x_t x_L L_{tube}} = \frac{\pi D_i}{x_t x_L}$$

$$\frac{A_i}{A_o} \approx \frac{\left(\dfrac{A_i}{\Omega} \right)}{\left(\dfrac{A_o}{\Omega} \right)} \approx \frac{\dfrac{\pi D_i}{x_t x_L}}{\alpha} = \frac{\pi D_i}{\alpha x_t x_L}$$

$$\frac{A_i}{A_o} \approx \frac{\pi \, (0.596 \text{ in.})}{\left(169 \text{ ft}^2/\text{ft}^3 \right) (1.50 \text{ in.}) (1.75 \text{ in.})} \times \frac{12 \text{ in.}}{1 \text{ ft}} = 0.0507.$$

The overall heat transfer coefficient is

$$\frac{1}{U_{FT}} = \frac{1}{\left(1183 \text{ Btu/(h ft}^2 \, {}^\circ\text{F)} \right) (0.0507)} + \left(\frac{0.00114}{0.0507} + 0.00199 \right) \text{ (h ft}^2 \, {}^\circ\text{F)/Btu}$$

$$+ \frac{1}{\left(17.7 \text{ (h ft}^2 \, {}^\circ\text{F)/Btu} \right) (0.737)}$$

$$\frac{1}{U} = 0.118 \text{ (h ft}^2 \, {}^\circ\text{F)/Btu}$$

$$U = 8.49 \text{ Btu/(h ft}^2 \, {}^\circ\text{F)}$$

Determine the configuration, number of coils, and dimensions of the heat exchanger. Find the total surface area of the tubes by using the ε-NTU method:

$$A_o = \frac{C_{min}\,NTU}{U}.$$

From Example 4.2,

$$C_a = C_c = C_{min} = 35.5 \text{ Btu/(min °F)}$$

$$C_w = C_h = C_{max} = 177.5 \text{ Btu/(min °F)}$$

$$c = 0.20$$

$$\varepsilon = 0.50$$

$$\dot{Q} = 1775 \text{ Btu/min} = 106500 \text{ Btu/h}$$

$$NTU \approx 0.75.$$

The heat transfer surface area is

$$A_o = \frac{(35.5 \text{ Btu/(min °F)})(0.75)}{8.49 \text{ Btu/(h ft}^2 \text{ °F)}} \times \frac{60 \text{ min}}{1 \text{ h}}$$

$$A_o = 188 \text{ ft}^2.$$

This value for the heat transfer surface area includes the extended surface area of the fins. The total heat transfer volume is found from the ratio of the heat transfer area to the total volume:

$$\alpha = \frac{A_o}{\Omega}.$$

Therefore,

$$\Omega = \frac{A_o}{\alpha} = \frac{188 \text{ ft}^2}{169 \text{ ft}^2/\text{ft}^3} = 1.11 \text{ ft}^3.$$

The depth of the heat exchanger system (dimension in the direction of airflow) is

$$W = \frac{\Omega}{A_f}.$$

A_f is the face area normal to the airflow direction, and is equal to 2 ft². Therefore,

$$W = \frac{1.11 \text{ ft}^3}{2 \text{ ft}^2} \times \frac{12 \text{ in.}}{1 \text{ ft}}$$

$$W = 6.7 \text{ in.} \approx 7 \text{ in.}$$

From the j-factor versus Re_{G_m} chart, the longitudinal tube pitch is 1.75 in. The number of tube rows is

$$N_r = \frac{W}{x_L} = \frac{6.7 \text{ in.}}{1.75 \text{ in.}}$$

$$N_r = 3.8 \approx 4 \text{ tube rows}.$$

Determine the number of tubes per row. From Example 4.2,

$$N_{tube} = \frac{4C_h}{c_{p,w}\rho_w \overline{V}_w \pi D_i^2}$$

$$N_{tube} = \frac{4\,(177.5 \text{ Btu/(min °F)})}{(1.0 \text{ Btu/(lb °F)})\left(61 \text{ lb/ft}^3\right)(4 \text{ ft/s})\,\pi\,(0.596 \text{ in.})^2} \times \left(\frac{12 \text{ in.}}{1 \text{ ft}}\right)^2 \times \frac{1 \text{ min}}{60 \text{ s}}$$

$$N_{tube} = 6.2 \approx 6 \text{ tubes per row}.$$

The total number of tubes along the length of the heat exchanger will be 24 tubes (4 rows × 6 tubes per row). If each row has six tubes, the height of the heating coil is

$$H = N_{tube}x_t = (6.2)(1.50 \text{ in.})$$

$$H = 9.3 \text{ in.} \approx 10 \text{ in.}$$

The length of the heat exchanger is

$$L = \frac{A_f}{H} = \frac{2 \text{ ft}^2}{9.3 \text{ in.}} \times \left(\frac{12 \text{ in.}}{1 \text{ ft}}\right)^2$$

$$L = 30.96 \text{ in.} \approx 31 \text{ in.}$$

The total length of the tubes in the heat exchanger is 62 ft (31 in. per tube × 24 tubes).

Pressure loss of air across the tube coils

The pressure drop of the air across the tube coil bank is given by

$$\Delta P_{bank} = \frac{G_m^2}{2\rho_{a,i}}\left[(1+\sigma^2)\left(\frac{\rho_{a,i}}{\rho_{a,o}}-1\right)+f\frac{A_T}{A_c}\frac{\rho_{a,i}}{\rho_{mean}}\right].$$

Remember: $G_m = 2.56 \text{ lb/s ft}^2$ and $Re_{G_m} \approx 2.3 \times 10^3$.
From the j-factor versus Re_{G_m} chart, the friction factor (f) is

$$f \approx 0.016.$$

The area ratio is

$$\frac{A_T}{A_c} = \frac{4W}{D_h} = \frac{4\,(6.7\ \text{in.})}{0.0114\ \text{ft}} \times \frac{1\ \text{ft}}{12\ \text{in.}} = 195.9.$$

From Example 4.2, the mean density (ρ_{mean}) is $\rho_{\text{mean}} = 0.075\ \text{lb/ft}^3$. From the j-factor versus Re_{G_m} chart, $\sigma = 0.481$.
Therefore,

$$\Delta P_{\text{bank}} = \frac{(2.56\ \text{lb/(s ft}^2))^2}{2(0.078\ \text{lb/ft}^3)} \left[(1 + 0.481^2) \left(\frac{0.078\ \text{lb/ft}^3}{0.071\ \text{lb/ft}^3} - 1 \right) + (0.016)\,(195.9)\,\frac{0.078\ \text{lb/ft}^3}{0.075\ \text{lb/ft}^3} \right]$$

$$\Delta P_{\text{bank}} = 142\ \text{lb/ft s}^2 \times \frac{1\ \text{lbf}}{32.2\ \text{lb ft/s}^2} = 4.4\ \text{lbf/ft}^2 = 4.4\ \text{psf.}$$

In practice, the pressure drop is reported in inches of water. Therefore,

$$\Delta P_{\text{bank}} = 142\ \text{lb/ft s}^2 \times \frac{1}{0.075\ \text{lb/ft}^3} \times \frac{1}{32.2\ \text{ft/s}^2} = 58.8\ \text{ft of air}$$

$$\Delta P_{\text{bank}} = 58.8\ \text{ft of air} \times SG_{\text{air,75°F}} = 58.8\ \text{ft of air} \times \frac{0.075\ \text{lb/ft}^3}{62\ \text{lb/ft}^3} \times \frac{12\ \text{in.}}{1\ \text{ft}}$$

$$\Delta P_{\text{bank}} = 0.85\ \text{in. of water} = 0.85\ \text{in. wg.}$$

Pressure loss of water in the tube coils

Determination of the pressure loss of the water in the tubes is needed to find the pump power required to move the fluid through the heat exchanger system. The total head loss is

$$H_{\text{IT}} = \left(f\frac{L_{\text{tube}}}{D_i} + K \right) \frac{\bar{V}_w^2}{2g}.$$

From the Moody chart, for $\text{Re}_{D_i} = 41789$ and for copper tubes with relative roughness, $\frac{\varepsilon}{D_i} = 0.0001$, the friction factor is $f \approx 0.022$.
   For the primary piping circuitry through the heat exchanger, there are four tube rows, including 1 supply run and 1 return run, each 31 in. long.
   Therefore,

$$L_{\text{tube}} = (4\ \text{rows})(31\ \text{in. per row}) = 124\ \text{in.} = 10.3\ \text{ft.}$$

For the minor losses, $K = 2.0$ for the soldered/brazed 180° return bends.
Hence,

$$H_{\text{IT}} = \left((0.022)\frac{124\ \text{in.}}{0.596\ \text{in.}} + (3)\,(2.0) \right) \frac{(4\ \text{ft/s})^2}{2\left(32.2\ \text{ft/s}^2 \right)}$$

$$H_{\text{IT}} = 2.6\ \text{ft wg.}$$

Drawings

A general schematic drawing of the heating coil is shown.

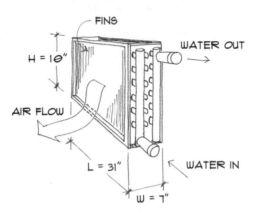

A detailed schematic drawing that shows an exploded view of the heat exchanger is also shown. It is not typical to show, in routine design problems, this type of schematic drawing with an exploded view. However, it will be required for fabrication of the coil. Additional information on dimensions would also be required.

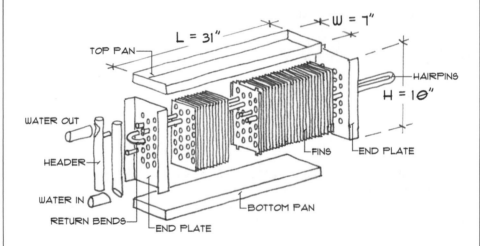

Conclusions

A heating coil heat exchanger complete with finned tubes has been designed. Improvements in the performance have been achieved by using finned tubes. The following points should be noted:

(i) G_m was lowered by choosing larger tube spacings, given the high airflow rate.
(ii) The number of rows and the number of tubes per row decreased compared with the heat exchanger with bare tubes. In addition, this heat exchanger is more compact with $\alpha = 169$ ft^2/ft^3 compared with 64.4 ft^2/ft^3 for the heat exchanger with bare tubes. The

total length of tube required decreased from 410 ft in the bare tube heat exchanger design to 62 ft in the finned-tube heat exchanger.
(iii) The calculated value of N_r was 3.8. It was assumed that N_r was equal to 4. However, the calculated value of N_{tube} was 6.2. It was assumed that N_{tube} was equal to 6. While it is typical to increase to the next whole number for more conservative designs, the compromise between N_r and N_{tube} should mitigate any significant error.
(iv) Lower pressure drop of the air across the tube bank was achieved. In particular, the pressure drop was on the order of 1 in. wg.
(v) The use of headers in this design resulted in low total head loss in the heating coil piping.

A heat exchanger design data sheet is shown in the following table.

Heat Exchanger Design Data Sheet

Type	Counter Flow
Section: Tube Bank	
Working fluid	Air
Volume flow rate	2000 cfm
Inlet temperature	50°F
Outlet temperature	100°F
Pressure drop	0.85 in. wg
Section: Tube	
Tube material	Copper
Working fluid	Water
Velocity	4 ft/s
Tube inner diameter	0.596 in.
Tube outer diameter	0.676 in.
Number of tube rows	4
Number of tubes per row	6
Tube spacing ($x_t \times x_L$)	1.50 in. × 1.75 in.
Total tube length	62 ft
Inlet temperature	150°F
Outlet temperature	140°F
Head loss	2.6 ft wg
Fin material	Aluminum
Fin pitch	7.75 fins per in.
Fin thickness	0.016 in.
Heat Exchanger Parameters	
Thermal capacity	106500 Btu/h
Effectiveness	0.50
Capacity ratio	0.20
Overall heat transfer coefficient	8.49 Btu/(h ft² °F)
Number of transfer units (NTU)	0.75
Heat exchanger dimensions ($L \times H \times W$)	31 in. × 10 in. × 7 in.

Example 4.4   Performance of an Oil Cooler

An oil cooler has been designed for use in a factory. The cooler has one tube inlet and one tube outlet into and out of a tube bank, respectively, to form an oil recycling system. Clean, unused engine oil is circulated through $\frac{1}{2}$-in. tubes that are connected with continuous-plate fins with a fin pitch of 8 fins per in. Cool air is forced over the tubes. The system operates under the following conditions:

   (i) Four rows of tubes
  (ii) Six tubes per row
 (iii) Width is 26 in.
 (iv) Fin thickness is 0.006 in.
  (v) Entering air temperature is 65°F
 (vi) Coil face velocity is 650 ft/min
(vii) Entering engine oil temperature is 150°F

    The design engineer who designed the original system used the following chart to aid in the design:

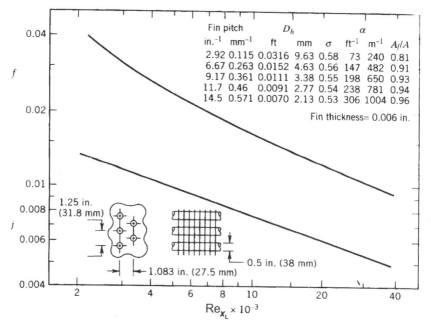

Source: ASHRAE Transactions, Vol. 79, Part II, 1973 (Reprinted with permission)

    Analyze the performance of the oil cooler and comment on the design.

Possible Solution

Definition

Analyze the performance of a finned-tube oil cooler (heat exchanger).

Preliminary Specifications and Constraints

   (i) The oil cooler is a heat exchanger with one tube inlet and one tube outlet.
  (ii) The working fluid in the tubes is clean, unused engine oil.
 (iii) The working fluid outside the tubes is cool air.
  (iv) The tubes have a nominal diameter of $1/2$ in.
   (v) This is a finned-tube heat exchanger with a fin pitch of 8 fins per in.
  (vi) Additional specifications are given in the problem preamble.

Detailed Design

Objective

To determine the thermal capacity, outlet temperatures, and pressure drops in a finned-tube oil cooler.

Data Given or Known

    (i) The working fluid in the tubes is clean, unused engine oil.
   (ii) The working fluid outside the tubes is cool air.
  (iii) The tubes have a nominal diameter of $1/2$ in.
   (iv) This is a finned-tube heat exchanger with a fin pitch of 8 fins per in. and fin thickness of 0.006 in.
    (v) There are four rows of tubes with six tubes per row.
   (vi) The width of the oil cooler is 26 in.
  (vii) The coil velocity (of air) is 650 fpm.
 (viii) The entering air temperature is 65°F.
   (ix) The entering engine oil temperature is 150°F.
    (x) A j-factor vs Re_{x_L} chart was provided.

Assumptions/Limitations/Constraints

    (i) Let the tube material be Type L copper. Copper has high heat transfer properties.
   (ii) Let the fin material be aluminum. This material is typically for the fabrication of fins for heating or cooling coils.
  (iii) Let the flow velocity of engine oil in the copper tubes be 3 fps. This velocity is lower than the erosion limit of copper with unused engine oil ($\approx$7 fps calculated for "Other Liquids" from Table A.13). A low tube velocity is desired since engine oil is highly viscous at lower temperatures. As the engine oil cools, its viscosity will increase, and it will require higher pump power to move through the coils.
   (iv) The tube thickness is approximately 0.04 in (see Table A.6). It will be assumed that the tube wall thickness is negligible compared to the inner and outer diameters.
    (v) The return bends will be soldered 180° return bends (regular).
   (vi) There is negligible elevation head. Assume that all the components are on the same level.
  (vii) Entrance and exit losses of the air over the coils will be negligible compared to other losses across the coils.
 (viii) This is a counter flow arrangement. The inlet and exit of the coils are on the same side of the cooler.

Sketch

A sketch is not needed for this performance analysis.

Performance Analysis

Determine the overall heat transfer coefficient (UFT)

The ε-NTU method will be used in this performance problem. To determine the effectiveness and NTU, the overall heat transfer coefficient (U_{FT}) must be determined first. The overall heat transfer coefficient for finned tubes is

$$\frac{1}{U_{FT}} = \frac{1}{h_i \left(A_i / A_o \right)} + \frac{R_{fi}}{\left(A_i / A_o \right)} + R_{fo} + \frac{1}{h_o \eta_{so}}.$$

Appropriate engine oil properties for the determination of Re_D and Pr will be found at 150°F. Note that the average temperature should be used, rather than the entering temperature. This will introduce a small error in the analysis. This will be verified. The inner diameter of the copper tube is approximately 0.545 in. (Table A.6).
    The Reynolds number is

$$Re_{D_i} = \frac{\rho \overline{V} D_i}{\mu} = \frac{\left(53.73 \text{ lb/ft}^3 \right) (3 \text{ ft/s}) (0.545 \text{ in.})}{3.833 \times 10^{-2} \text{ lb/(fts)}} \times \frac{1 \text{ ft}}{12 \text{ in.}} = 191.$$

The Prandtl number is

$$\text{Pr} = 848.3.$$

Assuming a constant surface heat flux from the tubes, and since the flow is laminar, the Nusselt number and heat transfer coefficient for flow in the tube is

$$\text{Nu} = \frac{\overline{h}_i D_i}{k} = 4.36$$

$$\overline{h}_i = \frac{4.36k}{D_i}$$

$$\overline{h}_i = \frac{4.36 \left(0.08046 \text{ Btu/(h ft R)} \right)}{0.545 \text{ in.}} \times \frac{12 \text{ in.}}{1 \text{ ft}} = 7.724 \text{ Btu/(h ft R)} = 7.724 \text{ Btu/(h ft °F)}.$$

The fouling resistances can be found from Table C.3.

For clean, unused engine oil: $R_{fi} = 0$.
Assume compressed air: $R_{fo} = 0.00199 \ (\text{h ft}^2 \ °\text{F})/\text{Btu}$.

The average heat transfer coefficient for airflow over the finned-tube bank is found from charts and appropriate correlation equations:

$$\overline{h}_o = \frac{j G_m c_p}{\text{Pr}^{2/3}} \quad \text{for air at 65°F,}$$

and the Chilton–Colburn j-factor is

$$j = \frac{\overline{h}_o}{G_m c_p} \text{Pr}^{2/3}.$$

The j-factor versus Re_{G_m} chart provided by the design engineer will be used. Note that the longitudinal pitch is 1.083 in. and the traverse pitch is 1.25 in.

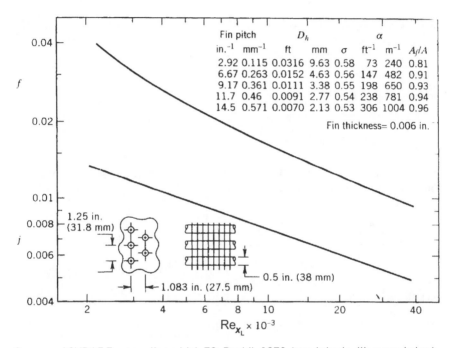

Source: ASHRAE Transactions, Vol. 79, Part II, 1973 (reprinted with permission).

Remember that air properties are found at 65°F. So, G_m is

$$G_m = \frac{\rho_a \overline{V}_a}{\sigma} = \frac{\left(0.07561\ \text{lb/ft}^3\right)(650\ \text{ft/min})}{0.555} \times \frac{1\ \text{min}}{60\ \text{s}} = 1.48\ \text{lb/(s ft}^2).$$

The Reynolds number based on G_m and x_L is

$$\text{Re}_{x_L} = \frac{G_m x_L}{\mu_a} = \frac{\left(1.48\ \text{lb/(s ft}^2)\right)(0.09025\ \text{ft})}{1.221 \times 10^{-5}\ \text{lb/(ft s)}} = 10939 \approx 11 \times 10^3.$$

Note that the values of ρ_a, σ, D_h, and μ were determined through interpolation. From the chart, $j \approx 0.0073$.

Therefore,

$$\bar{h}_o = \frac{jG_m c_p}{Pr^{2/3}}$$

$$\bar{h}_o = \frac{(0.0073)\left(1.48 \text{ lb/(s ft}^2)\right)(0.2404 \text{ Btu/(lb °F)})}{(0.73)^{2/3}} \times \frac{3600 \text{ s}}{1 \text{ h}}$$

$$\bar{h}_o = 11.5 \text{ Btu/(h ft}^2 \text{ °F)}.$$

The fin surface effectiveness is

$$\eta_{so} = 1 - \frac{A_{fin}}{A}(1 - \eta_o).$$

The fin efficiency is

$$\eta_o = \frac{\tan h\,(mr\phi)}{mr\phi}.$$

For aluminum, $k_{fin} = 100$ Btu/(hr ft °F). For the fins, the thickness is $t = 0.006$ in., as obtained from the j-factor versus Re_{x_L} chart. Therefore,

$$m = \frac{(2\bar{h}_o)^{1/2}}{(k_{fin}t)^{1/2}} = \left[\frac{2\left(22.1 \text{ Btu/(h ft}^2 \text{ °F)}\right)}{(100 \text{ Btu/(h ft °F)})(0.006 \text{ in.})\left(\dfrac{1 \text{ ft}}{12 \text{ in.}}\right)}\right]^{1/2} = 29.7 \text{ ft}^{-1}$$

The tube outer radius is

$$r = \frac{0.625 \text{ in.}}{2} \times \frac{1 \text{ ft}}{12 \text{ in.}} = 0.0260 \text{ ft.}$$

Find ϕ.

$$\phi = \left(\frac{R_{EQ}}{r} - 1\right)\left[1 + 0.35 \ln\left(\frac{R_{EQ}}{r}\right)\right]$$

$$\frac{R_{EQ}}{r} = 1.27\psi\,(\beta - 0.3)^{1/2}$$

$$\psi = \frac{M}{r} = \frac{x_t}{2} * \frac{1}{r} = \frac{x_t}{D} = X_t = \frac{1.25 \text{ in.}}{0.625 \text{ in.}} = 2.0$$

$$\beta = \frac{L}{M}$$

$$M = \frac{x_t}{2} = \frac{1.25 \text{ in.}}{2} = 0.625 \text{ in.}$$

$$L = \frac{\left[\left(\dfrac{x_t}{2}\right)^2 + x_L^2\right]^{1/2}}{2} = \frac{\left[\left(\dfrac{1.25 \text{ in.}}{2}\right)^2 + (1.083 \text{ in.})^2\right]^{1/2}}{2} = 0.625 \text{ in.}$$

Therefore,

$$\beta = \frac{0.625 \text{ in.}}{0.625 \text{ in.}} = 1.0.$$

Note that $L \geq M$ and $\beta \geq 1$, as observed in practice.
Thus,

$$\frac{R_{EQ}}{r} = 1.27\,(2.0)\,(1.0 - 0.3)^{1/2} = 2.13.$$

$$\phi = (2.13 - 1)\,[1 + 0.35\,\ln(2.13)] = 1.43.$$

The fin efficiency is

$$\eta_o = \frac{\tanh\left(29.7 \text{ ft}^{-1} \times 0.0260 \text{ ft} \times 1.43\right)}{29.7 \text{ ft}^{-1} \times 0.0260 \text{ ft} \times 1.43} = 0.727.$$

Therefore, the fin effectiveness is

$$\eta_{so} = 1 - 0.921\,(1 - 0.727) = 0.749.$$

Note that $\frac{A_{fin}}{A}$ was interpolated from data shown in the chart provided by the design engineer.

Find the $\frac{A_i}{A_o}$ ratio. Remember that $\alpha = \frac{A_o}{\Omega}$. From the j-factor versus Re_{x_L} chart and interpolation, $\alpha = 174 \text{ ft}^2/\text{ft}^3$. Assume that the tubes fill the volume bounded by $x_t x_L L_{tube}$.
Therefore,

$$\frac{A_i}{\Omega} \approx \frac{\pi D_i L_{tube}}{x_t x_L L_{tube}} = \frac{\pi D_i}{x_t x_L}$$

$$\frac{A_i}{A_o} \approx \frac{\left(\dfrac{A_i}{\Omega}\right)}{\left(\dfrac{A_o}{\Omega}\right)} \approx \frac{\dfrac{\pi D_i}{x_t x_L}}{\alpha} = \frac{\pi D_i}{\alpha x_t x_L}$$

$$\frac{A_i}{A_o} \approx \frac{\pi\,(0.545 \text{ in.})}{\left(174 \text{ ft}^2/\text{ft}^3\right)(1.25 \text{ in.})(1.083 \text{ in.})} \times \frac{12 \text{ in.}}{1 \text{ ft}} = 0.0872.$$

The overall heat transfer coefficient is

$$\frac{1}{U_{FT}} = \frac{1}{h_i\,(A_i/A_o)} + \frac{R_{fi}}{(A_i/A_o)} + R_{fo} + \frac{1}{h_o \eta_{so}}$$

$$\frac{1}{U_{FT}} = \frac{1}{(7.724 \text{ Btu}/(\text{h ft}^2 \, ^\circ\text{F}))\,(0.0872)} + 0.00199 \text{ (h ft}^2 \, ^\circ\text{F})/\text{Btu} + \frac{1}{(11.5(\text{h ft}^2 \, ^\circ\text{F})/\text{Btu})\,(0.749)}$$

$$\frac{1}{U} = 1.60 \text{ (h ft}^2 \, ^\circ\text{F})/\text{Btu}$$

$$U = 0.62 \text{ Btu}/(\text{h ft}^2 \, ^\circ\text{F}).$$

Determine the NTU of the heat exchanger.

The NTU is

$$\text{NTU} = \frac{UA_o}{C_{min}}.$$

The heat transfer surface area (A) is needed.

$$\alpha = \frac{A_o}{\Omega} = 174 \text{ ft}^2/\text{ft}^3$$

$$A_o = \left(174 \text{ ft}^2/\text{ft}^3\right)\Omega,$$

where Ω is the heat exchanger volume.
The depth of the oil cooler is

$$W = \frac{\Omega}{A_f}, \text{ and}$$

$$N_r = \frac{W}{x_L}, \quad H = N_{tube}\,x_t, \quad A_f = LH.$$

Hence,

$$W = N_r x_L = (4)\,(1.083 \text{ in.}) = 4.332 \text{ in.} \approx 5 \text{ in.}$$

$$H = N_{tube}\,x_t = (6)(1.25 \text{ in.}) = 7.5 \text{ in.} \approx 8 \text{ in.}$$

$$A_f = LH = (26 \text{ in.})(7.5 \text{ in.}) = 195 \text{ in.}^2 = 1.35 \text{ ft}^2$$

$$\Omega = WA_f = (4.332 \text{ in.})\left(195 \text{ in.}^2\right) \times \left(\frac{1 \text{ ft}}{12 \text{ in.}}\right)^3 = 0.49 \text{ ft}^3.$$

Thus,

$$A_o = \left(174 \text{ ft}^2/\text{ft}^3\right)\left(0.49 \text{ ft}^3\right) = 85 \text{ ft}^2.$$

Determine the heat capacities. For air,

$$C_a = \dot{m}_a c_{pa} = \rho_a \dot{V}_a c_{pa} = \rho_a V_a A_f c_{pa}$$

$$C_a = \left(0.07561 \text{ lb/ft}^3\right)(650 \text{ ft/min})\left(1.35 \text{ ft}^2\right)(0.2404 \text{ Btu/lb °F}) = 16.0 \text{ Btu/(min °F)}.$$

Since the outlet temperatures are unknown, the law of conservation energy cannot be used to find the heat capacity of the hot engine oil. The definition of the heat capacity for the engine oil is

$$C_e = \dot{m}_e c_{pe} = \rho_e \overline{V}_e c_{pe} A_{tube} = \rho_e \overline{V}_e c_{pe} \frac{\pi D_i^2}{4}$$

$$C_e = \left(53.73 \text{ lb/ft}^3\right)(3 \text{ ft/s})(0.4946 \text{ Btu/(lb °F)})\frac{\pi\,(0.545 \text{ in.})^2}{4} \times \left(\frac{1 \text{ ft}}{12 \text{ in.}}\right)^2 \times \frac{60 \text{ s}}{1 \text{ min}}$$

$$= 7.75 \text{ Btu/(min °F)}.$$

Therefore, $C_a = C_c = C_{max}$ and $C_e = C_h = C_{min}$.

It should be noted that since there is only one inlet tube connection in this heat exchanger, it was not necessary to include N_{tube} in the calculation of A_{tube}.

The NTU is

$$\text{NTU} = \frac{\left(0.62 \text{ Btu/(h ft}^2 \,^\circ\text{F)}\right)\left(85 \text{ ft}^2\right)}{7.75 \text{ Btu/(min }^\circ\text{F)}} \times \frac{1 \text{ h}}{60 \text{ min}}$$

$$\text{NTU} = 0.113.$$

Determine the thermal capacity and outlet temperatures

The thermal capacity and the outlet temperatures can be determined after determination of the effectiveness of the oil cooler.

The capacity ratio is

$$c = \frac{C_{min}}{C_{max}} = \frac{7.75 \text{ Btu/(min }^\circ\text{F)}}{16.0 \text{ Btu/(min }^\circ\text{F)}} = 0.484$$

$$c = 0.484.$$

Therefore,

$$\varepsilon = \frac{1 - \exp\left[-\text{NTU}\,(1-c)\right]}{1 - c \exp\left[-\text{NTU}\,(1-c)\right]}$$

$$\varepsilon = \frac{1 - \exp\left[-(0.113)\,(1-0.484)\right]}{1 - (0.484)\exp\left[-(0.113)\,(1-0.484)\right]}$$

$$\varepsilon = 0.104 = 10.4\%$$

The outlet temperature of the air can be determined:

$$\varepsilon = \frac{C_a}{C_{min}} \frac{(T_{a,o} - T_{a,i})}{(T_{e,i} - T_{a,i})}$$

$$T_{a,o} = T_{a,i} + \varepsilon \frac{C_{min}}{C_a}(T_{e,i} - T_{a,i}) = 65^\circ\text{F} + (0.104)\frac{7.75 \text{ Btu/(min }^\circ\text{F)}}{16 \text{ Btu/(min }^\circ\text{F)}}(150-65)\,^\circ\text{F}$$

$$T_{a,o} = 69.3^\circ\text{F} = 70^\circ\text{F}.$$

The thermal capacity of the oil cooler is

$$\dot{Q} = \dot{m}_a c_{pa}(T_{a,o} - T_{a,i}) = \rho_a V_a A_f c_{pa}(T_{a,o} - T_{a,i})$$

$$\dot{Q} = (0.07561 \text{ lb/ft}^3)(650 \text{ ft/min})(1.35 \text{ ft}^2)(0.2404 \text{ Btu/(lb }^\circ\text{F)})(70-65)^\circ\text{F} \times \frac{60 \text{ min}}{1 \text{ h}}$$

$$\dot{Q} = 4785 \text{ Btu/h}.$$

The outlet temperature of the engine oil can be determined:

$$\dot{Q} = \dot{m}_e c_{pe}(T_{e,i} - T_{e,o}) = C_e(T_{e,i} - T_{e,o})$$

$$T_{e,o} = T_{e,i} - \frac{\dot{Q}}{C_e} = 150^\circ\text{F} - \frac{4785 \text{ Btu/h}}{7.75 \text{ Btu/(min }^\circ\text{F)}} \times \frac{1 \text{ h}}{60 \text{ min}}$$

$$T_{e,o} = 139.7^\circ\text{F} = 140^\circ\text{F}.$$

The following will be needed to complete a heat exchanger design data sheet to show the results of the performance analysis:

Airflow rate: $\dot{V}_a = V_a A_f = (650 \text{ ft/min})(1.35 \text{ ft}^2) = 878 \text{ cfm}$;
Total tube length: $L = (4 \text{ rows})(6 \text{ tubes per row})(26 \text{ in. per tube}) = 624 \text{ in.} = 52 \text{ ft}$.

Pressure loss of air across the tube coils

The pressure drop of the air across the tube coil bank is given by

$$\Delta P_{\text{bank}} = \frac{G_m^2}{2\rho_{a,i}} \left[(1 + \sigma^2) \left(\frac{\rho_{a,i}}{\rho_{a,o}} - 1 \right) + f \frac{A_T}{A_c} \frac{\rho_{a,i}}{\rho_{\text{mean}}} \right].$$

Remember: $G_m = 1.48 \text{ lb/(s ft}^2)$ and $\text{Re}_{x_L} \approx 11 \times 10^3$.
From the j-factor versus Re_{x_L} chart, the friction factor (f) is

$$f \approx 0.016.$$

The area ratio is

$$\frac{A_T}{A_c} = \frac{4W}{D_h} = \frac{4\,(4.332 \text{ in.})}{0.0127 \text{ ft}} \times \frac{1 \text{ ft}}{12 \text{ in.}} = 113.7.$$

The mean density (ρ_{mean}) is

$$\rho_{\text{mean}} = \frac{\rho_{a,i} + \rho_{a,o}}{2} = \frac{(0.07561 + 0.07489) \text{ lb/ft}^3}{2} = 0.07525 \text{ lb/ft}^3.$$

From the j-factor versus Re_{x_L} chart, $\sigma = 0.555$.
Therefore,

$$\Delta P_{\text{bank}} = \frac{\left(1.48 \text{ lb/(s ft}^2)\right)^2}{2\left(0.07561 \text{ lb/ft}^3\right)} \left[(1 + 0.555^2) \left(\frac{0.07561 \text{ lb/ft}^3}{0.07489 \text{ lb/ft}^3} - 1 \right) \right.$$
$$\left. + (0.016)(113.7) \frac{0.07561 \text{ lb/ft}^3}{0.07525 \text{ lb/ft}^3} \right]$$

$$\Delta P_{\text{bank}} = 26.7 \text{ lb/(ft s}^2) \times \frac{1 \text{ lbf}}{32.2 \text{ (lb ft)/s}^2} = 0.828 \text{ lbf/ft}^2 = 0.828 \text{ psf}.$$

In practice, the pressure drop is reported in inches of water. Thus,

$$\Delta P_{\text{bank}} = 26.7 \text{ lb/(ft s}^2) \times \frac{1}{0.07525 \text{ lb/ft}^3} \times \frac{1}{32.2 \text{ ft/s}^2} = 11.1 \text{ ft of air}$$

$$\Delta P_{\text{bank}} = 11.1 \text{ ft of air} \times SG_{\text{air},68°F} = 11.1 \text{ ft of air} \times \frac{0.07525 \text{ lb/ft}^3}{62 \text{ lb/ft}^3} \times \frac{12 \text{ in.}}{1 \text{ ft}}$$

$$\Delta P_{\text{bank}} = 0.16 \text{ in. of water} = 0.16 \text{ in. wg.}$$

Pressure loss of water in the tube coils

Determination of the pressure loss of the engine oil in the tubes is needed to find the pump power required to move the fluid through the oil cooler system. The total head loss is

$$H_{IT} = \left(f \frac{L_{tube}}{D_i} + K \right) \frac{\overline{V}_w^2}{2g}.$$

For laminar flow in tubes,

$$f = \frac{64}{Re_{D_i}} = \frac{64}{191} = 0.335.$$

For the piping circuitry through the heat exchanger (excluding the hairpins), there are four tube rows, including one supply run and one return run, each 26 in. long. Therefore,

$$L_{tube} = (4 \text{ rows})(6 \text{ tubes per row})(26 \text{ in. per tube}) = 624 \text{ in.} = 52 \text{ ft.}$$

For the minor losses, $K = 2.0$ for the soldered/brazed $180°$ return bends. Thus,

$$H_{IT} = \left((0.335) \frac{624 \text{ in.}}{0.545 \text{ in.}} + (3)(6)(2.0) \right) \frac{(3 \text{ ft/s})^2}{2 \left(32.2 \text{ ft/s}^2 \right)}$$

$$H_{IT} = 58.6 \text{ ft wg.}$$

Drawings

No drawings are required for this performance analysis.

Conclusions

(i) A data sheet for this heat exchanger (oil cooler) is shown in the following table. This was a performance analysis problem. The following points should be noted:

(ii) The design engineer has designed an oil cooler that may not be capable of sufficiently cooling the engine oil. The temperature drop of the oil was approximately $10°F$. The engineer does not have control over the entering air temperature. Therefore, they should increase the size of the heat exchanger to increase the NTU, effectiveness, and the thermal capacity of the cooler.

(iii) Due to the viscosity of the engine oil and the assumed flow velocity (3 fps), the major head loss in the tube is very high. A large pump may be required, probably on the order of 1-½ hp. It would be recommended to decrease the flow velocity to reduce the pump size. However, the tube-side heat transfer coefficient would decrease further, resulting in lower heat transfer performance. A compromise may be needed. Another alternative could be to use inlet and outlet headers on the tubes, rather than having one tube inlet and one tube outlet. Use of the headers would reduce the longest run of piping in the heat exchanger to 104 in. (4 rows at 26 in. per tube), instead of 624 in. and the number of $180°$ return bends to 3, in lieu of 18. This reduction in length and number of bends would reduce the major head loss to 9.8 ft wg from 58.6 ft wg.

(iv) The pressure drop of the air over the oil cooler coils is low (less than 1 in. wg).

(v) The entering temperatures were used to find the fluid properties throughout the analysis. Since the temperature changes of the air and engine oil were small (less than 8%), the error introduced would be small.

A heat exchanger design data sheet is shown in the following table.

Heat Exchanger Design Data Sheet	
Type	Counter Flow
Section: Tube Bank	
Working fluid	Air
Volume flow rate	878 cfm
Inlet temperature	65°F
Outlet temperature	70°F
Pressure drop	0.16 in. wg
Section: Tube	
Tube material	Copper
Working fluid	Engine oil
Velocity	3 ft/s
Tube inner diameter	0.545 in.
Tube outer diameter	0.625 in.
Number of tube rows	4
Number of tubes per row	6
Tube spacing ($x_t \times x_L$)	1.083 in. × 1.25 in.
Total tube length	52 ft
Inlet temperature	150°F
Outlet temperature	140°F
Head loss	58.6 ft wg
Fin material	Aluminum
Fin pitch	8 fins per in.
Fin thickness	0.006 in.
Heat Exchanger Parameters	
Thermal capacity	4785 Btu/h
Effectiveness	0.104
Capacity ratio	0.484
Overall heat transfer coefficient	0.62 Btu/(h ft^2 °F)
Number of transfer units (NTU)	0.113
Heat exchanger dimensions ($L \times H \times W$)	26 in. × 8 in. × 5 in.

4.8 Manufacturer's Catalog Sheets for Heat Exchanger Selection

The material covered in most of this chapter focuses on developing an in-depth understanding of the fundamental design of heat exchangers. Given that the design of heat exchangers is time-consuming and laborious, in practice, only a select group of engineers will design heat exchangers as per the procedures outlined in this chapter.

The Unico System®

Bulletin 20-20.4 / June 2007

ENGINEERING SPECIFICATIONS

M SERIES HEATING MODULE

Packing List

Carton contains:
- 1 – Cabinet
- 1 – Hook Flange
- 2 – Latch keepers
- 2 – Latches
- 1 – Hot water heating coil *(Optional)*

Gasket as needed.

Applications

Unico System designed and built heating units can be easily installed with the matching blower and cooling modules. For matchups see table below. The heating module can be matched to a blower module for a heating only system or it can be matched with both a blower and a cooling module for a system that heats and cools. The slide-in hot water/glycol heating coil is supplied separately. If potable water is used, refer to Technote 112 for disinfection procedures.

Note: The MH2430 replaces the MH2436 and the MH3660 replaces the MH4260. Add HW to Part Number to include the coil, Example MH2430HW (coil included).

Table 1. Compatible Modules

Heating Module	Matching Unit	
	Blower Module	Cooling Module
MH2430	MB2430L	MC2430(C,H,W)
MH3660	MB3642L MB4860L	MC3642(C,H,W) MC4860(C,H,W)

Figure 1.   Heating Module

Cabinet Construction

The cabinet is constructed of 22 gauge galvanized steel. It has a removable panel to insert a hot water coil. The cabinet is fully lined with closed cell insulation. Easy snap latches are included for quick field assembly with the matching modules.

Coil Construction

Unico designed and fabricated hot water coils are constructed of evenly spaced corrugated aluminum fins mechanically bonded to copper tubes. The tubes are ½-in. diameter on staggered centers. The fins have full collars to provide greater tube-fin contact for excellent heat transfer. The coil is pressure tested at the factory. Bleed and drain valves are provided on the headers outside the cabinet. The coil is sold separately or with the cabinet.

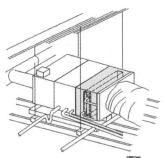

Typical Horizontal Installation with Unico System
Blower Module and Cooling Module

Certified to UL Standard 1995
Conforms to CAN/CSA Standard C22.2 NO. 236

 Unico products comply with the European regulations that guarantee the safety of the product.

Figure 4.24   M series heating coil from Unico, Inc.: (a) page 1 of the M series heating coil from Unico, Inc. (Unico, Inc., reprinted with permission); (b) page 2 of the M series heating coil from Unico, Inc. (Unico, Inc.; reprinted with permission); (c) page 3 of the M series heating coil from Unico, Inc. (Unico, Inc., reprinted with permission); (d) page 4 of the M series heating coil from Unico, Inc. (Unico, Inc.; reprinted with permission)

Model No.			MH2430	MH3660
Heating Coil	Coil Model		HW-2430	HW-3660
	Net Face Area, sq. ft. (m^2)		2.08 (0.20)	3.43 (0.32)
	Tube Diameter, in. (mm)		1/2 (12.7)	1/2 (12.7)
	Number of Rows		4	4
	Fins per inch (m)		12 (472)	12 (472)
	Connection Size, in. (mm) sweat		7/8 (22.2)	7/8 (22.2)
	Coil-only Shipping weight, lb. (kg)		33 (15)	47 (21)
Design Pressure, psi (kPa)			150 (1034)	150 (1034)
Dimensions, in. (mm)		A	25 (635)	38 (965)
		B	23 (584)	36 (914)
Shipping weight (without coil), lb. (kg)			20 (9)	28 (13)
Coil Water Volume, gal. (liters)			0.4 (1.8)	0.7 (3.2)

Module Dimensions

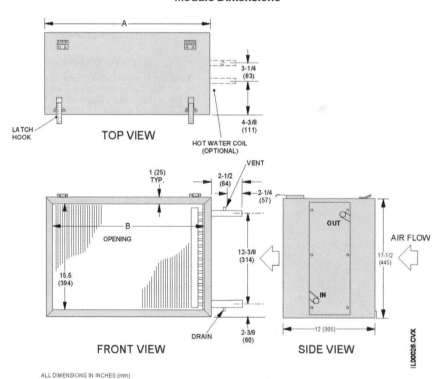

Figure 4.24 *(Continued)*

Hot Water Coil Performance

Capacity*, MBH (kW)

HW-2430

600 CFM (0.28 m³/s) — 18 Outlets minimum

Water Flow, GPM (L/s)		Entering Water Temperature, °F (°C)			
		120 (49)	140 (60)	160 (71)	180 (82)
4	(0.25)	26.7 (7.8)	37.5 (11.0)	48.4 (14.2)	59.4 (17.4)
6	(0.38)	27.7 (8.1)	38.9 (11.4)	50.1 (14.7)	61.4 (18.0)
8	(0.50)	28.2 (8.3)	39.5 (11.6)	50.9 (14.9)	62.4 (18.3)

500 CFM (0.24 m³/s) — 15 Outlets minimum

Water Flow, GPM (L/s)		Entering Water Temperature, °F (°C)			
		120 (49)	140 (60)	160 (71)	180 (82)
4	(0.25)	23.0 (6.7)	32.4 (9.5)	41.8 (12.2)	51.2 (15.0)
6	(0.38)	23.8 (7.0)	33.4 (9.8)	43.0 (12.6)	52.7 (15.4)
8	(0.50)	24.1 (7.1)	33.8 (9.9)	43.6 (12.8)	53.4 (15.6)

400 CFM (0.19 m³/s) — 12 Outlets minimum

Water Flow, GPM (L/s)		Entering Water Temperature, °F (°C)			
		120 (49)	140 (60)	160 (71)	180 (82)
4	(0.25)	19.3 (5.7)	27.1 (7.9)	34.9 (10.2)	42.8 (12.5)
6	(0.38)	19.7 (5.8)	27.7 (8.1)	35.6 (10.4)	43.6 (12.8)
8	(0.50)	19.9 (5.8)	27.9 (8.2)	36.0 (10.5)	44.0 (12.9)

EQUATIONS

The general equation of the sensible heat capacity, q, is:

$$q = \rho \dot{Q} c_p (\Delta T) \tag{1}$$

where ρ is density,

$\dot{Q}$ is the volumetric flow rate,

c_p is the specific heat capacity constant,

and ΔT is temperature difference through the coil.

The temperature difference is expressed differently depending on whether the fluid is being heated or cooled. It is expressed in the following way

Heated fluid: $\Delta T = T_{out} - T_{in}$ $\tag{2}$

Cooled fluid: $\Delta T = T_{in} - T_{out}$ $\tag{3}$

where T_{in} is the inlet temperature of the fluid,
and T_{out} is the outlet temperature of the fluid.
The fluid is either air or water.

HW-3660

1250 CFM (0.59 m³/s) — 37 Outlets minimum

Water Flow, GPM (l/s)		Entering Water Temperature, °F (°C)							
		120	(49)	140	(60)	160	(71)	180	(82)
4	(0.25)	45.0	13.2	63.4	18.6	81.8	24.0	100.4	29.4
6	(0.38)	49.9	14.6	70.2	20.6	90.5	26.5	111	32.6
8	(0.50)	52.5	15.4	73.8	21.6	95.2	27.9	117	34.3
10	(0.63)	54.1	15.9	76.0	22.3	98.0	28.7	120	35.2

1100 CFM (0.52 m³/s) — 33 Outlets minimum

Water Flow, GPM (l/s)		Entering Water Temperature, °F (°C)							
		120	(49)	140	(60)	160	(71)	180	(82)
4	(0.25)	42.1	12.3	59.2	17.4	76.5	22.4	93.8	27.5
6	(0.38)	46.1	13.5	64.8	19.0	83.6	24.5	102.5	30.1
8	(0.50)	48.2	14.1	67.7	19.9	87.3	25.6	107	31.4
10	(0.63)	49.5	14.5	69.5	20.4	90.0	26.4	110	32.3

1000 CFM (0.47 m³/s) — 30 Outlets minimum

Water Flow, GPM (l/s)		Entering Water Temperature, °F (°C)							
		120	(49)	140	(60)	160	(71)	180	(82)
4	(0.25)	39.9	11.7	56.2	16.5	72.5	21.3	88.9	26.1
6	(0.38)	43.4	12.7	61	17.9	78.6	23.1	96.3	28.2
8	(0.50)	45.2	13.3	63.4	18.6	81.8	24.0	100	29.3
10	(0.63)	46.2	13.6	64.9	19.0	83.6	24.5	102.3	30.0

900 CFM (0.42 m³/s) — 27 Outlets minimum

Water Flow, GPM (l/s)		Entering Water Temperature, °F (°C)							
		120	(49)	140	(60)	160	(71)	180	(82)
4	(0.25)	37.6	11.0	52.9	15.5	68.3	20.0	83.7	24.5
6	(0.38)	40.5	11.9	56.9	16.7	73.4	21.5	90	26.4
8	(0.50)	42	12.3	58.9	17.3	76	22.3	93	27.3
10	(0.63)	42.8	12.6	60.1	17.6	77.5	22.7	94.8	27.8

800 CFM (0.38 m³/s) — 24 Outlets minimum

Water Flow, GPM (l/s)		Entering Water Temperature, °F (°C)							
		120	(49)	140	(60)	160	(71)	180	(82)
4	(0.25)	35	10.3	49.3	14.5	63.5	18.6	77.9	22.8
6	(0.38)	37.3	10.9	52.5	15.4	67.6	19.8	82.8	24.3
8	(0.50)	38.5	11.3	54.1	15.9	69.6	20.4	85.3	25.0
10	(0.63)	39.2	11.5	55.0	16.1	70.8	20.8	86.7	25.4

700 CFM (0.33 m³/s) — 21 Outlets minimum

Water Flow, GPM (l/s)		Entering Water Temperature, °F (°C)						
		120 (49)		140 (60)		160 (71)		180 (82)
4	(0.25)	32.1	9.4	45.1	13.2	58.2	17.1	71.3	20.9
6	(0.38)	33.9	9.9	48.8	14.3	61.3	18.0	75.1	22.0
8	(0.50)	34.8	10.2	49.5	14.5	62.8	18.4	76.9	22.6
10	(0.63)	35.3	10.4	45.1	13.2	63.7	18.7	77.9	22.8

600 CFM (0.28 m³/s) — 18 Outlets minimum

Water Flow, GPM (l/s)		Entering Water Temperature, °F (°C)						
		120 (49)		140 (60)		160 (71)		180 (82)
4	(0.25)	28.8	8.4	40.5	11.9	52.2	15.3	63.9	18.7
6	(0.38)	30.1	8.8	42.2	12.4	54.4	16.0	66.6	19.5
8	(0.50)	30.7	9.0	43.1	12.6	55.4	16.2	67.8	19.9
10	(0.63)	31	9.1	43.5	12.8	56	16.4	68.5	20.1

* Capacity is based on 70°F (21°C) return air temperature (T_{in})
Conversion Factors: MBH = 1000 Btu/hr, 1 kW = 3413 Btu/hr

Figure 4.24 *(Continued)*

Equation 1 can be simplified by using standard density and specific heat. If you are at a high altitude please refer to Tech Note 103, *High Altitude Applications,* for more detailed information about effects of air density. Otherwise, use the following equations to find the leaving fluid temperature.

For air:

$$q = 1.08 \text{ (CFM)} \, \Delta T \ \text{ Btu/hr} \quad (\Delta T \text{ is in } °F) \quad (4)$$

$$q = 1.21 \text{ (L/s)} \, \Delta T \ \text{ Watts} \quad (\Delta T \text{ is in } °C) \quad (5)$$

For water:

$$q = 500 \text{ (GPM)} \, \Delta T \ \text{ Btu/hr} \quad (\Delta T \text{ is in } °F) \quad (4)$$

$$q = 4.15 \text{ (L/s)} \, \Delta T \ \text{ kW} \quad (\Delta T \text{ is in } °C) \quad (5)$$

Example. Consider a MH2430 with 6 GPM (38 L/s) at 140 °F (60 °C) and 600 CFM (280 L/s). The capacity from the table is 38.9 MBH (11.6 kW). Therefore, the leaving air temperature (LAT) is as follows:

$$\text{LAT} = 70 + \frac{38.9 \times 1000}{1.08 \times 600} = 130 \quad °F$$

$$\text{LAT} = 21 + \frac{11.4 \times 1000}{1.21 \times 280} = 54.6 \quad °C$$

Likewise, determine the Leaving Water Temperature (LWT) by using one of the following equations:

$$\text{LWT} = 140 - \frac{38.9 \times 1000}{500 \times 6} = 127 \quad °F$$

$$\text{LWT} = 60 - \frac{11.4}{4.15 \times .38} = 52.8 \quad °C$$

Coil Pressure Drop

Air Pressure Drop

Air Flow rate, CFM (m³/s)		ΔP, in. water (kPa)			
		HW2430		HW3660	
400	(0.19)	0.07	(0.017)	-	
500	(0.24)	0.10	(0.025)	-	
600	(0.28)	0.12	(0.030)	0.06	(0.015)
700	(0.33)	-		0.08	(0.020)
800	(0.38)	-		0.09	(0.022)
900	(0.42)	-		0.11	(0.027)
1000	(0.47)	-		0.13	(0.032)
1100	(0.52)	-		0.15	(0.037)
1250	(0.59)	-		0.19	(0.047)

Water Pressure Drop

Water Flow rate, GPM (L/s)		ΔPw, ft. water (kPa)			
		HW2430		HW3660	
4	(0.25)	4.3	(12.9)	2.6	(7.8)
6	(0.38)	9.3	(27.8)	5.5	(16.4)
8	(0.50)	16.1	(48.1)	9.6	(28.7)
10	(0.63)	-	-	14.6	(43.6)

Entering Water Temperature, °F (°C)	120 (49)	140 (60)	160 (71)	180 (82)	200 (93)
F1	1.046	1.000	0.959	0.921	0.888

Water Pressure drop = $\Delta P_w \times F1$

Figure 4.24 *(Continued)*

Most design engineers will use **manufacturer's catalog sheets** to select a heat exchanger. The designer would calculate the required thermal capacity, and with other parameters, an appropriate heat exchanger for the application would be selected. In practice, the size (physical dimensions) of the heat exchanger is typically a major factor that drives the final selection of a unit.

Figure 4.24 shows four sheets of a heat exchanger selection bulletin for a heating coil model used in high-velocity duct systems. In the bulletin, the manufacturer presents an overview of their M series heating coil heat exchanger (Figure 4.24a). In this overview, the manufacturer describes the construction of the unit, provides a list of the parts, and shows a drawing of the unit. Detailed drawings and specifications on the heating coil and the construction are also provided (Figure 4.24b). The manufacturer understands that each design will require different thermal capacities, and has provided three different models of hot water heating coils, HW-2430 and HW-3660 (Figure 4.24c). Each model is subdivided based on the flow rate of air across the heating coils and the minimum number of outlets from the duct system. The design engineer should note that to determine the heat transfer performance (i.e., thermal capacity) of the unit, the water flow rate and entering water temperature will need to be known. The final page of the bulletin (Figure 4.24d) shows pressure drops in and across the coils for different pipe velocities and airflow rates, respectively.

Problems

4.1. The design of a simple refrigerant condenser has been proposed for development. The condenser will accept hot, high-pressure, saturated refrigerant type R-134a from a compressor to reject its heat to ambient air at standard atmospheric conditions. The cooled refrigerant is drained back to an evaporator through an expansion valve to complete a standard refrigeration cycle. The design concept will be to use bare $^3/_8$-in. outer diameter tubes arranged in a staggered tube bank. To simplify the design, there will be only one tube inlet and one tube outlet into and out of the bank, respectively. No front end or rear end headers will be considered. It is hoped that the pressure drop across the tube bank will be kept as low as possible to ensure the smallest possible fan size. The system will operate under the following conditions:

   (i) Four rows of tubes
   (ii) Sixteen tubes per row
   (iii) Width is 32 in.
   (iv) Entering air temperature is 95°F
   (v) Coil face velocity is not to exceed 700 ft/min
   (vi) R-134a enters at a saturation pressure of 200 psia (assume liquid phase)

   Determine the heat rejected by the condenser and the temperature of the exiting air.

4.2. The Canadian Biosolids Partnership is considering the use of a small gas turbine with air as the working fluid in a cogeneration system to produce electricity for one of their infrastructures and heat for the bacterial decomposition of fecal matter in a water-based mixture. The pH of the water-based mixture is 7 (neutral; neither acidic or basic). The mixture is moved gently to maintain homogeneity of properties. Of interest is the use of fecal coliform bacteria to decompose human fecal matter in a large uninsulated concrete tank covered and located outside. After decomposition of the fecal matter, the solution must be heated to 170°F. The concrete tank has sides that are 10 ft long and 5 ft high. The construction is 8 in. thick concrete blocks. The cover has similar construction, and the tank is not pressurized. A large mechanical scrubber unit ensures that solid fecal matter never adheres to the concrete wall surface. A high-level switch ensures that the water level in the tank never exceeds $3^1/_2$ ft. The option exists to run a coiled piping system through the exhaust duct of the gas-turbine subsystem. The system would be directly connected to the concrete tank. High efficiency, fine mesh filters could be used at the concrete wall-to-pipe connection points. High-temperature resistant, high-efficiency HEPA filters will be installed near the exhaust of the gas turbine. The total distance between the concrete tank and the gas turbine exhaust duct is limited to 20 ft. Use of this tank will be restricted to the autumn months where the average outdoor temperature is approximately 55°F and wind speeds are low. Ground temperatures are between 40 and 48°F. The given schematic drawing provides additional information. Not all accessories are shown. Design an appropriate heat exchanger subsystem in this cogeneration system to provide enough energy to heat the fecal water to the required temperature.

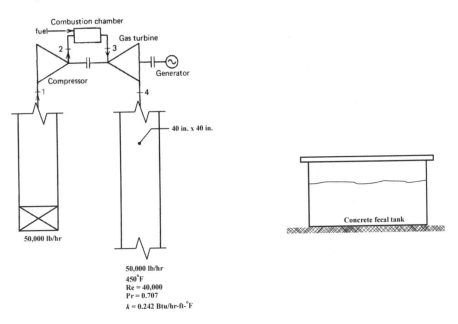

4.3. Rinnai US has recently developed a new hydronic air handling unit (AHU Series 37AHA) to heat air in residential buildings. It is expected that hot water from a tankless water heater will be pumped through a finned-tube heating coil installed in the main branch (20 in. × 18 in.) of a ductwork system. A homeowner has calculated that 96000 Btu/h will be needed to heat their home as desired. Air will enter the duct at approximately 70°F and should leave at 150°F. Water will enter the coil system at 180°F. Design the hot water heating coil for this application.

4.4. A *superheater* is a counter flow heat exchanger used in power plant systems to transfer heat from hot exhaust flue gases to saturated steam to increase its temperature before entrance to a steam turbine. The tubes containing saturated steam are usually arranged in-line and may be fabricated from low carbon steel, chrome-moly, stainless steel, super-alloys or other types of heat-resistant alloys. The tubes tend to be devoid of extended surfaces. A superheater will be required to produce steam at a rate of 1000000 lb/h at 1900 psia and 1000°F. The hot exhaust flue gas enters the superheater at 2000°F at a rate of 1230000 lb/h. Estimate the superheater size. In other words, estimate the heat transfer surface area, the number of tubes, and the number of tube rows for the following design conditions:

Tube nominal diameter	$2\frac{1}{2}$ in.
Tube center to center spacing	7 in.
Typical superheater length	12 ft
Typical superheater width	Variable
Estimated overall heat	
Transfer coefficient	8.8 Btu/(h ft^2 °F)

Further Information: A consideration of enthalpies may be useful.

4.5. Public Service Enterprise Group (PSEG) is a publicly traded, diversified energy company headquartered in New Jersey. A junior mechanical engineer working with PSEG wishes to size a reheater for use in a plant in Irvington, NJ. The reheater will receive steam at 744 psia and 600°F and release it at 1000°F. Due to losses in the tubing, there is a 7% pressure drop in the steam. The steam flow rate is 4000000 lb/h. Hot flue gases from a burner enter the reheater at 1600°F and 5250000 lb/h to flow over the steam tubes that are arranged in-line. The overall heat transfer coefficient is approximately 8.5 Btu/(h ft^2 °F). Of interest are the estimation of the reheater surface area and the temperature of the flue gas leaving the reheater.
   Further Information: A consideration of enthalpies may be useful.

4.6. Fuel oils such as #2 fuel oil may be used in boiler burners to provide energy to the working fluid in the boiler tubes. Due to the high viscosity of the fuel oil, a heater is usually needed in the oil storage tank to warm the oil to facilitate pumping. Heating the oil to at least 150°F will maintain good combustion in the burner. A design engineer wishes to reduce the total power required by the oil storage tank heater. They have devised a system in which hot exhaust gases from the boiler burner will be directed through an 18 in. × 20 in. rectangular ductwork to a heat exchanger for the purposes of providing additional energy to heat the oil. The average flue gas temperature from the burner is on the order of 650°F, and to avoid metal corrosion and proper operation of pollution control equipment, the temperature can never be lower than 480°F. The oil pumping system available to the engineer will fail if the oil temperature is lower than 50°F. Design a heat exchanger that could be used in the engineer's system.
   Properties of #2 fuel oil are: $T_{freezing} = -22°F$; $T_{boiling} = 374\text{--}689°F$; $SG = 0.86$; $v = 3.66 \times 10^{-5}$ ft^2/s; $c_p = 0.44$ Btu/lb-°F; $k = 0.0797$ Btu/(h ft °F).

4.7. A counter flow evaporator based on R-134a refrigerant has been designed with a capacity of 12 tons of refrigeration. The evaporator is used to cool

50 vol% ethylene glycol solution ($c_p = 0.775$ Btu/(lb °F) that enters the heat exchanger at 55°F with a mass flow rate of 20000 lb/h. The R-134a enters the heat exchanger at 35°F with a quality of 0.1 and leaves with a quality of 0.65. Estimate the exit temperature of the ethylene glycol solution, the mass flow rate of the refrigerant, and the increase in efficiency of the heat exchanger if the surface area of the evaporator were doubled.

Further Information: The R-134a refrigerant has experienced a phase change.

4.8. A low-velocity residential air duct is to be equipped with a manufacturer's heating module. It is expected that the module will temper dry air by heating it from 40°F to 80°F. Hot water from a boiler package is available. Due to inefficiencies in the boiler, the exit water temperatures will range from 140°F to 180°F. Select a manufacturer's heating module to heat the air and prepare an equipment schedule for the client's contract documents. The air duct size will depend on the size of the module unit that is selected by the mechanical engineer.

References and Further Reading

[1] Lee, H. (2010) *Thermal Design: Heat Sinks, Thermoelectrics, Heat Pipes, Compact Heat Exchangers, and Solar Cells*, John Wiley & Sons, Inc., Hoboken.

[2] Kays, W. and London, A. (1964) *Compact Heat Exchangers*, 2nd edn, McGraw-Hill, Inc., New York.

[3] Çengel, Y. (2007) *Heat and Mass Transfer: A Practical Approach*, 3rd edn, McGraw-Hill, Inc., New York, p. 469.

[4] Edwards, D., Denny, V., and Mills, A. (1979) *Transfer Processes*, 2nd edn, Hemisphere Publishing Corp., New York.

[5] Sieder, E. and Tate, G. (1936) Heat transfer and pressure drop of liquids in tubes, *Industrial Engineering Chemistry*, **28**, 1429–1435.

[6] Dittus, F. and Boelter, L. (1930) *University of California Publications on Engineering*, **2**, 433.

[7] Gnielinski, V. (1976) New equations for heat and mass transfer in turbulent pipe and channel flow, *International Chemical Engineering*, **16**, 359–368.

[8] Churchill, S. and Bernstein, M. (1977) A correlating equation for forced convection from gases and liquids to a circular cylinder in cross flow, *Journal of Heat Transfer*, **99**, 300–306.

[9] Schmidt, T. (1945–1946) La production calorifique des surfaces munies d'ailettes, *Annexe du Bulletin de L'Institut International du Froid*, Annexe G-5.

[10] McQuiston, F., Parker, J., and Spitler, J. (2000) *Heating, Ventilating, and Air Conditioning: Analysis and Design*, 5th edn, John Wiley & Sons, Inc., New York, pp. 489–503.

[11] Wolverine Tube Inc. (2009) *Wolverine Tube Heat Transfer Data Book*, Wolverine Tube Inc., Huntsville, pp. 45–56.

5

Applications of Heat Exchangers in Systems

The fundamental design of heat exchangers can be a complex and laborious process, requiring the use of multiple charts and correlation equations. To that end, equipment manufacturers have invested millions of dollars into conducting performance tests and securing appropriate government certification of their equipment. This performance data is usually available in catalogs from the manufacturers. This greatly reduces the work required by design engineers during larger system designs that include heat exchangers.

This chapter focuses on the practical applications of heat exchangers in thermo-fluids systems. Particular attention will be given to a cooling system for a high-temperature plasma spray torch and hot water heating systems.

5.1 Operation of a Heat Exchanger in a Plasma Spraying System

Plasma spraying is a high-temperature process used to melt and accelerate powdered particles in order to fabricate protective coatings on machine and industrial parts. Figure 5.1 shows a picture of a plasma torch in operation, the small plasma jet shown, can reach temperatures between 10000 and 20000 K. As a consequence of the high plasma jet temperatures, a system is needed to cool the torch electrodes and housing to prevent overheating.

Figure 5.2 shows a heat exchanger that could be used to provide chilled water to the torch. In addition, Figure 5.3 shows a **functional diagram** of the heat exchanger. The functional diagram shows all the functional parts of the heat exchanger. Of special interest is the distilled/deionized water storage tank (label 7 in Figure 5.3). In this application, it is paramount to have chilled deionized water in contact with the components of the torch. If the cooling water is hard or ionized, electricity from the torch could flow through the cooling water, with dangerous consequences. A

Introduction to Thermo-Fluids Systems Design, First Edition. André G. McDonald and Hugh L. Magande.
© 2012 André G. McDonald and Hugh L. Magande. Published 2012 by John Wiley & Sons, Ltd.

Figure 5.1   A Praxair SG-100 plasma spray torch in operation

heat exchanger module (label 3 in Figure 5.3) is needed for the exchange of heat between the heated deionized water and the cold plant/building or chiller water, which typically contains ions or other chemical contaminants.

A **flow diagram** is usually presented to show the flow of water (the working fluid) in the cooling system that includes the heat exchanger. Figure 5.4 shows a flow diagram

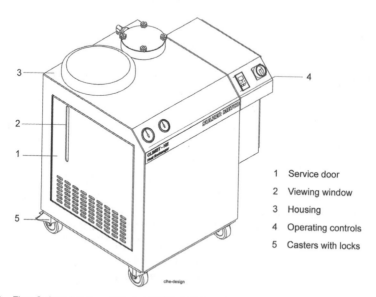

1   Service door
2   Viewing window
3   Housing
4   Operating controls
5   Casters with locks

Figure 5.2   The Sulzer Metco Climet-HE™-200 heat exchanger (Sulzer Metco, Product Manual MAN 41292 EN 05; reprinted with permission)

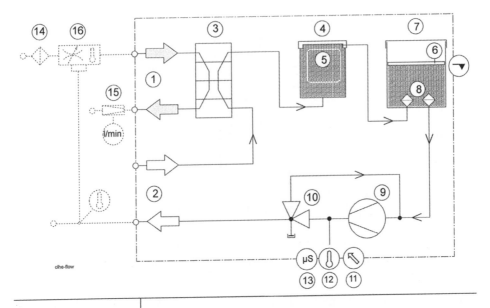

1   Primary cooling circuit (plant water)

2   Secondary cooling circuit (spray gun cooling water)

3   Heat exchanger module

4   Conditioning tank

5   Conditioning chemical element

6   Floating disk

7   Storage tank

8   Screen filters

9   Regenerative turbine pump

10 Pressure adjustment valve

11 Pressure indicator

12 Temperature indicator

13 Conductivity indicator

14 Basket strainer (option)

15 Flow indicator (option)

16 Temperature control valve (option)

Figure 5.3 Functional diagram for the Sulzer Metco Climet-HE^{TM}-200 (Sulzer Metco, Product Manual MAN 41292 EN 05; reprinted with permission)

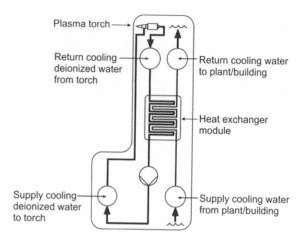

Figure 5.4 Flow diagram for cooling a typical plasma torch (modified from Sulzer Metco, Product Manual MAN 41292 EN 05; reprinted with permission)

for cooling a typical plasma torch. Note that the deionized water line to and from the plasma torch is a closed-loop circuit; on the other hand, the flexible tube circuit, which is in series with the chilled water lines to and from the plant/building is an open loop.

5.2 Components and General Operation of a Hot Water Heating System

Hot water heating systems or **hydronic heating systems** are multicomponent systems. Some of the components include

 (i) a **boiler** or **water heater** to heat water (the working fluid). This is the heat generator;
 (ii) water in an open- or closed-loop piping system;
 (iii) a **circulator** (a pump) to move fluid through the piping system;
 (iv) an **expansion tank** to facilitate volumetric expansion of the heated water;
 (v) an **air purger** or **vent** to allow the release of entrapped air;
 (vi) valves and other appurtenances, as needed;
 (vii) **baseboard heaters** (or other **terminal units**) to provide heat to the space or process.

Figure 5.5 shows a schematic of a closed-loop hydronic heating system c/w a boiler. The following points should be noted regarding this typical system:

 (i) This is a closed system under pressure.
 (ii) The boiler serves to generate hot water.

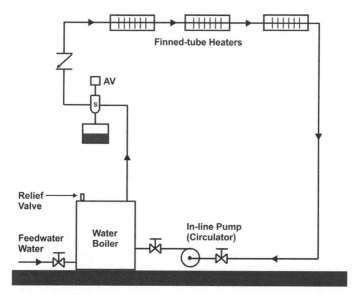

Figure 5.5   Schematic of a closed-loop hydronic heating system complete with a boiler

(iii) The circulator pumps water through the system piping. The circulator may be placed on the inlet or the discharge side of the boiler.

(iv) The hot water will lose heat to the space or the process through the baseboard heaters or other terminal units.

(v) Other heat losses may occur through the piping system if they are not well insulated.

(vi) The **pressure relief valve** on the boiler serves to protect it from excessive pressure buildup and possible rupture.

(vii) The check valve prevents backflow of water into the boiler when the pump is not energized. Other valves may be included in the system, as required (for isolation, control, etc.).

5.3 Boilers for Water

5.3.1 Types of Boilers

A **boiler** for water is a pressure vessel designed to transfer heat to cold water to generate steam or hot water. All boilers are constructed to meet the ASME (American Society of Mechanical Engineers) boiler and pressure vessel code, Section IV.

There are two types of water boilers that will be encountered frequently in industry:

(i) *Low-pressure boilers*: In these boilers, the maximum working pressure is approximately 15 psig for steam and 160 psig for hot water. The temperature in the hot water boilers is limited to 250°F (operating temperature).

(ii) *Medium- and high-pressure boilers*: These boilers will operate above 15 psig for steam and above 160 psig, but less than 250°F for hot water.

5.3.2 Operation and Components of a Typical Boiler

Figure 5.6 shows a schematic of a typical gas-fired hot water boiler. Note that the arrangement of the components can vary.

Of interest are the following additional components that expand the operation of a typical boiler:

(i) A drain valve complete with a **hose bib** or **pipe cock**. This is used to drain the boiler to allow for maintenance/repairs or to prevent flooding.

(ii) Water feed line. This line supplies fresh plant or building water to the boiler. For homes or small commercial buildings, the **feedwater** may be supplied by the local municipality.

(iii) **Burners** c/w with a gas valve. The burners supply the heat required. Open flames will heat the water contained in the boiler tubes (water tube) or storage tank (fire tube). Natural gas, oil, wood, or coal are examples of **fuels** that may be used.

(iv) **Aquastat**. The aquastat is a thermostat that is used to measure the water temperature and control the burners by modulating the fuel flow rate (through the gas valve).

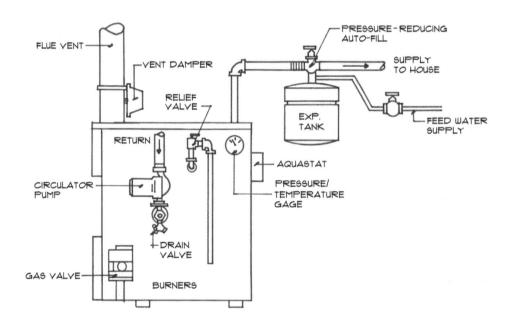

GAS-FIRED HOT WATER HEATING BOILER
NTS

Figure 5.6 A typical gas-fired hot water boiler

(v) **Vent** or **chimney**. This serves to channel combustion **flue gases** from the burners to the atmosphere. As the burners (enclosed in the combustion chamber) heat the fluid inside the heat exchanger, combustion gases are formed. The vent is therefore connected to the combustion chamber to facilitate routing of the flue gases to atmosphere.

Practical Note 5.1   Condensing Boilers

The typical boiler system described in Section 5.3.2 is a conventional, noncondensing boiler. In this type of boiler, the combustion flue gases are exhausted from the combustion chamber at high temperatures on the order of 350–400°F. This results in significant waste of energy. Currently, condensing boilers are being specified by design engineers and design-build contractors, in lieu of conventional, noncondensing boilers. In a condensing boiler, the temperature of the flue gases may be lowered to 140°F through the use of a secondary heat exchanger to preheat the return water. A larger primary heat exchanger, which provides a larger condensing surface area, may be used instead of a secondary heat exchanger to reduce the flue gas temperature (fully condensing boiler). The design engineer should note that reducing the flue gas temperature to 140°F results in the formation of condensate. This could pose a serious corrosion issue if not properly drained from the system. Additionally, the condensate should be neutralized prior to disposal into the building's plumbing drain system. The abovementioned issue should be considered in all designs, selections, and consultations. Regarding efficiency of the condensing boiler, it is increased with lower return water temperatures to the boiler heat exchanger. The water return temperature to the boiler should not exceed 113–122°F [1] to improve efficiency.

Water heaters are also used extensively in industry to provide hot water (not steam), especially in residential buildings or small process applications. The operation of gas-fired water heaters is similar to that of boilers. Figure 5.7 shows a schematic of the internal sections of typical water heaters. The figure shows that the water heaters can either be electric or gas fired. In the electric water heater, heating elements are placed at different levels in the tank. In both water heater configurations, **dip tubes** are used to supply cold feedwater to the tank. The dip tube is placed near the bottom of the tank since colder water has a higher density than heated water, and will thus fall to the bottom of the tank. In addition, the dip tube has several small orifices. As the water exits the orifices at high velocity, this promotes mixing in the tank to enhance heating and uniform tank water temperature. The hot water outlet that supplies hot water to the building or process will be installed close to the top of the tank. This will ensure that only hot water will be supplied.

Tankless water heaters are also being used to supply hot water for a variety of different residential and commercial applications. In some of these systems (condensing), direct heat from a burner flame and flue gases are used to heat water flowing over finned copper tubes in a counterflow heat exchanger arrangement. The high efficiency of the heat exchanger design eliminates the need for large water tanks. This reduces

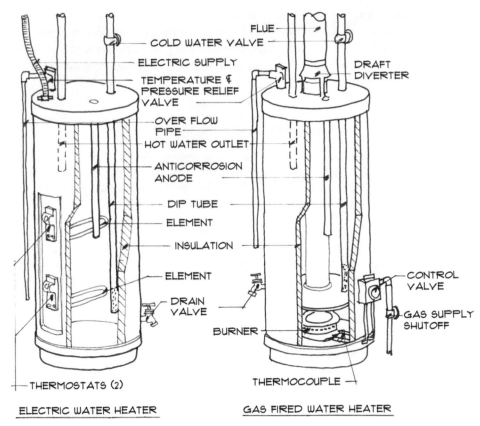

FLUE
COLD WATER VALVE
ELECTRIC SUPPLY
TEMPERATURE &
PRESSURE RELIEF
VALVE
DRAFT
DIVERTER
OVER FLOW
PIPE
HOT WATER OUTLET
ANTICORROSION
ANODE
DIP TUBE
ELEMENT
INSULATION
ELEMENT
CONTROL
VALVE
DRAIN
VALVE
GAS SUPPLY
SHUTOFF
BURNER
THERMOSTATS (2)
THERMOCOUPLE

ELECTRIC WATER HEATER   GAS FIRED WATER HEATER

Figure 5.7 Schematic of the internal section of typical water heaters

energy consumption since water is only heated as needed, rather than being heated and stored in a tank. Systems of this type provide continuous hot water. Figure 5.8 shows images and a typical schematic diagram of a tankless water heater from Rinnai.

5.3.3 Water Boiler Sizing

Boiler (or water heater) sizing refers to the determination of the amount of heat that a boiler will need to provide for a given process. During boiler sizing, the physical size of the boiler or water heater tank must also be determined.

Boiler sizes or **heating capacity** are rated in **boiler horsepower**. 1 boiler horsepower is the amount of heat needed to evaporate approximately 34.5 lbs of water to steam at 212°F or higher.

Therefore,

$$1 \text{ hp (boiler)} = 33475 \text{ Btu/h} = 33.475 \text{ MBH}, \tag{5.1}$$

where MBH = 1000 Btu/h.

Small- to medium-sized boilers are rated below 10000000 Btu/h or 10000 MBH.

Non-Condensing tankless water heater

(*a*)

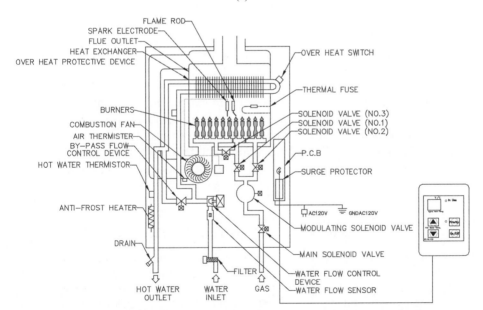

SCHEMATIC OF RINNAI NON-CONDENSING TANKLESS WATER HEATER

(*b*)

Figure 5.8   (a) A Rinnai noncondensing tankless water heater. (b) Schematic of Rinnai noncondensing tankless water heater (reprinted with permission)

To determine the heating capacity of a boiler or water heater, the amount of heat required by the application **must** first be determined. That will determine the size of the boiler or water heater. For example, consider the heating of a space (residential or commercial) and domestic water. The **thermal size** of the boiler or water heater is found from

$$\dot{Q}_{boiler} = \dot{Q}_{space} \times \text{OSF} + \frac{G_h \rho c_p (T_{out} - T_{in})}{\eta}, \tag{5.2}$$

where

$\dot{Q}_{boiler}$ = thermal size of the boiler;
$\dot{Q}_{space}$ = energy required to maintain the space at a set temperature. This will be the total heat transferred through the baseboard heaters;
OSF = **oversizing factor**. Consider this as a safety factor, which ranges between 1.05 and 1.25;
G_h = **recovery rate**. The recovery rate is the volume flow rate of hot water that is required to ensure that a sufficient amount of hot water is available at the required outlet temperatures for domestic use. Charts are used to find G_h;
T_{out} = outlet temperature of the hot water from the boiler or water heater;
T_{in} = inlet temperature of cold water from the plant/building or from the municipal line that enters the boiler or water heater; and
η = efficiency of the boiler or water heater. Boiler efficiencies may range between 70% and 85%. For fully condensing boilers, the efficiency may be above 90%.

It is important to note that in Equation (5.2), the total thermal size of the boiler is the total heat required to heat the space and to supply hot water for domestic use. For other applications, the onus is on the design engineer to determine the amount of heat required by the application or process in order to size the boiler appropriately. Equation (5.2) may be modified to determine the thermal size of a boiler or water heater that will provide hot water for an industrial process. In this case, $\dot{Q}_{space} = 0$ and $G_h = \dot{V}$.

Practical Note 5.2   Typical OSF Values

Residential houses and apartments are typically heavily occupied (more people per unit area who are cooking and constantly using hot water.) and will require OSF values from 1.05 to 1.1. Commercial buildings or schools that are lightly occupied or unoccupied (less people per unit area) will require OSF values from 1.1 to 1.25. Higher OSF values enable faster achievement of the set temperatures because the water boiler and/or water heater will have excess capacity to respond to demand faster. When no domestic water is required, the OSF may be considered as the efficiency of the boiler.

Practical Note 5.3   Domestic Water Data for Edmonton, Alberta, Canada

Water supplied by the City of Edmonton has temperatures on the order of 40–48°F (4–9°C). The water pressure will range between 35 and 100 psig (240–700 kPa gage). Typical water consumption is on the order of 60 gal per person per day (225 L per person per day) [2].

The physical size of the boiler or water heater is represented as the volume of the hot water tank. The tank size (for domestic use) is given by

$$\Psi_{tank} = \frac{\Psi_{capacity}}{F_{usable}} N, \tag{5.3}$$

where

Ψ_{tank} = required size of the boiler or water heater tank (volume);

$\Psi_{capacity}$ = **usable storage capacity** required per person (L/person or gal/person). This is the amount of hot water that the tank should provide per person or other unit, such as an apartment. Charts are used to find $\Psi_{capacity}$;

F_{usable} = **fraction of the tank water that has the minimum required temperature.** Normally, F_{usable} ranges between 60% and 80%. Remember that heating of cold water may occur from the bottom of the tank (as in the case of gas-fired boilers or water heaters). Heat will be conducted from the bottom to the top of the tank, which will not occur instantaneously. Therefore, the water temperature will not be uniform at the set temperature throughout the tank volume; and

N = number of units (e.g., persons, beds, apartments, etc.).

Practical Note 5.4   Hot Water Temperatures from Faucets

Division B, Section 7.2.6.1 of the 2006 Alberta Building Code [2] requires that hot water be supplied at temperatures higher than 113°F (45°C). To prevent burning (scalding), Section 7.2.6.7 requires that hot water from faucets must be limited to 130°F (54°C). For domestic use, hot water boilers and water heaters should not produce water with temperatures higher than 140°F (Section 501.6 of the International Plumbing Code). For other jurisdictions, the appropriate codes should be consulted. To limit the possibility of legionella growth, tank water should be kept at a minimum of 140°F. If lower water temperatures are required, a thermostatic mixing valve should be employed.

Table 5.1 shows some typical values of the minimum recovery rates and the corresponding minimum usable storage capacities for a variety of residential and industrial applications.

Table 5.1   Minimum recovery rates and minimum usable storage capacities

Type of Building	Minimum Recovery ptRates (G_h)	Corresponding Minimum ptUsable Storage Capacity pt($\Psi_{capacity}$)
Dormitories		
Men's dormitories	0.85 gal/h per student	10 gal per student
Women's dormitories	1.10 gal/h per student	12 gal per student
Motels: number of units		
20 or less	1.50 gal/h per unit	16 gal per unit
60	1.25 gal/h per unit	14 gal per unit
100 or more	1.00 gal/h per unit	12 gal per unit
Nursing homes	1.25 gal/h per bed	12 gal per bed
Office building	0.10 gal/h per person	1.6 gal per person
Food service		
Type A: full meal restaurant	0.45 gal per maximum number of meals per hour	7.5 gal per maximum number of meals per hour
Type B: drive-ins, luncheonettes	0.25 gal per maximum number of meals per hour	2.0 gal per maximum number of meals per hour
Apartment houses		
20 or less	3.40 gal/h per apartment	42 gal per apartment
50	3.00 gal/h per apartment	38 gal per apartment
75	2.75 gal/h per apartment	34 gal per apartment
100	2.40 gal/h per apartment	28 gal per apartment
130 or more	2.10 gal/h per apartment	24 gal per apartment
Elementary schools	0.06 gal/h per student	1.5 gal per student
Junior and senior high schools	0.15 gal/h per student	3.0 gal per student

Source: Bobenhausen [3].

5.3.4 Boiler Capacity Ratings

Once the thermal size of the boiler or water heater has been determined, an appropriate unit is selected from manufacturers' catalogs. Figure 5.9 shows an excerpt from a page taken from a Smith gas-fired residential boiler (GB100 series) technical brochure that shows typical specifications for a line of gas-fired boilers.

The following points are noted regarding the specifications presented in Figure 5.9:

(i) Three different **heating capacity ratings** are presented. These ratings are determined through different types of testing procedures by the manufacturer, and may be conducted with different types of burner fuels such as natural gas or liquified petroleum (LP) gas:

   (a) The **AGA** heating capacity rating is based on standards developed by the American Gas Association. This rating was phased out in 1997.

GB100 series

GAS FIRED RESIDENTIAL BOILER

A.G.A. Certified heating capacity ratings from 50,000 – 275,000 MBH

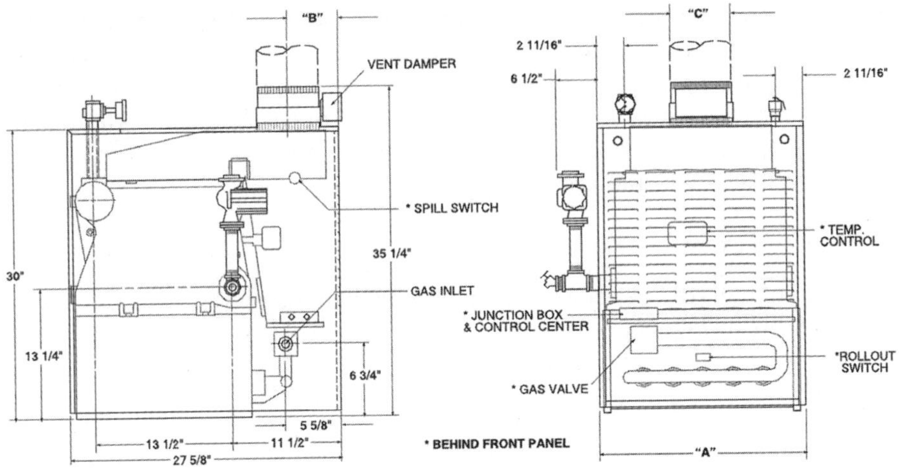

Supply and Return pipe size 1¼" NPT

Note: For closet, alcove or combustible floor installations, add 3" to the height dimensions.

GB100 Specifications

Boiler Model Number	Natural/L.P. AGA Input MBH	Natural/L.P. D.O.E Htg. Cap. MBH	Natural/L.P. Net I=B=R Ratings MBH	A.F.U.E.① Standing Pilot	A.F.U.E.① Hot Surface Ignition	Dimensions A	Dimensions B	Dimensions C	Gas Connection, Natural Gas: Standing Pilot	Gas Connection, Natural Gas: Spark Pilot	Gas Connection, Natural Gas: Hot Surface Ignition	Gas Connection: L.P. Gas
GB100-3	50.0	42	36	81.5%	84.3%	10⅞"	4½"	5"	½"	¾"	½"	¾"
GB100-4	75.0	63	55	82.0%	84.0%	13⅝"	4½"	5"	¾"	¾"	½"	¾"
GB100-5	100.0	83	72	82.0%	83.5%	16⅜"	5"	6"	¾"	¾"	½"	¾"
GB100-6	125.0	104	90	82.0%	83.2%	19⅛"	5"	6"	¾"	¾"	½"	¾"
GB100-7	150.0	125	109	82.3%	83.3%	21⅞"	5½"	7"	¾"	¾"	½"	¾"
GB100-8	175.0	146	127	82.1%	83.2%	24⅝"	5½"	7"	¾"	¾"	¾"	¾"
GB100-9	200.0	167	145	82.1%	83.2%	27⅜"	6"	8"	¾"	¾"	¾"	¾"
GB100-10	225.0	187	163	82.1%	83.2%	30⅛"	6"	8"	¾"	¾"	¾"	¾"
GB100-11	250.0	208	181	82.1%	83.2%	32⅞"	6"	9"	¾"	¾"	¾"	¾"
GB100-12	275.0	229	199	82.1%	83.2%	35⅝"	6"	9"	¾"	¾"	¾"	¾"

① With Stack Damper

In the interest of product improvement, we reserve the right to make changes without notice.

Figure 5.9 Brochure showing specifications for a line of gas-fired boilers (Smith Cast Iron Boilers, GB100 series technical brochure; reprinted with permission)

(b) The US Department of Energy (**DOE**) heating capacity rating was developed by the US Federal Government. This rating tests and rates boilers with capacities up to 300 MBH.

(c) The net **I=B=R rating system** is the most conservative rating system. It is based on a 1.15 allowance on the DOE rating system. This rating is used for larger capacity boilers and water heaters. It may, however, be used for all boilers regardless of thermal size. As a design engineer, it is recommended to base designs and contractor specifications on this rating system, unless advised otherwise by a senior professional engineer. Further, this rating should be used when the boiler and piping are outside of the space that will be heated.

(ii) **AFUE** is the **annual fuel utilization efficiency** of the boiler. The AFUE accounts for the combustion efficiency of the burner fuel and includes other losses such as heat losses to the unheated boiler room, start-up/cool-down losses, vent losses in the chimney, heat exchanger performance losses, standing pilot losses, etc. Typically, AFUE values for most conventional oil or gas-fired boilers are 78–85%.

(iii) The physical boiler sizes and the sizes of the gas connections are typically presented in the specifications chart.

5.3.5 Burner Fuels

Several types of fuels can be used to heat the water in boilers and water heaters. Some types of fuels are as follows:

Electricity
Natural gas
Propane (LP)
Oil
Wood

Combustion of these fuels occurs in the burner chamber to provide heating for the heat exchanger component of the system. The typical **energy content** or **heating value** of the respective fuels is usually available in tabulated form as shown in Table 5.2. The heating value of a fuel is the amount of energy released when the fuel is completely burned in oxygen (combustion) to form carbon dioxide and water.

The heating values of fuels for use in boilers, water heaters, and **furnaces** are based on the **higher heating values** (HHV) of the fuels. HHV is the amount of

Table 5.2 Approximate heating value of fuels

Fuel Type	Heat Content
Natural gas (methane)	1030 Btu/ft^3
Propane	2500 Btu/ft^3
#2 Oil	140000 Btu/gal
#4 Oil	145000 Btu/gal
#6 Oil	153000 Btu/gal
Electricity	3413 Btu/h per kW
Anthracite (hard) coal	14000 Btu/lb
Bituminous (soft) coal	12000 Btu/lb
Hardwoods	24000000 Btu/cord
Softwoods	15000000 Btu/cord

Source: Bobenhausen [3].

energy released after complete combustion of fuel and complete condensation of water. **Lower heating value** (LHV) is the amount of energy released after complete combustion of the fuel, but with water as a vapor.

5.4 Design of Hydronic Heating Systems c/w Baseboards or Finned-Tube Heaters

5.4.1 Zoning and Types of Systems

Space heating in a residential or commercial building may be accomplished with the use of piping system arrangements, which include a water boiler or water heater, circulators, valves, expansion tanks, and baseboard heaters. These types of **hydronic heating systems** may include multiple-pipe circuits to provide heating to different zones of the space or building. A **zone** is an area of the space or building that is controlled by a single thermostat.

5.4.2 One-Pipe Series Loop System

Figure 5.10 shows a schematic diagram of a one-pipe series loop system. As the figure shows, this is a closed-loop system with a single pipeline routed from and to the boiler. The line also includes several heaters.

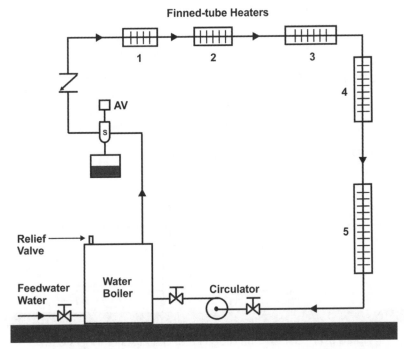

Figure 5.10 Schematic diagram of a one-pipe series loop system

This type of system works best in small buildings or residences where the amount of heat loss from the space will be small. It is important to note that the water cools as it flows through each successive heater in the system. This is because the leaving water temperature from heater 1 is the entering water temperature to heater 2 and this continues consecutively to the other heaters. Therefore, due to progressive heat loss in the system, each sequential heater size **must** be increased to provide sufficient surface to facilitate the requisite heating capacity as the water cools towards the end of the loop.

The one-pipe series loop system may be improved by modifying the arrangement of the piping system. Following is a discussion of some of those modifications.

Split Series Loop

This type of design arrangement is most useful for applications with larger design heat loss. The main hot water supply branch from the boiler or water heater is split into multiple supply loops. Figure 5.11 shows a schematic of this type of system. Note that the heaters (heat exchangers) must be sized larger towards the end of each loop.

One-Pipe "Monoflow" Series Loop with Diversion Fittings

In a one-pipe "monoflow" series loop system, the water flow from the main supply line is diverted to the individual heaters through **diversion fittings**. This helps to eliminate the need to increase the lengths of the heaters towards the end of the loop by ensuring that hotter water enters the downstream heaters. Figure 5.12 shows a schematic of this type of system.

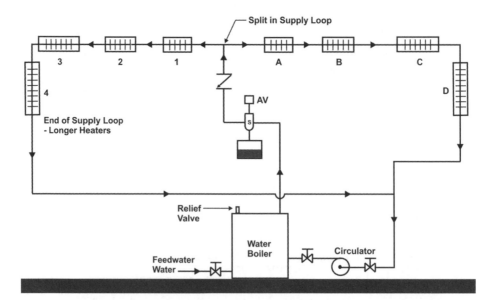

Figure 5.11   Schematic diagram of a split series loop system

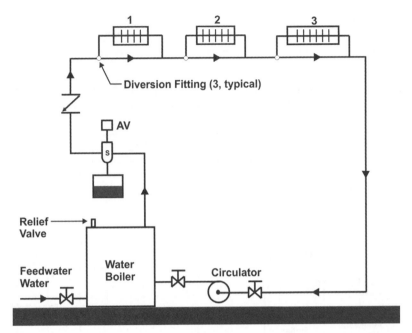

Figure 5.12   Schematic of a one-pipe "monoflow" series loop system

Multiple one-pipe series circuits can be used to provide heat to different zones of a large space or building. To improve the performance of the system, the supply line for each circuit is manifold to the main hot water supply line. In addition, diversion fittings may be used with each heater. Figure 5.13 shows a schematic of a large multizone heating system of one-pipe series loops.

5.4.3 Two-Pipe Systems

Two-pipe systems were developed to eliminate the need for larger downstream heaters due to water temperature drops. In these types of systems, each heater possesses its own return line that is connected to a main return line. There are two main types of two-pipe systems.

Two-Pipe Direct Return System

Figure 5.14 shows a schematic of this type of system. In this system, return water from each heater is returned directly to a main return line that is piped back to the boiler.

Use of this system minimizes the amount of piping required. However, for the downstream heaters, pipe pressure losses will be greater. The fluid may not flow through the heaters, and the heaters will be **starved** of heat. This is an example of an **unbalanced flow system**.

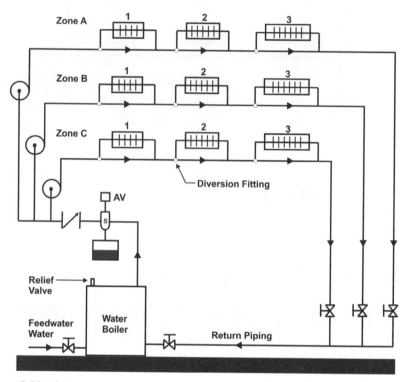

Figure 5.13   Schematic diagram of a multizone system of one-pipe series loops

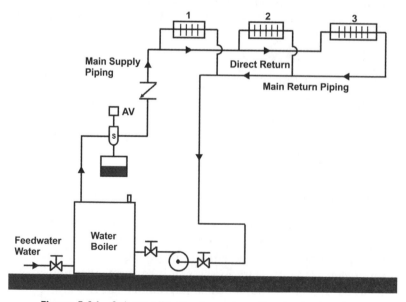

Figure 5.14   Schematic of a two-pipe direct return system

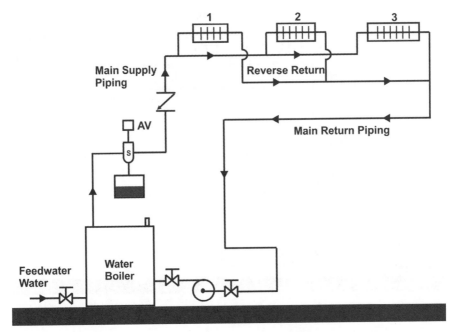

Figure 5.15 Schematic of a two-pipe reverse return system

Two-Pipe Reverse Return System

Figure 5.15 shows a schematic of this type of two-pipe system. In this system, the heater return lines are connected downstream, resulting in one larger pipe connected to the main return line. For this system, more piping and larger circulators are required. As a result of the longer return piping lines for each heater, this type of system is easier to balance since the amount of pressure drop in each heater line are equivalent.

In some cases, unbalanced flow and starvation of the heaters is a serious concern. **Orifice plates** or **balancing valves** may be used to improve flow balance in the system. Consider the unbalanced flow through a long two-pipe direct return system shown in Fig. 5.16. On this long run of piping system, the resistance (ft water column (wc)) through the heaters vary, increasing progressively downstream. That is, the resistance in heater P is six times larger than that in heater X, which is upstream. This is a very unbalanced system, and heater P may experience heat starvation. A solution to this problem could be the installation of orifice plates or balancing valves. Consider the same system, shown in Fig. 5.17, with orifice plates at each heater. Each orifice plate carries a resistance of 3 ft wc. Though the installation of the orifice plates have increased the system resistance (and the circulator size), the resistance in heater P is now only two times larger than that in heater X. This is a more balanced system, and starvation of heater P is more likely to be avoided.

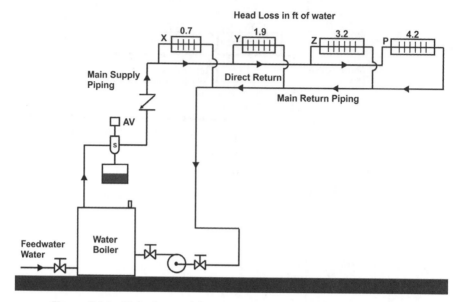

Figure 5.16   Unbalanced flow in a two-pipe direct return system

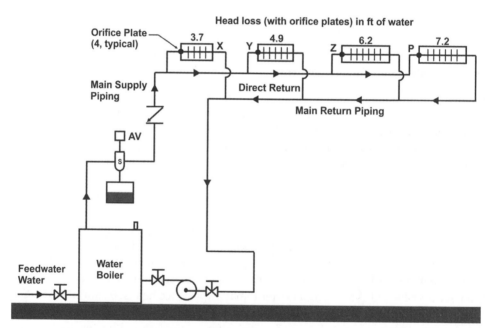

Figure 5.17   Improved balance in a two-pipe direct return system

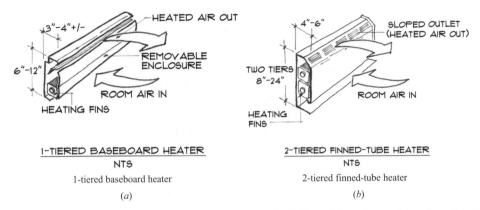

Figure 5.18 Diagrams of baseboard heaters. (a) 1-tiered baseboard heater; (b) 2-tiered finned-tube heater

5.4.4 Baseboard and Finned-Tube Heaters

Baseboard and finned-tube heaters are finned-tube heat exchangers used to heat spaces through buoyancy-driven free convection mechanisms. These heaters typically consist of a copper water pipe with aluminum fins attached to increase the heat transfer surface area. Most units will have 25–60 fins per foot of piping. The depth of the heaters will range from 3 to 6 in. and the height may range from 6 to 24 in. Baseboard heaters are small in height, ranging from 6 to 12 in. Finned-tube heaters are larger than baseboard heaters, with heights of 8–24 in. Baseboard heaters are typically **1-tiered**, with one finned-tube pipe. Finned-tube heaters are typically at least **2-tiered**, with two rows of finned-tube pipes. Figure 5.18 shows two types of heaters, baseboard and finned-tube.

Manufacturer's catalogs **must** be consulted to select the heaters that will be installed by the mechanical contractor. Tables 5.3 and 5.4 provide ratings and specifications for

Table 5.3   Baseboard heater rated outputs at 1 gpm water flow rate

Average Water Temperature (°F)	Rated Output (Btu/h) per foot of pipe
220	840
215	810
210	770
200	710
190	640
180	580
170	510
160	450

Source: Slant/ Fin Corporation.

finned-tube heaters. Table 5.3 shows heat outputs per foot of $^3/_4$ in. copper pipe as a function of water temperature for baseboard heaters. Table 5.4 is extended to provide more detail for a wider variety of pipe sizes for copper and schedule 40 steel pipes. Details on the fins and enclosure heights are also provided. Though the data presented in Table 5.4 applies to finned-tube heaters, the data presented for 1-tiered heaters could be used to approximate the heat output for a 1-tiered baseboard heater.

The ratings provided in Table 5.4 are based on steam or hot water at 215°F, a water velocity of 3 fps, and free convection to air at 65°F. For other temperatures

Table 5.4   "Front outlet" finned-tube heater ratings for Trane heaters

Tube		Fins			Actual Enclosure Height (in.)	Rating (Btu/h/ft)
Size (in.)	Material[a]	H×W[b]	Per foot	Tiers		
$^3/_4$	C	$3^1/_4$	35	1	10	1080
$^3/_4$	C	$3^1/_4$	35	1	14	1140
$^3/_4$	C	$3^1/_4$	35	1	18	1170
$^3/_4$	C	$3^1/_4$	50	1	10	1210
$^3/_4$	C	$3^1/_4$	50	1	14	1350
$^3/_4$	C	$3^1/_4$	50	1	18	1450
$^3/_4$	C	$3^1/_4$	58	1	10	1280
$^3/_4$	C	$3^1/_4$	58	1	14	1460
$^3/_4$	C	$3^1/_4$	58	1	18	1560
1	C	$3^1/_4$	35	1	10	1090
1	C	$3^1/_4$	35	1	14	1150
1	C	$3^1/_4$	35	1	18	1190
1	C	$3^1/_4$	50	1	10	1200
1	C	$3^1/_4$	50	1	14	1350
1	C	$2^1/_4$	50	1	18	1460
1	C	$3^1/_4$	58	1	10	1260
1	C	$3^1/_4$	58	1	14	1450
1	C	$3^1/_4$	58	1	18	1600
$1^1/_4$	C	$3^1/_4$	35	1	10	1100
$1^1/_4$	C	$3^1/_4$	50	1	10	1200
$1^1/_4$	C	$3^1/_4$	58	1	10	1250
$1^1/_4$	S	$2^1/_2 \times 5^1/_8$	40	1	16	1300
$1^1/_4$	S	$2^1/_2 \times 5^1/_8$	40	2	16	1980
$1^1/_4$	S	$2^1/_2 \times 5^1/_{16}$	52	1	16	1510
$1^3/_4$	S	$2^1/_2 \times 5^3/_{16}$	52	2	16	2060

Source: The Hydronics Institute, *I=B=R Ratings for Boilers, Baseboard Radiation, and Finned Tube (Commercial) Radiation, 1993.*
[a]C, copper; S, schedule 40 steel.
[b]Unless otherwise indicated, fin dimensions are $3^1/_4$ in.×$3^1/_4$ in.

Table 5.5 Flow rate correction factors for water
velocities less than 3 fps

Water Velocity (fps)	Correction Factor (C_F)
3.0	1.000
2.5	0.992
2.0	0.984
1.5	0.973
1.0	0.957
0.5	0.931
0.25	0.905

Source: The Hydronics Institute, *I=B=R Ratings for Boilers, Baseboard Radiation, and Finned Tube (Commercial) Radiation, 1993.*

and/or velocities, correction factors are used to determine the actual heat output of the heaters. The actual heat output is

$$\dot{Q}_{actual} = \dot{Q}_{rating} \times C_T \times C_F,\tag{5.4}$$

where

$\dot{Q}_{actual}$ = actual heat transferred from the heater at water temperatures other than 215°F and velocities other than 3 fps;
$\dot{Q}_{rating}$ = heat output rating obtained from Table 5.4;
C_T = temperature correction factor. This is found from tables or charts;
C_F = velocity correction factor. This is found from tables or charts.

Tables 5.5 and 5.6 give C_F and C_T values, respectively. For the C_F values presented in Table 5.5, if the flow velocities are larger than 3 fps, $C_F = 1.00$.

Practical Note 5.5   Temperature Data for Sizing Finned-Tube Heaters

Some manufacturers may provide data on the inlet temperatures to the heaters that will give a specific heat output. As the inlet temperatures decrease, their data will show a decrease in heat output. In addition, the temperature change (ΔT) may be provided to determine the outlet temperature.

The length of the heater is found from

$$L_{heater} = \frac{\dot{Q}_{design}}{\dot{Q}_{actual}} = \frac{\dot{Q}_{design}}{\dot{Q}_{rating} \times C_T \times C_F},\tag{5.5}$$

where $\dot{Q}_{design}$ is the total amount of heat that needs to be supplied by the heater.

Table 5.6   Temperature correction factors for hot water ratings

Average Heater Temperature (°F)	Correction Factor (C_T)
100	0.15
120	0.26
150	0.45
170	0.61
175	0.65
180	0.69
185	0.73
190	0.78
195	0.82
200	0.86
205	0.91
210	0.95
215	1.00
220	1.05
230	1.14
240	1.25

Source: The Hydronics Institute, *I=B=R Ratings for Boilers, Baseboard Radiation, and Finned Tube (Commercial) Radiation, 1993.*

5.5 Design Considerations for Hot Water Heating Systems

The following are several points that should be borne in mind during the design of hot water heating systems:

   (i) Study the architectural drawings or other relevant information for the space or process that requires hot water. For architectural drawings, these should be furnished by the architect.
  (ii) Obtain the heat requirement (**heating loads**) for each space for building heating applications. A Heating, Ventilating, and Air-conditioning (HVAC) engineer should provide the heating loads for each space. For processes, determine the heat requirement by using appropriate formula.
 (iii) Size the boiler or hot water heater (heat capacity and tank volume or volume flow rate). In some jurisdictions (e.g., California and Vermont), the use of a single boiler or hot water heater for domestic water and space heating is strictly prohibited. In that case, a **dual use hot water boiler** (combi system) may be considered. In that case, the boiler/water heater must have two distinct sections for heating water, one for domestic water use and one for space heating.
  (iv) The International Plumbing Code (Section 501.2) requires the maximum hot water temperature for domestic use (drinking, showers) to be 140°F. Authorities having juridiction may specify minimum hot water temperatures for domestic

use (e.g., The Alberta Building Code (Division B, Section 7.2.6) requires that hot water for domestic use (drinking, showers) be at least 113°F to avoid the growth of microorganisms [4]). Industry standard recommends that hot water for space heating be about 180°F for houses and about 200–220°F for larger building applications. Lower values can be used where energy conservation is an issue or when a forced air hydronic system is employed.

(v) Size and select the heaters for space heating, if needed.
(vi) Layout and size the piping system. Multiple piping arrangements are available.
(vii) Size the circulator (pump). Some boilers or water heaters may be available as packaged units, complete with a circulator.
(viii) Where applicable, show drawings and prepare specifications for selected equipment.

Example 5.1   Sizing a Boiler and Heaters for a Hostel

A design engineer has calculated the heat losses from several zoned spaces in a hostel located in Edmonton, Alberta, Canada. The zones include three rooms and a large corridor that connects the three rooms. To calculate the heat losses from the zones, the engineer considered the required indoor temperature, the average outdoor temperature, the resistance of the wall material to heat transfer, the number of people and heat-generating equipment in the spaces, among other parameters.

Analysis produced the following zone heat losses:

Zone	Heat Losses (Btu/h)	Zone Dimensions
Corridor and recreational area	18125	25 ft×30 ft
Room 131	11565	15 ft×18 ft
Room 132	8700	12 ft×12 ft
Room 133	7500	10 ft×12 ft

Size and select the boiler and specify the total length of finned-tube heaters required for each room. A dedicated hot water boiler should be available.

Possible Solution

Definition

Size a boiler and heaters to provide heat and hot water to three rooms and a corridor in a hostel.

Preliminary Specifications and Constraints

(i) The area of interest has three rooms and a corridor in a hostel.
(ii) The dimensions of the zones are specified.

(iii) The heat losses from the spaces were specified by the design engineer.
(iv) Limited space is available for mechanical equipment and units.

Detailed Design

Objective

Size a boiler to provide hot water and space heating to a three-room/corridor space in a hostel. Specify the boiler and the lengths of the heaters for space heating.

Data Given or Known

(i) Three rooms in a hostel
(ii) One large corridor
(iii) Dimensions are given in the following table:

Zone	Zone Dimensions
Corridor	25 ft×30 ft
Room 131	15 ft×18 ft
Room 132	12 ft×12 ft
Room 133	10 ft×12 ft

(iv) Heat losses from the spaces are provided in the following table:

Zone	Heat Losses (Btu/h)
Corridor	18125
Room 131	11565
Room 132	8700
Room 133	7500

Assumptions/Limitations/Constraints

(i) Assume that the spaces will be heavily occupied. Therefore, the boiler OSF is 1.10.
(ii) The average outlet temperature of water from the boiler will be 200°F. Industry standard recommends that hot water for space heating be about 180°F for houses and about 200–220°F for larger building applications. The assumed outlet temperature is slightly higher than that recommended for houses or hostels.
(iii) The average inlet temperature of the water into the boiler is 40°F. For Edmonton, inlet water temperature is about 40–48°F.
(iv) The boiler efficiency will be at least 80%. Selection of a manufacturer's unit will give a precise value.
(v) The heater tube size will be $3/4$ in. The tubes will be Type L copper to promote heat transfer.

(vi) The flow velocity in the tube will not exceed 6 fps to prevent erosion of the copper tubes. Assume a water velocity of 1.5 fps. This low velocity should increase heat transfer to the space, while keeping the circulator size small.

(vii) All properties of the water used for sizing the boiler will be taken at an average temperature.

(viii) The pipelines that connect the heaters in each room will be properly insulated by the mechanical contractor. The design engineer should provide details in the Specifications section of the contract documents.

Sketch

No sketch is needed.

Analysis

Heater Sizing

Finned-tube heaters will be used in this application. The architect has specified that space is limited for mechanical equipment and units. Therefore, heaters with the smallest enclosure heights will be chosen. Therefore, choose the Trane 1-tier finned-tube heaters with 58 fins per foot of tube, an enclosure height of 10 in., and a rating of 1280 Btu/(h ft). Tube diameter is $3/4$ in., and the material is copper.

For this application, a simple one-pipe series loop system will be chosen. A two-pipe direct return system could have been chosen.

The heater length is given by

$$L_{\text{heater}} = \frac{\dot{Q}_{\text{design}}}{\dot{Q}_{\text{rating}} \times C_T \times C_F}.$$

Each space will need to be considered individually when sizing the heaters because there will be a temperature drop in the water as it proceeds from one heater to the next in the series loop system. This decrease in inlet temperatures to the heaters will result in a change in the temperature correction factors (C_T). Since the corridor is the space with the largest heat load requirement, the heaters in this space will be the first to receive hot water from the boiler.

For the corridor:

For a hot water temperature of 200°F, $C_T = 0.86$. For tube velocities of 1.5 fps, $C_F = 0.973$. Therefore,

$$L_{\text{corridor}} = \frac{18125 \text{ Btu/h}}{(1280 \text{ Btu/(h ft)}) (0.86) (0.973)} = 16.9 \text{ ft} \approx 17 \text{ ft}.$$

Since the corridor is large, install **three heaters each with 6-ft length** around the perimeter of the corridor.

The exit temperature of the water from the heater is found from

$$\dot{Q} = \rho_{\text{water}} \dot{V} c_{\text{p,water}} (T_{\text{in}} - T_{\text{out}}) = \rho_{\text{water}} \frac{\pi D_i^2 V_{\text{water}}}{4} c_{\text{p,water}} (T_{\text{in}} - T_{\text{out}}).$$

Hence, assuming that the water properties are at 200°F,

$$T_{out,corridor} = T_{in,corridor} - \frac{4\dot{Q}}{\pi D_i^2 V_{water} \rho_{water} c_{p,water}},$$

$$T_{out,corridor} = 200°F - \frac{4\,(18125\ \text{Btu/h})}{\pi\,(0.785\ \text{in.})^2\,(1.5\ \text{ft/s})\left(60.12\ \text{lb/ft}^3\right)(1.005\ \text{Btu/(lb\,°F)})}$$

$$\times \left(\frac{12\ \text{in.}}{1\ \text{ft}}\right)^2 \times \frac{1\ \text{h}}{3600\ \text{s}} = 184°F.$$

Note that the inner diameter of a $^3/_4$ in. nominal Type L copper pipe is 0.785 in.

For Room 131
For a hot water temperature of 184°F, $C_T = 0.73$. For tube velocities of 1.5 fps, $C_F = 0.973$.
Therefore,

$$L_{131} = \frac{11565\ \text{Btu/h}}{(1280\ \text{Btu/(h ft)})\,(0.73)\,(0.973)} = 12.7\ \text{ft} \approx 13\ \text{ft}.$$

Install **two heaters each with 7-ft length** around the perimeter of the room.
Hence, assuming that the water properties are at 184°F,

$$T_{out,131} = T_{in,131} - \frac{4\dot{Q}}{\pi D_i^2 V_{water} \rho_{water} c_{p,water}},$$

$$T_{out,131} = 184°F - \frac{4\,(11565\ \text{Btu/h})}{\pi\,(0.785\ \text{in.})^2\,(1.5\ \text{ft/s})\left(60.4\ \text{lb/ft}^3\right)(1.003\ \text{Btu/(lb\,°F)})}$$

$$\times \left(\frac{12\ \text{in.}}{1\ \text{ft}}\right)^2 \times \frac{1\ \text{h}}{3600\ \text{s}} = 174°F.$$

For Room 132
For a hot water temperature of 174°F, $C_T = 0.65$. For tube velocities of 1.5 fps, $C_F = 0.973$.
Therefore,

$$L_{131} = \frac{8700\ \text{Btu/h}}{(1280\ \text{Btu/(h ft)})\,(0.65)\,(0.973)} = 10.8\ \text{ft} \approx 11\ \text{ft}.$$

Install **three heaters each with 4-ft length** around the perimeter of the room.
Hence, assuming that the water properties are at 174°F,

$$T_{out,132} = T_{in,132} - \frac{4\dot{Q}}{\pi D_i^2 V_{water} \rho_{water} c_{p,water}},$$

$$T_{out,132} = 174°F - \frac{4\,(8700\ \text{Btu/h})}{\pi\,(0.785\ \text{in.})^2\,(1.5\ \text{ft/s})\left(60.6\ \text{lb/ft}^3\right)(1.001\ \text{Btu/(lb\,°F)})}$$

$$\times \left(\frac{12\ \text{in.}}{1\ \text{ft}}\right)^2 \times \frac{1\ \text{h}}{3600\ \text{s}} = 166°F.$$

For Room 133

For a hot water temperature of 166°F, $C_T = 0.61$. For tube velocities of 1.5 fps, $C_F = 0.973$. Therefore,

$$L_{131} = \frac{7500 \text{ Btu/h}}{(1280 \text{ Btu/(h ft))} (0.61) (0.973)} = 9.9 \text{ ft} \approx 10 \text{ ft.}$$

Install **two heaters each with 5-ft length** around the perimeter of the room. Hence, assuming that the water properties are at 166°F,

$$T_{out,133} = T_{in,133} - \frac{4\dot{Q}}{\pi D_i^2 V_{water} \rho_{water} C_{p,water}},$$

$$T_{out,133} = 166°F - \frac{4(7500 \text{ Btu/h})}{\pi (0.785 \text{ in.})^2 (1.5 \text{ ft/s}) \left(60.8 \text{ lb/ft}^3\right)(1.001 \text{ Btu/(lb°F))}}$$

$$\times \left(\frac{12 \text{ in.}}{1 \text{ ft}}\right)^2 \times \frac{1 \text{ h}}{3600 \text{ s}} = 159°F.$$

The water will return to the boiler at a temperature of approximately 159°F.

Boiler Sizing

In this application, the boiler or water heater is not required to provide domestic hot water, only hot water for space heating. A Rinnai condensing boiler can be considered for selection. The thermal size of the boiler is given by

$$\dot{Q}_{boiler} = \dot{Q}_{space} \times OSF.$$

The total space heat load is

$$\dot{Q}_{space} = (18125 + 11565 + 8700 + 7500) \text{ Btu/h} = 45890 \text{ Btu/h.}$$

Therefore,

$$\dot{Q}_{boiler} = 45890 \text{ Btu/h} \times 1.10 = 50479 \text{ Btu/h,}$$
$$\dot{Q}_{boiler} = 50479 \text{ Btu/h} = 51 \text{ MBH} = 1.5 \text{ hp (boiler).}$$

The fuel will be natural gas. The maximum total fuel flow rate is

$$\dot{V}_{fuel} = \frac{\dot{Q}}{HV \times \eta} = \frac{50479 \text{ Btu/h}}{\left(1030 \text{ Btu/ft}^3\right)(0.965)} \times \frac{1 \text{ h}}{60 \text{ min}} = 0.85 \text{ ft}^3/\text{min} = 0.85 \text{ cfm.}$$

A Rinnai condensing wall-mounted gas boiler model Q75C will be specified for this application. The AFUE is specified as 96.5%. Below is an excerpt from the technical catalog for this boiler. This information will be used to prepare the boiler schedule for use by the mechanical contractor.

Rinnai.

E75C / E110 Condensing Boiler

Standard Features

- Onboard outdoor reset control system with sensor standard
- Prebuilt plumbing kit with insulated Low Loss Header
- Priority DHW standard
- Built in DHW plate heat exchanger
- Domestic hot water plate warming
- Modulating Ceramic premix burner
- 5:1 turn down ratio
- Spark ignition
- Exceeds SCAQMD 1146.2 Low NO_x requirements
- Stainless Steel Water tube condensing heat exchanger
- Direct Vent sealed combustion
- Concentric and twin pipe venting adapters included
- Approved for room and closet installations

Optional Accessories

- RS100 Single zone controller
- Flue Gas temperature sensor
- Room Air Filter
- NG to LP conversion Kit
- LP to NG Conversion Kit

SPECIFICATIONS

General Specifications

	Units	Boiler Model E75C	Boiler Model E110C
Water content	gal	0.9	1.3
Max. supply boiler temperature	°F	176	176
Max operating pressure	psi	45	45
Relief valve rating	MBH	375	375
Relief valve pressure rating	psi	30	30
Dry weight	lbs	91	101
Min inlet gas pressure NG	"W.C.	4.0"	4.0"
Max inlet gas pressure NG	"W.C.	10.5"	10.5"
Min inlet gas pressure LP	"W.C.	8.0"	8.0"
Max inlet gas pressure LP	"W.C.	14.0"	14.0"
Max equivalent exhaust vent length	ft	100	100
Max equivalent combustion air vent length	ft	100	100
Approved venting materials		Polypropylene, Stainless Steel, PVC, CPVC	

Performance Specifications

	Units	Boiler Model E75C	Boiler Model E110C
Fuel Type		NG/LP	NG/LP
Input	MBH	75	110
Heating Capacity	MBH	69	101
AFUE (I=B=R)	%	96.5	96.1
Part Load Efficiency (EN677)	%	98.8	99.0
DHW flow rate 75°F ΔT	gpm	2.1	3.2

Electrical Specifications

	Units	Boiler Model E75C	Boiler Model E110C
Electrical Voltage Mains	V/Hz	120/60	120/60
Electrical Voltage Controls	V	24	24
Power Consumption Max Load	W	145	145
Power Consumption Stand by Load	W	14	14
Recommended Circuit breaker rating	A	15	15

Rinnai.   E75C / E110C

DIMENSIONS

		E75C	E110C
		inch / mm	
A	Height	25.6 / 650	
B	Height with expansion tank	NA	34.3 / 870
C	Width	19.7 / 500	
D	Depth	15.6 / 395	
E	Left side / vent	13.2 / 335	
F	Center to center / vent and air supply	4.7 / 120	
G	Back / vent	10.6 / 270	
H	Left side / gas pipe	9.8 / 250	
J	Left side / supply pipe	5.9 / 150	
K	Left side / return pipe	13.8 / 350	
L	Left side / condensate pipe	15.9 / 405	
N	Left side / cold water pipe	11.2 / 285	
P	Left side / hot water pipe	8.5 / 215	
Q	Pipe length of g	0.7 / 19	8.5 / 215
R	Pipe length of c	1.6 / 40	
S	Pipe length of f, k, r, and w	2 / 50	6.3 / 160
T	Back / center of pipe c, k, and w	1 / 26	
U	Back / Center of pipe f, g, and r	2 / 50	
V	Pipe length vent co-axial / Pipe length vent parallel	3.7 / 95 / 7 / 177	

Clearances

	Minimum required clearance to combustibles and non-combustibles	Recommended service clearances
	inch / mm	inch / mm
Top	2 / 50	10 / 250
Back	0	0
Front	6 / 150	24 / 600
Left side	2 / 50	2 / 50
Right side	2 / 50	2 / 50
Floor / Ground	12 / 300	30 / 762
Vent	0	0

Rinnai Corporation • 103 International Drive • Peachtree City, GA 30269 • Toll-Free: 1-800-621-9419 • Fax: 678-364-8643 • www.rinnai.us

Source: Rinnai America Corp. (reprinted with permission)

Drawings

Not needed.

Conclusions

The boiler has been sized. The lengths of the heaters have also been specified. The lengths of the heaters are smaller than the total length of the walls.

While a condensing boiler was chosen, a conventional boiler could have been selected instead. The designer will need to make a final selection based on the preferences of the client, need for routine equipment maintenance, and costs (fixed and operating).

The water is returned to the boiler at 159°F. Industry standard recommends that hot water for space heating be about 180°F for houses. For some of the heaters, the return temperature is lower than the minimum temperature required. Therefore, the heaters may need to be larger to provide the required amount of heat.

It should be noted that while the AFUE is specified as 96.5%, this efficiency will decrease when higher temperature water returns to the condensing boiler. Larger ΔT between the boiler supply water temperature and return water temperature will improve condensing boiler efficiency. It is well-known that in order to extract more heat from the flue gases in the boiler heat exchanger and increase the boiler efficiency, the water return temperature should not exceed 113–122°F [1]. To that end, a return water temperature of 110°F could be chosen. In that case, the water temperature at the outlet of the finned-tube heater that serves Room 133 will be 110°F. Therefore, the water that exits the boiler will have a temperature of approximately 152°F. For energy efficient buildings, the heat loads are usually smaller. Therefore, it may not be necessary that water for space heating be 180°F or higher.

With smaller loads, the flow rate of the fuel supplied to the boiler can be reduced after initial start of the system. Since the water will return to the boiler at 159°F, a lower amount of energy will be required to heat it to 200°F. Further, given that the heating capacity of the Rinnai model Q75C (69000 Btu/h with natural gas) condensing boiler is much higher than the energy requirement for this application (50479 Btu/h), the fuel flow rate can be modulated to meet the lower requirements. In less sosphisticated systems, a control system may be installed to modulate the fuel flow; but since the Q75C is a modulating boiler, such adjustment will be automatic. Another option could be to purchase more than one smaller boilers and schedule their operation in order to meet the heating requirements of the client.

It should be noted that the boiler will be slightly oversized in this application. The OSF used was 1.10 (equivalent to 91% efficiency). However, the AFUE of the selected boiler is about 96%.

Orifice plates or balancing valves may be required to balance the losses through this system. This will ensure that the heaters in Room 133 will not be "starved" of hot water due to large pressure drops.

A two-pipe direct return system could have been specified for the finned-tube heaters instead of the one-pipe series loop system that was selected. In this case, there would be no need to verify the inlet temperatures to the heaters, assuming no heat loss from the pipes outside the finned-tube heater units.

Data Sheets and Equipment Schedules

Below are a data sheet and a schedule of the major equipment.

1-Tiered Finned-Tube Heater Data

Zone	Tube Size (in.)	Tube Material	Fins H×W (in.×in.)	Fins Per Foot	Height (in.)	Total Length (ft)
Corridor	$^3/_4$	copper	$3^1/_4 \times 3^1/_4$	58	10	17
Room 131	$^3/_4$	copper	$3^1/_4 \times 3^1/_4$	58	10	13
Room 132	$^3/_4$	copper	$3^1/_4 \times 3^1/_4$	58	10	11
Room 133	$^3/_4$	copper	$3^1/_4 \times 3^1/_4$	58	10	10

Boiler Schedule

Tag	Manufacturer and Model Number	Boiler Thermal Size	Boiler Type	Fuel Flow Rate (cfm)	Fuel Type	Fuel HHV (Btu/ft³)	Dimensions Width	Dimensions Height	Dimensions Depth
B-1	Rinnai model Q75C, or equal	69000 Btu/h (capacity)	Condensing; wall-mounted	0.85	Natural gas	1030	19.7″	25.6″	15.6″

Example 5.2   Design of a Water Heating System for a Chemical Plant

A chemical engineer would like to use warm, liquid ethyl alcohol (CH_3CH_2OH) to convert 2-bromo-3-methylbutane ($CH_3CHBrCH(CH_3)_2$) to 2-ethoxy-2-methylbutane ($CH_3CH_2C(CH_3)_2OCH_2CH_3$). For this process, the concentration of the ethyl alcohol should be 40% vol. and the temperature should be 100°F. Pure ethyl alcohol (at 68°F) from a storage tank will be mixed in mixing tank 1 (see sketch) with water from a hot water heater or boiler. The warm, dilute ethyl alcohol will be mixed with 2-bromo-3-methylbutane in mixing tank 2 (see sketch) to produce 2-ethoxy-2-methylbutane. A requirement of the chemical process sets the ethyl alcohol storage tank at 4 ft or less from the mixing tanks. An unlimited, constant supply of pure ethyl alcohol is available for the storage tank. The following sketch provides additional information on the existing system and the chemical process. Not all accessories are shown.

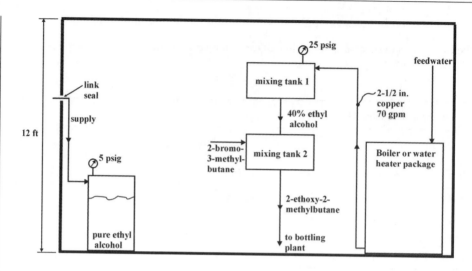

The pressure in mixing tank 1 will be fixed to 25 psig. Pure liquid ethyl alcohol can be volatile. So, for safety reasons, the pressure in the pure ethyl alcohol storage tank will be 5 psig to ensure limited vaporization of the liquid. Design the piping system between the ethyl alcohol storage tank and mixing tank 1 and specify all appropriate equipment to meet the requirements of the chemical engineer.

Possible Solution

Definition

Design a system for delivery of warm ethyl alcohol to mixing tank 1 from the storage tank.

Preliminary Specifications and Constraints

  (i) The working fluid is pure ethyl alcohol at 68°F.
 (ii) The final concentration of the ethyl alcohol in mixing tank 1 should be 40% vol. and the temperature should be 100°F.
(iii) Only hot water is available to dilute the pure ethyl alcohol.
 (iv) The height of the ethyl alcohol tank will be constrained to 12 ft or less.
  (v) The pressure in mixing tank 1 is fixed at 25 psig; the pressure in the ethyl alcohol storage tank will be at least 5 psig for safety reasons.
 (vi) The ethyl alcohol storage tank cannot be more than 4 ft from the mixing tanks.

Detailed Design

Objective

To size a pipe and to size and select an appropriate pump for delivery of pure ethyl alcohol to a mixing tank in a chemical process plant. The minimum thermal size of the hot water heater or boiler will be specified. An appropriate packaged boiler unit will be selected.

Data Given or Known

(i) Chemicals in the process are known. Of interest is the pure ethyl alcohol.

(ii) The pressures in the ethyl alcohol storage tank and mixing tank 1 are 5 psig and 25 psig, respectively.

(iii) The ethyl alcohol must be mixed with hot water to a final concentration of 40% vol. and 100°F.

(iv) Pure ethyl alcohol in the storage tank is available at 68°F.

(v) A hot water heater or boiler will be available to provide 70 gpm of hot water.

(vi) The hot water pipe size is $2\frac{1}{2}$ in. and the pipe material is copper.

(vii) Pure liquid ethyl alcohol can be volatile.

(viii) The ethyl alcohol storage tank cannot be more than 4 ft from the mixing tanks.

Assumptions/Limitations/Constraints

(i) This system uses volatile chemicals. Assume that this system is located in a chemical plant and noise from pipe flow may not be an issue.

(ii) For transportation of ethyl alcohol between the storage tank and mixing tank 1, assume that the piping material is **PEX (cross-linked polyethylene)**. This should ensure that limited reactions occur between the piping material and the working ethyl alcohol fluid.

(iii) Assume that the sizes of the PEX piping is equivalent to that of Schedule 40 PVC plastic piping.

(iv) For the purpose of sizing the PEX piping, the friction loss chart for Schedule 40 steel will be used. If available, the chart for plastic piping can be used.

(v) Let the flow velocity be about 6 fps or less. Without information on the erosion limit of PEX, a velocity lower than the erosion limit of soft metals such as aluminum is chosen. This is a conservative choice to avoid system failure.

(vi) Limit pipe frictional losses to 7 ft of water per 100 ft of pipe. This is more than the standard of 3 ft of water per 100 ft of pipe. Larger head losses are acceptable since the system will be located in a plant and the length of pipe will be short (no more than 12 ft). The viscosity of ethyl alcohol is larger than that of water, but the two viscosities are on the same order of magnitude for a given temperature.

(vii) All fittings will be screwed.

(viii) Assume that the ethyl alcohol storage tank height is 4 ft. This will ensure proper maintenance by a 5 ft 6 in. person. In addition, this height is well below the maximum height of the building of 12 ft.

(ix) Since the supply flow rate of the pure ethyl alcohol to the storage tank is unknown, assume that the storage tank is always nearly full (a constant, unlimited supply is available). To prevent spills, assume that the tank is always full to a height of 3 ft. A high-level limit switch could be installed to control the fill height.

(x) To permit easy access, assume that the discharge pipe connection to the storage tank is 6 in. from the finished floor.

(xi) The pump will be an in-line pump mounted on brackets and mounted to the back wall.

(xii) Ball valves will be used to isolate the pump to facilitate maintenance.

(xiii) The pipe connection to mixing tank 1 will be above the free surface of the liquid in the tank. This will eliminate minor losses at the pipe exit.

(xiv) Mixing tank 1 is 8 ft above the finished floor. This will provide a 4 ft clearance between the top of the mixing tank and the ceiling.

(xv) In Edmonton, the inlet temperature of the feed water to the boiler is about 40°F.

Sketch

The following drawing shows a possible layout of the piping between the ethyl alcohol storage tank and mixing tank 1. Note the locations of the pipe connections to the tanks and the valves.

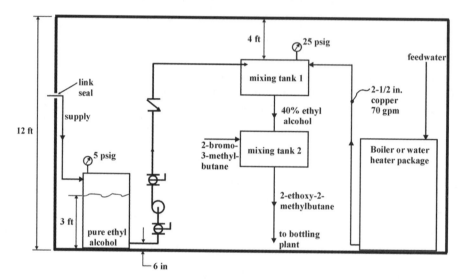

Analysis

Determine the Pipe Size Required

The pipe size will be determined by considering the flow rate of pure ethyl alcohol that is required and the maximum allowed head loss (7 ft of water per 100 ft of pipe).

Use the volume flow rate of water into mixing tank 1 and the final concentration (based on volume) of ethyl alcohol to determine the volume flow rate of ethyl alcohol into mixing tank 1.

Therefore,

$$\frac{\dot{V}_{ethyl}}{\dot{V}_{ethyl} + \dot{V}_{water}} = 0.40.$$

Thus,

$$\dot{V}_{ethyl} = \frac{0.40}{0.60} \dot{V}_{water} = \frac{0.40}{0.60}(70\ \text{gpm}) = 47\ \text{gpm}.$$

Use the friction loss chart for Schedule 40 steel for open piping systems (Figure A.4) with a flow rate of 47 gpm and a maximum head loss of 7 ft of water per 100 ft of pipe to estimate the pipe size.

Therefore,

$$D_{pipe} = 2 \text{ in. nominal.}$$

For this pipe diameter, the head loss is approximately 6.9 ft of water per 100 ft of pipe and the pipe velocity is approximately 4.8 fps. The head loss and pipe velocity are below the maximum values of the design constraints. If the friction loss chart for Schedule 80 plastic piping was used (Figure 3.3), a nominal 2 in. pipe that transports 47 gpm of alcohol would generate head loss of approximately 5.5 ft of water per 100 ft of pipe.

Determine the Minimum Pump Size Required

The energy equation will be used to determine the pump head required:

$$h_{pump} = H_{IT} + \left[\frac{p_2}{\rho g} + \alpha_2 \frac{V_{ave,2}^2}{2g} + z_2 \right] - \left[\frac{p_1}{\rho g} + \alpha_1 \frac{V_{ave,1}^2}{2g} + z_1 \right].$$

Let point 1 be at the inlet to the pipe from the ethyl alcohol storage tank and point 2 be in the fluid jet at the exit of the pipe in mixing tank 1. Therefore, $V_{ave,1} = V_{ave,2} = 4.8$ fps. It is assumed that the ethyl alcohol fluid jet into mixing tank 1 has the same diameter as the pipe to ensure that the jet velocity is equal to the pipe velocity.

At point 1

$p_1 = p_{s\text{-tank}} + \rho g h$, where h is the height of fluid in the tank above the pipe connection
$h = 3 \text{ ft} - 6 \text{ in.} = 2 \text{ ft } 6 \text{ in.}$
$z_1 = 0.$

At point 2

$p_2 = p_{m\text{-tank}}$,
$z_2 = 12 \text{ ft} - 4 \text{ ft} - 6 \text{ in.} = 7 \text{ ft } 6 \text{ in.}$

The energy equation becomes

$$h_{pump} = H_{IT} + \left[\frac{p_2}{\rho g} + z_2 \right] - \left[\frac{p_1}{\rho g} + z_1 \right]$$

$$h_{pump} = H_{IT} + \left[\frac{p_{m-tank}}{\rho g} + z_2 \right] - \left[\frac{p_{s-tank} + \rho g h}{\rho g} \right]$$

$$h_{pump} = H_{IT} + \left[\frac{p_{m-tank}}{\rho g} + z_2 \right] - \left[\frac{p_{s-tank}}{\rho g} + h \right]$$

$$h_{pump} = H_{IT} + \left[\frac{p_{m-tank} - p_{s-tank}}{\rho g} \right] + [z_2 - h].$$

The total head loss is the sum of the major head loss and the minor head losses. The major head loss is

$$H_{\text{lM}} = \frac{6.9 \text{ ft water}}{100 \text{ ft pipe}} \times L_{\text{pipe}} = \frac{6.9 \text{ ft water}}{100 \text{ ft pipe}} \times (7.5 \text{ ft} + 4 \text{ ft}) = 0.79 \text{ ft water.}$$

For the minor head losses, the loss coefficients for the bends, fittings, and valves are given below for a 2 in. diameter pipe,

Ball valve: $K_{\text{ball}} = 0.05$
Check valve: $K_{\text{check}} = 2.1$
Screwed 90° regular bends: $K_{90 \text{ deg bend}} = 0.95$
Pipe entrance, sharp-edged: $K_{\text{entrance}} = 0.50$.

Remember that it was assumed that an in-line pump will be selected. No 90° bends at the pump are expected for this arrangement.
    Therefore,

$$H_{\text{lm}} = \sum K_L \frac{V_{\text{ave}}^2}{2g} = (0.50 + 2\,(0.95) + 2\,(0.05) + 2.1)\left[\frac{(4.8 \text{ ft/s})^2}{2\left(32.2 \text{ ft/s}^2\right)}\right] = 1.65 \text{ ft ethyl alcohol.}$$

Multiply by the specific gravity of ethyl alcohol ($SG_{\text{ethyl}} = 0.794$) to convert the units to ft of water.

$$H_{\text{lm}} = 1.65 \text{ ft ethyl alcohol} \times SG_{\text{ethyl}} = (1.65 \times 0.794) \text{ ft water} = 1.31 \text{ ft water.}$$

The total head loss is

$$H_{\text{lT}} = H_l + H_{\text{lm}} = 0.79 \text{ ft} + 1.31 \text{ ft} = 2.10 \text{ ft water.}$$

The pump head is

$$h_{\text{pump}} = 2.10 \text{ ft water} + \left[\frac{25 \text{ lbf/in.}^2 - 5 \text{ lbf/in.}^2}{\left(49.3 \text{ lbm/ft}^3\right)\left(32.2 \text{ ft/s}^2\right)}\right] \times \frac{(12 \text{ in.})^2}{1 \text{ ft}^2} \times \frac{32.2(\text{lbm ft})/\text{s}^2}{1 \text{ lbf}}$$

$$+ [7.5 \text{ ft} - 2.5 \text{ ft}] ,$$

$h_{\text{pump}} = 2.10 \text{ ft water} + 58.4 \text{ ft ethyl alcohol} + 5 \text{ ft ethyl alcohol}$
$h_{\text{pump}} = 2.10 \text{ ft water} + 58.4 \text{ ft ethyl alcohol} \times (SG_{\text{ethyl}}) + 5 \text{ ft ethyl alcohol} \times (SG_{\text{ethyl}}).$

In terms of ft of water, the pump head required is

$$h_{\text{pump}} = 53 \text{ ft water.}$$

The discharge of the pump should be 47 gpm. For this application, a pump that is rated 47 gpm at 53 ft of head is required. Readily available are the pump performance curves of Series 60 Bell & Gossett in-line centrifugal pumps. The master selection chart is shown below:

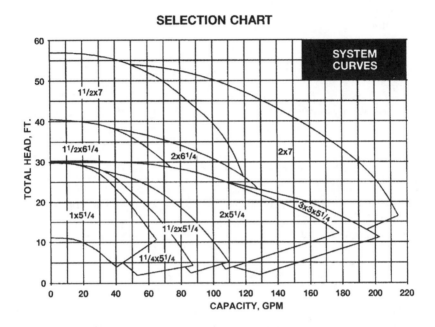

The operating point would fall in the contour of the $1\frac{1}{2}$ in.×7 in. pump casing. Choose the $1\frac{1}{2}$ in.×$1\frac{1}{2}$ in.×7 in. family of pumps.

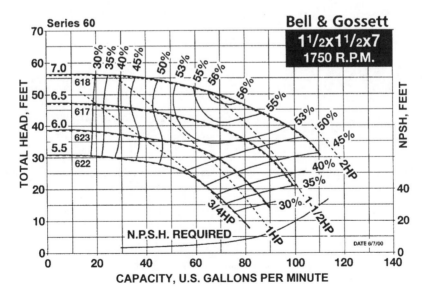

The final choice is a pump with the following operating parameters:

$1\frac{1}{2}$ in. $\times$ $1\frac{1}{2}$ in. $\times$ 7 in. pump casing
7.0 in. impeller diameter
$1\frac{1}{2}$ hp motor
1750 rpm.

Boiler Sizing

This boiler serves only to provide hot water to a chemical process. The thermal size of the boiler is

$$\dot{Q}_{\text{boiler}} = \dot{m}_{\text{water}} c_{\text{p,water}} \left(T_{\text{out,boiler}} - T_{\text{in,boiler}} \right).$$

The efficiency of the boiler is not considered at this point until an appropriate unit is finally selected.

In terms of volume flow rate,

$$\dot{Q}_{\text{boiler}} = \rho_{\text{water}} \dot{V}_{\text{water}} c_{\text{p,water}} \left(T_{\text{out,boiler}} - T_{\text{in,boiler}} \right).$$

The outlet temperature of the water from the boiler is still unknown. The heat transferred from the hot water will be equal to the heat absorbed by the ethyl alcohol in mixing tank 1 to achieve the tank temperature of 100°F. Therefore,

$$q = \dot{m}_{\text{water}} c_{\text{p,water}} \left(T_{\text{out,boiler}} - T_{\text{tank}} \right) = \dot{m}_{\text{ethyl}} c_{\text{p,ethyl}} \left(T_{\text{tank}} - T_{\text{ethyl,storage}} \right)$$

$$q = \rho_{\text{water}} \dot{V}_{\text{water}} c_{\text{p,water}} \left(T_{\text{out,boiler}} - T_{\text{tank}} \right) = \rho_{\text{ethyl}} \dot{V}_{\text{ethyl}} c_{\text{p,ethyl}} \left(T_{\text{tank}} - T_{\text{ethyl,storage}} \right).$$

Hence,

$$T_{\text{out,boiler}} = T_{\text{tank}} + \frac{\rho_{\text{ethyl}} \dot{V}_{\text{ethyl}} c_{\text{p,ethyl}} \left(T_{\text{tank}} - T_{\text{ethyl,storage}} \right)}{\rho_{\text{water}} \dot{V}_{\text{water}} c_{\text{p,water}}}.$$

Assume that the properties of ethyl alcohol are at 68°F and those of water are at 100°F. Then,

$T_{\text{out,boiler}} = 100°F$

$$+\frac{\left(49.3 \text{ lb/ft}^3\right)(47 \text{ gal/min})(0.678 \text{ Btu/(lb °F)})(100-68)°\text{F}\left(35.315 \text{ ft}^3/264.17 \text{ gal}\right)}{\left(62 \text{ lb/ft}^3\right)(70 \text{ gal/min})(0.999 \text{ Btu/(lb °F)})\left(35.315 \text{ ft}^3/264.17 \text{ gal}\right)}$$

$T_{\text{out,boiler}} = 112°F.$

The properties of water that will be used to size the boiler will be taken at an average temperature. For the boiler, the average temperature is

$$T_{\text{ave,boiler}} = \frac{T_{\text{out,boiler}} + T_{\text{in,boiler}}}{2}.$$

In Edmonton, the inlet temperature of the feed water to the boiler is about 40°F. Assume $T_{\text{in,boiler}} = 40°F$.
Thus,

$$T_{\text{ave,boiler}} = \frac{112°F + 100°F}{2} = 106°F.$$

Then,

$$\dot{Q}_{\text{boiler}} \approx \left(61.86 \text{ lb/ft}^3\right)(70 \text{ gal/min})(0.999 \text{ Btu/(lb °F)})(112-40)°\text{F} \times \frac{35.315 \text{ ft}^3}{264.17 \text{ gal}} \times \frac{60 \text{ min}}{1 \text{ h}}.$$

The minimum thermal size of the hot water boiler is

$$\dot{Q}_{\text{boiler}} \approx 2.29 \times 10^6 \text{ Btu/h} = 2290 \text{ MBH} = 68 \text{ hp (boiler)}.$$

A Cleaver-Brooks commercial package boiler will be selected for this application (www.cbboilers.com/commercial_boilers.htm). Sometimes, the efficiencies of the boilers may not be provided in the manufacturer's boiler book. However, Cleaver-Brooks uses a nominal boiler efficiency of 82% in fuel consumption calculations. Therefore,

$$\dot{Q}_{\text{boiler}} \approx 84 \text{ hp (boiler)}.$$

A Cleaver-Brooks model 4WG packaged hot water boiler will be specified for this application. Below is an excerpt from the technical catalog of the boiler book. This information will be used to prepare the boiler schedule for use by the mechanical contractor.

Model 4WG   **100-800 HP Boilers**

Table A3-3. 4WG Ratings - Hot Water

BOILER H.P.	100	125	150	200	250	300	350	400	500	600	700	800
BURNER MODEL	FP-3	FP-3	FP-3	FP-4	FP-4	FP-4	D145P	D175P	D210P	D252P	D300P	D378P
RATINGS - SEA LEVEL TO 700 FT.												
Btu Output (1000 Btu/hr)	3347	4184	5021	6694	8368	10042	11715	13389	16736	20083	23430	26778
APPROXIMATE FUEL CONSUMPTION AT RATED CAPACITY BASED ON NOMINAL 82% EFFICIENCY												
Light Oil gph (140,000 Btu/gal)	29.2	36.4	43.7	58.3	72.9	87.5	102.0	116.6	145.8	174.9	204.1	233.3
Gas CFH (1000 Btu)	4082	5102	6123	8164	10205	12246	14287	16328	20410	24492	28574	32656
Gas (Therm/hr)	40.8	51.0	61.2	81.6	102.0	122.5	142.9	163.3	204.1	244.9	285.7	326.6
POWER REQUIREMENTS - SEA LEVEL TO 700 FT. (60 HZ)												
Blower Motor hp	2	3	5	5	5	7-1/2	15	20	25	30	40	75
Circulating Oil Pump Motor hp (Oil only)	1/3*	3/4*	3/4*	1/2	1/2	3/4	3/4	3/4	3/4	3/4	1	1
Oil Metering Pump Motor hp (Oil only)	-	-	-	-	-	-	-	1/2	3/4	3/4	3/4	1
Integral Oil/Air Motor hp (Oil only)	-	-	-	-	-	-	2	-	-	-	-	-
Air Compressor Motor hp (Oil only)	**	**	**	3	3	3	-	5	5	7-1/2	7-1/2	15
BOILER DATA												
Heating Surface sq-ft. (Fire-side)	500	625	750	1000	1250	1500	1750	2000	2500	3000	3500	See Note "B"

* Base rail mounted oil pump will be 3-phase voltage.
** No air compressor required (pressure atomized)
NOTES:
A. All fractional hp motors will be single phase voltage except oil metering pump motor(3-phase); integral hp motors will be 3-phase voltage.
B. 800 hp boilers are available w/ 3500 or 4000 sq. ft. of heating surface

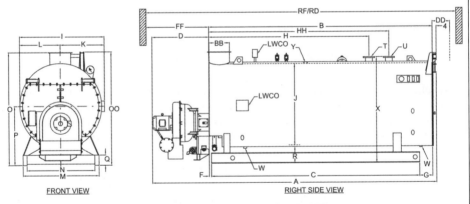

FRONT VIEW   RIGHT SIDE VIEW

Figure A3-2. 4WG - Hot Water 100-800 HP

Model 4WG   **100-800 HP Boilers**

Table A3-4. 4WG Dimensions - Hot Water

BOILER H.P.	DIM	100	125	150	200	250	300	350	400	500	600	700	800	*800 See Note "C"
LENGTHS														
Overall Length	A	174	198	186	227.5	220.25	244.25	245.25	263.25	276.75	311.75	296.5	304	331
Shell	B	131	155	143	177	172.5	196.5	189.75	207.75	213.75	248.75	232.75	232.75	259.75
Base Frame	C	110	124	122	156	150.125	174.125	167.25	185.25	188.25	223.25	207.25	207.25	234.25
Burner Extension	D	39	39	39	46.5	43.75	43.75	51.5	51.5	59	59.75	67.25	67.25
Smokebox to Base	F	0.5	0.5	0.5	0.5	0.5	0.5	0.5	0.5	0.5	0.5	0.5	0.5	0.5
Rear Ring Flange to Base	G	20.5	20.5	20.5	20.5	22	22	22	22	25	25	25	25	25
Smokebox to Return	H	78	99	87	121	113.5	137.5	130.75	148.75	143	151.75	146.25	146.75	173.75
Smokebox to Outlet	HH	103	124	112	146	139.5	163.5	156.75	174.75	179	187.75	182.25	182.75	209.75
WIDTHS														
Overall Width	I	70	70	77.5	77.5	88	88	95	95	106	106	116	116	116
I.D. Boiler	J	60	60	67	67	78	78	85	85	96	96	106	106	106
Center to LWCO Controller	K	37	37	40.75	40.75	46	46	49.5	49.5	55	55	60	60	60
Center to Lagging	L	33	33	36.75	36.75	42	42	45.5	45.5	51	51	56	56	56
Base Outside	M	52.5	52.5	51	51	64	64	60	60	71.88	71.88	74.75	74.75	74.75
Base Inside	N	44.5	44.5	43	43	56	56	47	47	58.88	58.88	61.75	61.75	61.75
HEIGHTS														
Base to Rear Davit	OO	86.12	86.12	92.75	92.75	98.88	98.88	112.75	112.75	125.12	125.12	134.25	134.25	134.25
Base to Vent Outlet	O	87	87	92.63	92.63	106	106	115	115	126	126	135.63	135.63	135.63
Base to Boiler Centerline	P	46	46	50	50	56	56	61	61	67	67	71	71	71
Height of Base Frame	Q	12	12	12	12	12	12	12	12	12	12	12	12	12
Base to Bottom of Boiler	R	15.63	15.63	16.13	16.13	16.5	16.5	18	18	18.5	18.5	17.5	17.5	17.5
Base to Return & Outlet	X	82.38	82.38	89.88	89.88	101.5	101.5	110	110	121.5	121.5	130.5	130.5	130.5
BOILER CONNECTIONS														
Water Return (See Note "A")	T	4	6	6	6	8	8	8	10	10	12	12	12	12
Water Outlet (See Notes "A & B")	U	4	6	6	6	8	8	8	10	10	12	12	12	12
Drain-Front & Rear	W	1.5	1.5	1.5	2	2	2	2	2	2	2	2	2	2
Air Vent	Y	1.5	1.5	1.5	1.5	1.5	1.5	1.5	2	2	2	2	2	2
VENT STACK														
Vent Stack Diameter (Flanged)	BB	16	16	16	16	20	20	24	24	24	24	24	24	24
MINIMUM CLEARANCES														
Rear Door Swing	DD	36	36	40	40	46	46	50	50	55	55	60	60	60
Tube Removal - Front Only	FF	96	120	108	142	132.5	156.5	148	166	169	204	188	188	215
MINIMUM BOILER ROOM LENGTH ALLOWING FOR DOOR SWING AND TUBE REMOVAL FROM:														
Thru Window or Door	RD	234	258	261	295	307.5	331.5	337	355	377	412	411	411	438
Front of Boiler	RF	263	311	291	359	351	399	388	424	438	508	481	481	535
WEIGHTS IN LBS														
Normal Water Weight	-	6,890	8,580	8,870	11,600	14,760	17,380	19,220	21,520	26,260	31,580	35,900	35,900	40,930
Approx. Shipping Weight - (30 psig)	-	10,860	12,080	13,090	15,260	19,110	21,050	24,760	27,640	33,295	38,150	42,320	42,320	46,300
Approx. Shipping Weight - (125 psig)	-	11,600	12,980	14,040	16,680	20,670	23,140	26,930	30,060	35,390	40,550	45,430	45,430	49,750

NOTES:
Accompanying dimensions, while sufficiently accurate for layout purposes, must be confirmed for construction by certified dimen sion diagram/drawing. All Connections are Threaded Unless Otherwise Indicated:
NOTE "A": ANSI 150 psig Flange
NOTE "B": Water Outlet includes 2" Dip Tube
NOTE "C": *800 hp w/ 4000 sq. ft. of heating surface

Section A3-8   Rev. 09-09

Source: Cleaver-Brooks, Inc. (reprinted with permission)

Drawings

Below is a schematic of the system c/w with all the pipe sizes and flow rates.

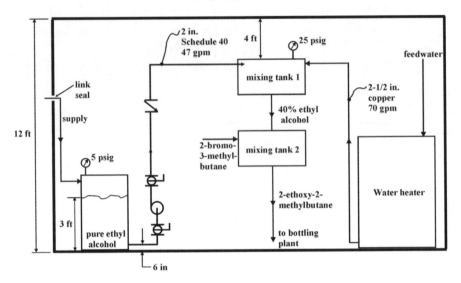

Conclusions

The piping system has been designed, complete with two ball valves, a check valve for backflow prevention, and a pump. The pipe size was determined to be 2 in. in diameter and the total pipe length is 11.5 ft.

For the Cleaver-Brooks boiler, the manufacturer provides information on the procedures to determine the efficiency of the boiler at full and part-load operation. The efficiency used in the analysis was based on a typical value chosen by the manufacturer. Low boiler efficiencies increase the boiler size. Hence, a control system may be required when lower thermal capacity is needed. The designer could have considered using smaller boilers (arranged in a cascade configuration) that would be scheduled to operate when increased capacity was required. This would add greater control to the system operation, and provide contingencies in the event of equipment failure or maintenance.

A conventional boiler was chosen. A condensing boiler could have been chosen. This type of boiler would probably be larger and cost as much as 50% more than a conventional boiler.

A check of the NPSHA could be conducted to ensure that cavitation will not occur and to determine the minimum height of fluid required in the tank to avoid cavitation and pump damage.

Schedules

Below are schedules of the major equipment.

Pump Schedule

				Fluid			Electrical		
Tag	Manufacturer and Model Number	Type	Construction	Flow Rate (gpm)	Working Fluid	Head Loss (ft)	Motor Size (hp)	Motor Speed (rpm)	Volt/ pH/ Hz
P-1	Bell & Gossett Series 60, or equal	Centrifugal, in-line mounted	Iron $1\frac{1}{2}'' \times 1\frac{1}{2}'' \times 7''$ casing, 7.0"ϕ	47	Ethyl alcohol	53	$1\frac{1}{2}$	1750	208/ 3/ 60

Boiler Schedule

				Fuel			Dimensions		
Tag	Manufacturer And Model Number	Boiler Thermal Size	Burner Model	Flow Rate (gph)	Fuel Type	HHV (Btu/gal)	A	X	I
B-1	Cleaver-Brooks model 4WG, or equal	100 HP (3347 MBH)	Cleaver-Brooks model FP-3	29.2	#2 OIL	140000	174"	82.4"	70"

Problems

5.1. The Ministry of Defense in Ottawa has rehired the same consulting firm that they hired initially to complete the piping design of Problem 3.7. The original piping design has served its purpose, and now the Ministry would like to reuse the piping system to heat a large space. It is estimated that an average of 20 persons could occupy the space at any given time in an office setting. The in-house engineers have provided a sketch of the piping system. The expansion tank is shown. The base-mounted pump will be replaced by an in-line pump (circulator) and the chiller, by a hot water boiler package. The units in the existing system (units a, b, c) were replaced with finned-tube heaters. All valves, orifice plates, and pipe lengths are to remain.

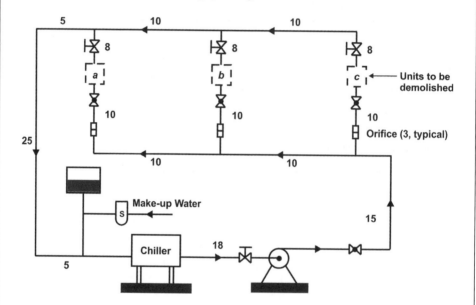

Values for the different types of heat losses from the space in winter were calculated, and are presented in the table below.

Types of Heat Losses	Values (Btu/h)
Wall transmission losses	4200
Window transmission losses	13800
Ceiling transmission losses	2700
Floor transmission losses	7080
Cold air infiltration losses	27000

Size and select the boiler and heaters for the modified system.

5.2. The City of Edmonton has solicited the services of a mechanical engineering consulting firm to design a heating system for a public pool located in the City. The pool will be 20 ft wide and 15 ft long. It will be 4 ft deep at one end, varying linearly along the length of the pool to a depth of 10 ft. The pool water is to be heated to 77°F in the month of May. The focus of the design will be to provide a system that can heat the water from an initial temperature, after winter closure. The 2006 Alberta Building Code, Division B, Section 7.3.3.9 requires that the slope of the bottom of public swimming pools be no more than 1:3. Section 7.3.3.39 of the same code also requires that the turnover period (recirculation period) of the pool water shall not exceed 4 hours. Design a heating system that is complete with one or more boiler packages. The maximum length of the piping system may be 25–50 ft from the pool. Reference to the 2006 Alberta Building Code may be useful.

5.3. A tankless water heater is required to heat water to 140°F for domestic use. The water flow rate should be fixed between 2 and 4 gpm. Select and specify (present a schedule) a manufacturer's tankless water heater. Determine the minimum amount of fuel that will be needed for this application.

5.4. Manufacturers of tankless water heaters will usually recommend that their units be vented vertically or horizontally over short runs of pipe. In vertical venting, the vent will protrude about 6 in. upward from the top of the unit and, through a regular 90° bend, will penetrate the interior wall to exhaust to the exterior. In the case of a client living in a condominium, this type of installation was not permitted by the bylaws of the condominium board. As a result, the length of the vent was extended so that venting could be accomplished through a wall that was a horizontal distance, L from the tankless water heater unit. The designer noted that one manufacturer's vent model was a concentric tubular venting system in which there was a $2\frac{1}{2}$ in. inner metal tube vent for hot exhaust gases and a 4 in. outer PVC tube for fresh air from outside for combustion. The fan installed with the unit moves 50 cfm of air and exhaust gases through the respective sections of the vent. The average outdoor air temperature is –22°F and the exhaust gases exit the finned copper tubes of the heat exchanger bank at about 200°F. Estimate the length of the vent if the air should enter the combustion chamber at a minimum temperature of 40°F.

5.5. Campbell, Thompson, and Stewart Architects Ltd. have provided the following floor plan to a mechanical consulting engineer.

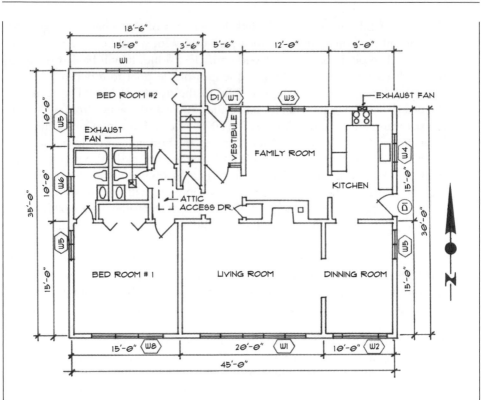

The engineer has conducted preliminary calculations to determine the room design heat loss values that are presented in the table below:

Room	Heat Loss (Btu/h)
Dining room	5480
Living room	8100
Bedroom #1	6440
Exterior bathroom	1340
Bedroom #2	4680
Vestibule	1460
Family room	2620
Kitchen	3030

The owner of the home ("the client") has decided that they would like baseboard or finned-tube heaters for the dining room, the living room, bedroom #1, and the exterior bathroom. All the other rooms would receive forced air heating from a dedicated furnace. Design and layout a hydronic heating system for the rooms that were specified by the client. Size and select a boiler or water heater to provide all the hot water needs of the house. The first floor (shown) is above a finished basement.

Further Information: Applicable codes and standards require different water temperatures for domestic use and space heating.

5.6. For most building design projects, the architectural trade tends to be the consultant (i.e., the lead consultant in the project) who hires the mechanical trade as a subconsultant on the project. In some cases, the mechanical engineering subconsultant has expertise in the design of ductwork to transport air for the purposes of heating and/or cooling an occupied space. A section of a second floor tenant plan of an office space has been designed and presented by an architect.

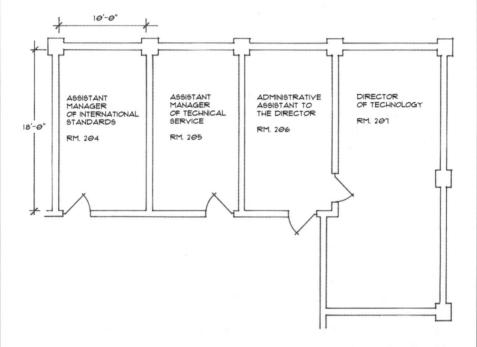

For the offices shown in the plan (complete with the occupant and work function), the architect has requested that the design temperature in each room be 75°F. If an air distribution system was used to heat the spaces, the following table shows the total amount of air that would be required. The exit temperature of the air from the diffusers would be 110°F in order to ensure the indoor design temperature under winter conditions.

Office Space	Heating Air Requirement
Office 204	330 cfm
Office 205	470 cfm
Office 206	190 cfm
Office 207	520 cfm

However, instead of an air distribution system, the client has requested that a finned-tube heating system, complete with a dedicated condensing boiler be designed and installed. Comment on any trade-offs between boiler efficiency and heater lengths.

Further Information: The length of a standard door is 3 ft.

5.7. WoodBridge ClimateCare, located in Toronto, has solicited the services of GTA Designs Inc. in Brampton, Ontario to design and layout the ductwork system that will provide heated air to the space shown in the drawing below. A dedicated furnace, complete with a fan will be purchased and installed below the space in an open-concept basement. The ducts will be hung from the ceiling of the basement, and appropriate flexible connections of short length will be made from the ducts to the diffusers shown in the drawing. Only supply air ducts will be required, since return air will be drawn into the fan from the open basement space. GTA Designs Inc. has conducted a heat load analysis, and the heat required to maintain each room/area in the space at 77°F is shown in the table below. Given the low winter outdoor air temperature in Toronto proper, the air will exit the diffuser grilles at 125°F to maintain the desired indoor space temperature. Design a supply air ductwork system, based on round ducts, to meet the most stringent demands for heated air. At this stage of the design, consideration of system balancing and furnace selection is not required.

Room/Area	Heating Load (Btu/h)
Eating area	12800
Stairs	1000
Private wash room (PWR)	1000
Family room	6600
Dining	6050
Foyer	1200
Living room	7050

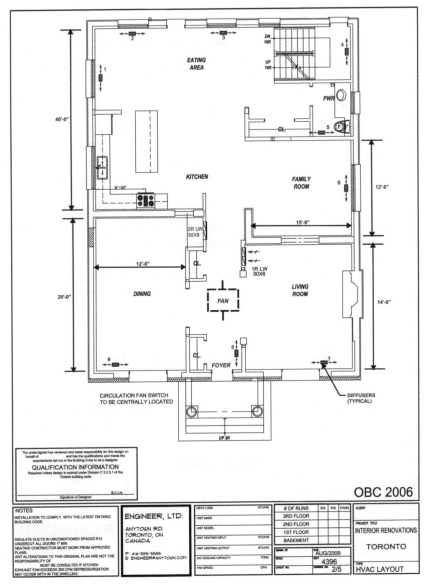

Modified from: GTA Designs, Inc. (reprinted with permission)

5.8. The design of a processing plant is being considered for development. In this plant, ten large centrifugal compressors will be used for water cooling. Refined lubricating oil will be required at a rate of 5 gpm for each compressor. The lubricating oil will be refined onsite to ensure that the content of unsaturated hydrocarbons in the oil will be low to prevent damage to the compressors. Due to the large number of compressors and auxiliary equipment required for the refinement of the oil, it has been proposed to use two separate buildings to house the compressors and oil refinement equipment. The sketch below shows the distance between the buildings. As an initial design concept, it has been suggested to transport the refined oil from an 11-ft-diameter vented storage tank in Building 1 to the main header line feeding the compressors in Building 2. To manage construction costs, the outdoor piping between the buildings will not be buried and will be mounted $1\frac{1}{4}$ ft above grade. For this location, the average outdoor temperature in the summer is 80°F and in the winter, it is –5°F. The average wind speed in winter is about 9 mph (miles per hour). For the lubricating oil, the *pour point* (temperature at which oil ceases to flow) is approximately 20°F and the *flash point* (temperature at which oil vapor will ignite when exposed to a flame) is 400°F. For improved operation of the compressors, the oil temperature and pressure at the inlet to the main header line shall be no less than 70°F and 20 psig, respectively. Design a piping system, up to the main header line feeding the compressors, and specify the information that will be required for the selection of equipment (including equipment for heating) needed to facilitate transportation of the oil from the storage tank in Building 1 to the main header line of the compressors in Building 2. Given the preliminary stage of this design, no equipment schedules or catalog sheets are required.

Useful Property Data for Standard Lubricating Oil: Density of 58 lb/ft³; kinematic viscosity of 3.8 cSt $= 4.09\times10^{-5}$ ft²/s; specific heat of 0.450 Btu/(lb R); thermal conductivity of 0.087 Btu/(h ft R); vapor pressure of 0.030 psi.

Further Information: Special heaters for oil are available that come complete with a vented tank and burner, only. Typical heater heights will be between 3 and 6 ft.

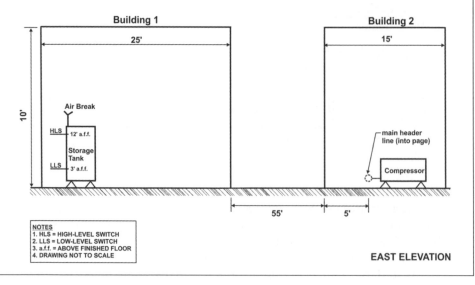

References and Further Reading

[1] Natural Resources Canada (2004) *Heating with Gas, Home Heating and Cooling Series*, Natural Resources Canada, Ottawa.
[2] Epcor, Corp. (2009) *Edmonton Water Utilities Statistics*, Epcor, Corp., Edmonton.
[3] Bobenhausen, W. (1994) *Simplified Design of HVAC Systems*, John Wiley and Sons, Inc., New York.
[4] National Research Council of Canada (2006) *Alberta Building Code*, Division B, vol. 2, part 7, Plumbing Services and Health, Ottawa, pp. 7-9–7-10.

6

Performance Analysis of Power Plant Systems

Power plant systems are used to generate electrical power and heat. Power plants are multicomponent systems with turbines, compressors, steam boilers (**steam generators**), condensers, other heat exchangers, pumps, and combustion burners, to name a few. Although this type of system operates on a **thermodynamic (power) cycle** to produce electrical power, other fundamental concepts from fluid mechanics (pipe and duct sizing, pump sizing) and heat exchanger design (design of condensers, feedwater heaters, reheaters) may be needed. Due to the multicomponent nature of power plant systems, their design can be laborious, especially when control system design is included. In addition, significant analysis must be conducted to determine the most economical power plant that will yield the greatest efficiency and performance to deliver a required amount of power. In performance analysis of power plant systems, the design engineer will analyze existing power plants and their components to make recommendations for improvement of the systems.

6.1 Thermodynamic Cycles for Power Generation—Brief Review

6.1.1 Types of Power Cycles

The thermodynamic (power) cycles are categorized into two groups: **gas cycles and vapor cycles**.

In a **gas (power) cycle**, the working fluid remains as a gas throughout the entire cycle. An example is a gas-turbine system based on the **Brayton cycle** with air or combustion gas as the working fluid.

In a **vapor (power) cycle**, the working fluid is a gas in one part of the cycle and is a liquid in another part of the system. An example is a steam-turbine system based

---

Introduction to Thermo-Fluids Systems Design, First Edition. André G. McDonald and Hugh L. Magande.

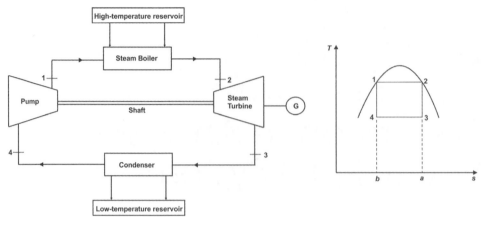

Figure 6.1   Ideal Carnot cycle

on the **Rankine cycle** with pressurized, superheated steam entering a steam turbine and liquid water leaving a condenser.

6.1.2 Vapor Power Cycles—Ideal Carnot Cycle

The Carnot cycle is an ideal reversible cycle that can operate between two constant temperature reservoirs. It is the most efficient power cycle, lacking any practical counterpart, and is used as a comparison for other power-generating cycles. A schematic of the ideal Carnot cycle is shown in Figure 6.1.

The following points should be noted regarding this cycle:

  (i) Process 1-2 is reversible and isothermal ($T_1 = T_2$). Liquid working fluid is converted to vapor in the boiler—a phase change occurs.
 (ii) Process 2-3 is reversible and adiabatic. There is no heat transfer (adiabatic), no changes in entropy (isentropic), and no losses (no irreversibilities). The vapor is expanded isentropically (see the *T-s* curve in Figure 6.1) in the turbine to produce useful work to drive the pump shaft and electrical generator.
(iii) Process 3-4 is reversible and isothermal ($T_3 = T_4$). The expanded vapor is condensed to liquid in the condenser—a phase change occurs.
(iv) Process 4-1 is reversible and adiabatic. The liquid working fluid is pressurized isentropically in a pump. There is no heat transfer (adiabatic), no changes in entropy (isentropic), and no losses (no irreversibilities).

6.1.3 Vapor Power Cycles—Ideal Rankine Cycle for Steam Power Plants

The ideal Rankine cycle for steam power plants is similar to the Carnot cycle, and includes at least a steam turbine, a condenser, a pump, and a steam boiler or steam generator. Figure 6.2 shows a schematic of the ideal Rankine cycle.

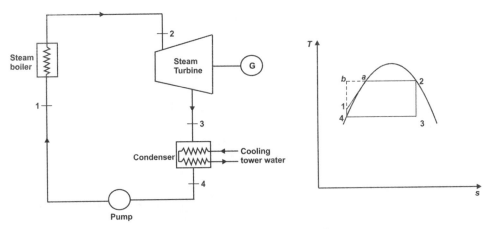

Figure 6.2 Ideal Rankine cycle

In the Rankine cycle for steam power plants, the condenser may be a large shell-and-tube heat exchanger supplied with cooling water to permit steam condensation. As shown in Figure 6.2, steam condensation terminates at the saturated liquid state (point 4 on the saturation curve), which is different from the Carnot cycle (see Figure 6.1). The steam generator (boiler) may include a burner in which fuel is combusted to provide the heat needed for steam generation. A simple liquid pump, with reduced pump power requirements, can be used. For smaller pumps, a dedicated motor could be used to drive them (motor-driven pump); for larger pumps, an auxiliary steam turbine could be used to drive them (turbine-driven pump). It will be shown later through second-law analysis that losses in the boiler are high (see the T-s diagram of Figure 6.2). This makes the Rankine cycle much less efficient than the Carnot cycle.

6.1.4 Vapor Power Cycles—Ideal Regenerative Rankine Cycle for Steam Power Plants

Regeneration can be included in the ideal Rankine cycle to preheat the liquid that enters the boiler. Hot steam is bled from an intermediate section of the steam turbine and used to heat the liquid water. The heating occurs in **feedwater heaters (regenerators)**. As a result of this preheating, less heat is lost at the condenser, and the overall cycle efficiency increases. Figure 6.3 shows schematics of a single-stage regenerative Rankine cycle and a four-stage regenerative Rankine cycle. In the single-stage regenerative Rankine cycle, one feedwater heater is present; in the four-stage regenerative Rankine cycle, four feedwater heaters are used.

Figure 6.3 shows two types of feedwater heaters: **open feedwater heaters** and **closed feedwater heaters**.

In **open feedwater heaters (contact feedwater heaters)**, direct mixing of steam with liquid water occurs. This will increase the efficiency of heating. However, proper control of heating is difficult and a special feedwater pump may be needed, which

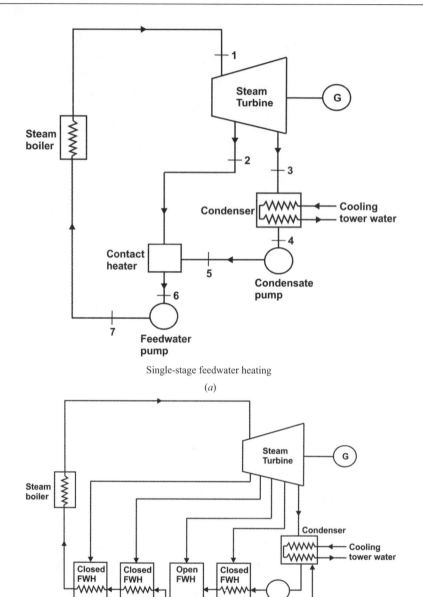

Single-stage feedwater heating

(*a*)

4-stage feedwater heating

(*b*)

Figure 6.3   Ideal regenerative Rankine cycles. (a) Single-stage feedwater heating; (b) four-stage feedwater heating

may increase first costs (installation/initial construction costs) and operation costs of the plant.

In **closed feedwater heaters** (**surface heaters**), there is no direct mixing of steam and liquid water. All heat transfer occurs though the heat transfer surface area (may be tubular). This type of feedwater heater may be more desirable for steam power plants because the heat exchanger can be designed (tube size, arrangement, number of fins, etc.) to provide some control over the transfer of heat. In addition, only the condensate pump may be required to move the liquid through the tube line that passes through the feedwater heaters.

Practical Note 6.1   Optimizing the Number of Feedwater Heaters

Many feedwater heaters could be included in a regenerative Rankine cycle-based steam power plant. It is the responsibility of the design engineer to optimize the number of feedwater heaters to produce the greatest system efficiency at the most acceptable and reasonable cost (installation and operating costs).

Fundamental Note: A reheat stage may be added to the cycle between two different stages of the steam turbine. Moisture in the steam can reduce the efficiency and undermine the performance of the turbine during expansion. In addition, with less steam to expand, the work output of the turbine will be reduced. Reheating partially expanded steam in a **reheater** (complete with a burner) to facilitate complete expansion in the intermediate-pressure (IP) or low-pressure (LP) stage of the cycle will increase the performance and life of the turbine.

6.2 Real Steam Power Plants—General Considerations

Real steam power plants based on the Rankine cycle will differ from the ideal Rankine cycle due to losses throughout the system. The following general points should be considered for real steam power plants:

(i) The actual cycle efficiency will depend on the efficiencies of the components in the system (turbine, pump, condenser, boiler/steam generator) and losses in the pipes and fittings. A second-law analysis may be more appropriate for this type of practical, multicomponent system.

(ii) There may be multiple feedwater heaters to improve the efficiency of the plant.

(iii) Feedwater pumps tend to be larger (larger pump horsepowers). So, they are typically driven by auxiliary steam-turbine units rather than electric motors.

(iv) Steam may be extracted to preheat combustion air that is used in the boiler burners or in supplementary burners for the feedwater heaters. Energy from the combustion gases that exit the boiler stack may also be used to preheat the combustion air.

6.3 Steam-Turbine Internal Efficiency and Expansion Lines

For an isentropic, adiabatic expansion in a turbine, the turbine **internal efficiency** would be 100%, and $s_{in} = s_{out}$. However, due to losses in real turbines, the expansion is not isentropic and $s_{in} \neq s_{out}$. So, the entropy of steam at the inlet to the turbine is not equal to the entropy of expanded steam that exhausts from the turbine.

The internal efficiency of the turbine is given as

$$\eta_i = \frac{h_{in} - h_{out}}{h_{in} - h_{out,s}}, \tag{6.1}$$

where $h_{out,s}$ = enthalpy of steam at the turbine exit with the same entropy as that of h_{in} at the turbine inlet.

If the internal efficiency of the turbine is known and the objective is the determination of the real work output of the turbine, the real enthalpy of the steam at the turbine exhaust (h_{out}) would be needed. Equation (6.1) could be used to find h_{out}.

In the case of steam bleeding for feedwater or process heating, the enthalpies of the steam bled off from the turbine at IPs would be needed to determine the turbine work. In the case of multistage feedwater heating, the determination of all the enthalpies of the bled steam could be time consuming if the quality of steam is first determined to find these enthalpies. Instead, it is typical to use the **Mollier diagram for water** to plot straight **expansion lines** between h_{in} and h_{out} to find the enthalpies of the steam bled from the turbine at IPs between the inlet and outlet pressures of the turbine. Figure 6.4 shows the Mollier diagram for water. The diagram presents entropy on the x-axis, enthalpy on the y-axis, and lines of pressure, temperature, and quality. The diagram only applies to superheated steam or steam in which the quality is greater than 50%. The Mollier diagram for water can become very busy. Care should be taken to read the diagram carefully.

Consider the simple plant with one stage of contact feedwater heating and one steam turbine that is shown in Figure 6.3a. The turbine internal efficiency is 85%. The inlet pressure to the turbine is $P_1 = 300$ psia (saturated vapor). Steam at 60 psia is bled from the turbine for heating in the contact (open) feedwater heater ($P_2 = 60$ psia). Steam is exhausted from the turbine at 1 psia to the condenser ($P_3 = 1$ psia ≈ 2.0 in. Hg. abs.). The Mollier diagram can be used to find the enthalpy at point 2.

The following steps may be employed:

(i) Find $h_{in} = h_1$ from appropriate thermodynamic tables for steam. At this point, only the pressure was given. Therefore, the Mollier diagram cannot be used to find h_1.

Therefore, $h_1 = 1203.3$ Btu/lb for $P_1 = 300$ psia and $T_1 = 417.4°F$. The entropy is $s_1 = 1.51$ Btu/(lb R).

(ii) Identify point 1 on the Mollier diagram (see Figure 6.5).

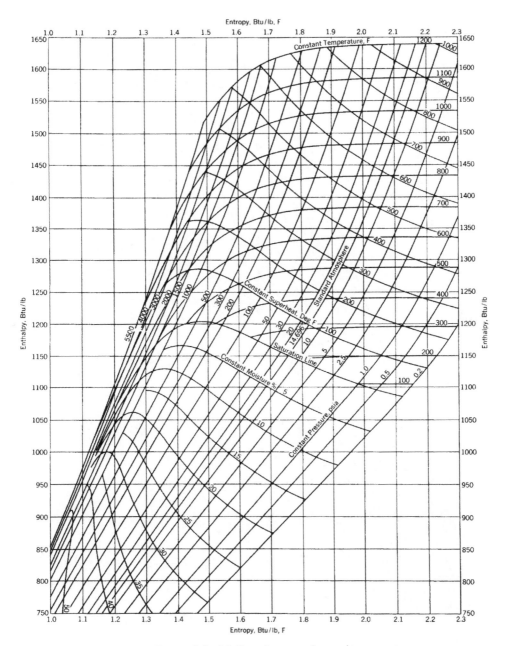

Figure 6.4 Mollier diagram for water

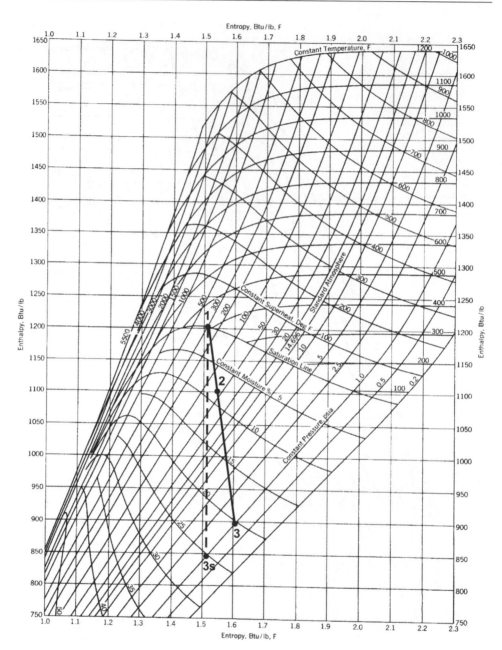

Figure 6.5 Mollier diagram for water showing an expansion line

(iii) Find h_{3s}. This is the enthalpy at the exhaust of the turbine for an isentropic expansion from 300 to 1 psia. On the Mollier diagram, follow the $s_1 = s_{3s} = 1.51$ Btu/(lb R) line down from 300 to 1 psia, and identify the enthalpy.

Hence, $h_{3s} \approx 850$ Btu/lb.

(iv) The turbine internal efficiency can be used to find the real enthalpy of the steam at the turbine exhaust (h_3).

$$\text{Remember:} \quad \eta_i = \frac{h_1 - h_3}{h_1 - h_{3s}}.$$

Therefore,

$$h_3 = h_1 - \eta_i \, (h_1 - h_{3\,s})$$
$$h_3 = 1203.3 \text{ Btu/lb} - (0.85)(1203.3 - 850) \text{ Btu/lb}$$
$$h_3 \approx 903 \text{ Btu/lb at 1 psia}$$

(v) Identify point 3 on the Mollier diagram (see Figure 6.5).
(vi) Connect a straight line between points 1 and 3.
(vii) Point 2 will lie along this straight, **expansion line** at 60 psia.
(viii) From the Mollier diagram, $h_2 \approx 1095$ Btu/lb at 60 psia.
(ix) Use the expansion line between the inlet (point 1) and the outlet (point 3) of the turbine to find the enthalpies of the extracted steam at the specified pressures for multiple feedwater heaters.

Example 6.1   First-Law Efficiencies of the Rankine Cycle

A steam power plant is being considered for development. It is expected that saturated steam at 300 psia will enter the turbine and exhaust at 1 psia to the condenser. It is proposed to extract some of the steam at 60 psia for the purpose of feedwater heating. Calculate the first-law cycle efficiencies of the plant with and without steam extraction.

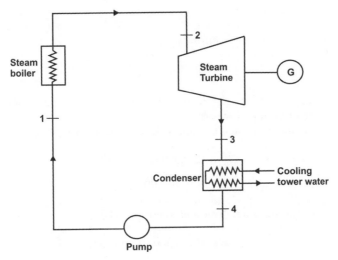

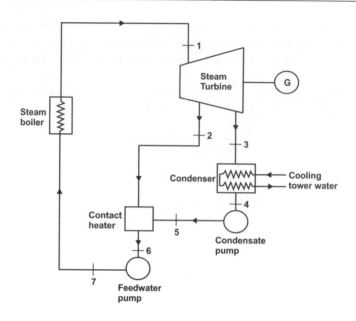

Solution

Without Steam Extraction

The first-law efficiency of the cycle is

$$\eta_{\text{cyc}} = \frac{w_{\text{net}}}{q_h}.$$

w_{net} is the net work output from the cycle and is

$$w_{\text{net}} = w_{\text{turbine}} - w_{\text{pump}}$$

$$w_{\text{net}} = (h_2 - h_3) - \int_4^1 v dp$$

$$w_{\text{net}} = (h_2 - h_3) - v_4 (p_1 - p_4),$$

where h_2 and h_3 are the enthalpies of steam at the turbine inlet and outlet, respectively, v is the specific volume of the liquid water.

From thermodynamic tables and for saturated steam at 300 psia,

$$h_2 = 1203.3 \text{ Btu/lb}$$

$$s_2 = 1.51 \text{ Btu/(lb °F)}.$$

Expansion in the turbine is adiabatic and isentropic. Therefore,

$$s_2 = s_3 = 1.51 \text{ Btu/(lb °F)}.$$

The Mollier diagram can be used to estimate h_3 at $p_3 = 1$ psia and $s_3 = 1.51$ Btu/(lb °F). From Figure 6.4,

$$h_3 \approx 850 \text{ Btu/lb.}$$

If a Mollier diagram is not readily available, h_3 can be determined by considering the quality of the expanded steam exiting the turbine. The quality of the steam is

$$x_{\text{quality}} = \frac{s_3 - s_{f3}}{s_{fg3}} = \frac{(1.51 - 0.133) \text{ Btu/(lb °F)}}{1.85 \text{ Btu/(lb °F)}} = 0.744.$$

With the quality known, the enthalpy at point 3 is

$h_3 = h_{f3} + x_{\text{quality}} h_{fg3}$
$h_3 = 69.72 \text{ Btu/lb} + (0.744)(1035.7 \text{ Btu/lb})$
$h_3 = 841 \text{ Btu/lb} \approx 850 \text{ Btu/lb.}$

Assuming that there are no losses in the system,

$p_4 = 1$ psia
$p_1 = 300$ psia
$v_4 = 0.01614 \text{ ft}^3/\text{lb}$ (obtained from thermodynamic tables)
$h_4 = 69.72 \text{ Btu/lb}$ (obtained from thermodynamic tables).

Therefore,

$$w_{\text{net}} = (1203.3 - 850) \text{ Btu/lb} - 0.01614 \frac{\text{ft}^3}{\text{lb}} \times (300 - 1) \frac{\text{lbf}}{\text{in}^2} \times \frac{32.2 \text{ lb/ft/s}^2}{1 \text{ lbf}}$$

$$\times \left(\frac{12 \text{ in.}}{1 \text{ ft}}\right)^2 \times \frac{1 \text{ Btu/lb}}{25037 \text{ ft}^2/\text{s}^2}$$

$$w_{\text{net}} = 352.4 \text{ Btu/lb.}$$

Heat (q_h) is supplied at the boiler to vaporize the liquid water to steam. This occurs between points 1 and 2. Thus,

$$q_h = h_2 - h_1.$$

The enthalpy of the liquid at the inlet to the boiler is

$$h_1 = h_4 - w_{\text{pump}} = h_4 - [-v_4(p_1 - p_4)].$$

Hence,

$$q_h = h_2 - h_4 - [v_4(p_1 - p_4)]$$

$$q_h = (1203.3 - 69.72) \text{ Btu/lb} - 0.01614\frac{\text{ft}^3}{\text{lb}} \times (300 - 1) \frac{\text{lbf}}{\text{in.}^2} \times \frac{32.2 \text{ lb/ft/s}^2}{1 \text{ lbf}}$$

$$\times \left(\frac{12 \text{ in.}}{1 \text{ ft}}\right)^2 \times \frac{1 \text{ Btu/lb}}{25037 \text{ ft}^2/\text{s}^2}$$

$$q_h = 1132.7 \text{ Btu/lb}.$$

The first-law cycle efficiency is

$$\eta_{\text{cyc}} = \frac{352.4 \text{ Btu/lb}}{1132.7 \text{ Btu/lb}}$$

$$\eta_{\text{cyc}} = 31.1\%.$$

With Steam Extraction

The first-law efficiency of the cycle is still

$$\eta_{\text{cyc}} = \frac{w_{\text{net}}}{q_h}.$$

The use of regenerative feedwater heaters makes the problem more complicated since steam is bled from the turbine at an intermediate stage. Each point in the cycle will be analyzed.

Point 1: Steam at 300 psia. Use tables.

$$h_1 = 1203.3 \text{ Btu/lb and } s_1 = 1.51 \text{ Btu/(lb °F)}.$$

Point 2: Steam at 60 psia after isentropic expansion. Use the Mollier diagram.

$$h_2 \approx 1075 \text{ Btu/lb and } s_2 = 1.51 \text{ Btu/(lb °F)}.$$

Point 3: Steam at 1 psia after isentropic expansion. Use the Mollier diagram.

$$h_3 \approx 850 \text{ Btu/lb and } s_3 = 1.51 \text{ Btu/(lb °F)}.$$

Point 4: Liquid water at 1 psia. Use tables.

$$h_4 = 69.72 \text{ Btu/lb, } s_4 = 0.133 \text{ Btu/(lb °F), and } v_4 = 0.01614 \text{ ft}^3/\text{lb}$$

Point 5: Liquid water at the condensate pump exit:

$$h_5 = h_4 - w_{con.pump} = h_4 - [-v_4(p_5 - p_4)]$$

$$h_5 = 69.72 \text{ Btu/lb} + 0.01614 \, \frac{\text{ft}^3}{\text{lb}} \times (60 - 1) \, \frac{\text{lbf}}{\text{in.}^2} \times \frac{32.2 \text{ lb/ft/s}^2}{1 \text{ lbf}}$$

$$\times \left(\frac{12 \text{ in.}}{1 \text{ ft}}\right)^2 \times \frac{1 \text{ Btu/lb}}{25037 \text{ ft}^2/\text{s}^2}$$

$$h_5 = 69.9 \text{ Btu/lb}.$$

Note that $p_5 = 60$ psia to match $p_2 = 60$ psia. This will prevent backflow of liquid into the discharge line of the condensate pump.

Point 6: Liquid water at the open feedwater heater exit. Use tables.

At this point, the pressure is 60 psia. The enthalpy, h_6, is a sum of the enthalpies of the fluid masses from the condensate pump and the bleed line (line 2) from the turbine. Therefore,

$$\Delta m_6 h_6 = \Delta m_2 h_2 + \Delta m_5 h_5.$$

The mass is $\Delta m_6 = \Delta m_2 + \Delta m_5$.

Then,

$$\Delta m_6 h_6 = \Delta m_2 h_2 + (\Delta m_6 - \Delta m_2) h_5$$

$$h_6 = \frac{\Delta m_2}{\Delta m_6} h_2 + \left(1 - \frac{\Delta m_2}{\Delta m_6}\right) h_5.$$

From tables, $h_6 = 262.2$ Btu/lb and $v_6 = 0.01738$ ft^3/lb.
The mass flow ratio, $\frac{\Delta m_2}{\Delta m_6}$, is

$$262.2 \text{ Btu/lb} = \frac{\Delta m_2}{\Delta m_6} (1075 \text{ Btu/lb}) + \left(1 - \frac{\Delta m_2}{\Delta m_6}\right) (69.9 \text{ Btu/lb})$$

$$\frac{\Delta m_2}{\Delta m_6} = 0.191.$$

Point 7: Liquid Water at the Feedwater Pump Exit:

$$h_7 = h_6 - w_{feed.pump} = h_6 - [-v_6(p_7 - p_1)]$$

$$h_7 = 262.2 \text{ Btu/lb} + 0.01738 \, \frac{\text{ft}^3}{\text{lb}} \times (300 - 60) \, \frac{\text{lbf}}{\text{in.}^2} \times \frac{32.2 \text{ lb/ft/s}^2}{1 \text{ lbf}}$$

$$\times \left(\frac{12 \text{ in.}}{1 \text{ ft}}\right)^2 \times \frac{1 \text{ Btu/lb}}{25037 \text{ ft}^2/\text{s}^2}$$

$$h_7 = 262.97 \text{ Btu/lb}.$$

All the enthalpies are known. The turbine work output is

$$w_{turbine} = h_1 - \frac{\Delta m_2}{\Delta m_6} h_2 - \left(1 - \frac{\Delta m_2}{\Delta m_6}\right) h_3$$

$$w_{turbine} = 1203.3 \text{ Btu/lb} - (0.191)(1075 \text{ Btu/lb}) - (1 - 0.191)(850 \text{ Btu/lb})$$

$$w_{turbine} = 310.3 \text{ Btu/lb}.$$

The total pump work is

$$w_{pump} = w_{con.pump} + w_{feed.pump}$$

$$w_{pump} = \left(1 - \frac{\Delta m_2}{\Delta m_6}\right)(h_5 - h_4) + (h_7 - h_6)$$

$$w_{pump} = (1 - 0.191)(69.9 - 69.72) \text{ Btu/lb} + (262.97 - 262.2) \text{ Btu/lb}$$

$$w_{pump} = -0.916 \text{ Btu/lb}.$$

Note that the negative sign indicates that the system inputs energy to drive the pumps. The heat supplied at the boiler is

$$q_h = h_1 - h_7 = (1203.3 - 262.97) \text{ Btu/lb} = 940.3 \text{ Btu/lb}.$$

The first-law cycle efficiency is

$$\eta_{cyc} = \frac{(310.3 - 0.916) \text{ Btu/lb}}{940.3 \text{ Btu/lb}}$$

$$\eta_{cyc} = 32.9\%.$$

With regeneration, the first-law cycle efficiency is increased.

6.4 Closed Feedwater Heaters (Surface Heaters)

In closed feedwater heaters (typically, shell-and-tube heat exchangers), condensate water (feedwater) flows through tubes in the heater, and energy is absorbed from condensing steam in the shell. The closed feedwater heater may have three main heat exchanger zones: **the desuperheating zone, the condensing zone,** and **the drain cooling zone**. In the desuperheating zone, superheated steam will be cooled to about 50°F above the saturation temperature before entering the condensing zone. The liquid condensate may be cooled in a **drain cooler** to recover additional energy. Figure 6.6 shows two types of drain disposals for closed feedwater heaters. Liquid water from the drain cooler may be pumped to the main water that is routed toward the boiler (Figure 6.6a) or it may be throttled back to a LP heater or to the condenser (Figure 6.6b). Some feedwater heater designs may be devoid of drain coolers. In which case, the liquid condensate will leave the feedwater at the saturation temperature.

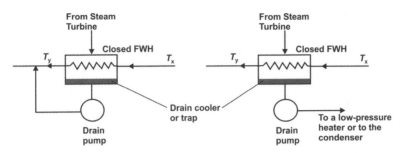

Figure 6.6 Drain disposals for closed feedwater heaters (surface heaters)

To analyze the plant system completely, the enthalpies of the water exiting the piping system and the feedwater heater will be needed. To find the enthalpies, the water temperatures will be needed. Figure 6.6 shows that the temperature of the inlet water to the feedwater pipe is denoted as T_x and for the exiting pipe water, the temperature is denoted as T_y.

Two methods are available to find T_x and T_y: the **drain cooler approach** (**DCA**) and the **terminal temperature difference** (**TTD**).

For the case of drain cooling, the DCA is

$$DCA = T_{\text{drain}} - T_x, \tag{6.2}$$

where

T_{drain} = temperature of the water leaving the feedwater heater drain cooler;
$\quad T_x$ = inlet temperature of the water to the feedwater heater pipe.

For the TTD,

$$TTD = T_{\text{sat}} - T_y, \tag{6.3}$$

where

T_{sat} = saturation temperature at the operating pressure of the condensing zone;
$\quad T_y$ = exit temperature of the water from the feedwater heater pipe.

Practical Note 6.2   DCA and TTD Values

The performance of a feedwater heater is specified by the values of *DCA* and *TTD*. Lower *DCA* and *TTD* indicate greater heat transfer. In practice, the *DCA* will range from 10°F to 20°F for an internal drain cooler. The *TTD* will range from −3°F to 10°F. Negative values of *TTD* are possible for feedwater heaters with desuperheating zones. As the superheated steam exchanges heat with the feedwater exiting the heater, the exit water temperature may exceed the saturation temperature at the operating pressure of the condensing zone to produce a negative *TTD* value. Typically, though, *TTD* values will be on the order of 2°F to 10°F, with 2°F being a practical lower limit for a feedwater heater without a desuperheating zone. A desuperheating zone may not be necessary for extraction steam with less than 100°F superheat.

6.5 The Steam Turbine

The steam turbine is the component of the plant that generates the power that is converted to electrical power by the generator. There are two types of steam turbines: **impulse** and **reaction** turbines.

In **impulse turbines**, steam expands in stationary nozzles to attain very high velocities, and then flows over the moving turbine blades. This converts kinetic energy to mechanical work. In **reaction turbines**, steam expands in both stationary nozzles and across the moving blades.

In practice, turbines used in power generation will have both impulse and reaction sections. Figure 6.7 shows diagrams of operation of an impulse and reaction turbine. Note that the steam pressure decreases across the turbine.

Steam turbines have many **stages**. Each stage generally consists of one row of stationary nozzles and one row of moving curved blades to convert a certain amount of thermal energy into mechanical work.

Practical Note 6.3   Stages of a Steam Turbine

The term "stages" are used to define sections of an individual steam turbine. However, a steam turbine may be subdivided into **HP, IP**, and **LP** sections. These may also be referred to as stages.

6.5.1 Steam-Turbine Internal Efficiency and Exhaust End Losses

The performance of a steam-turbine unit is usually measured by its internal efficiency. Equation (6.1) may be expanded to

$$\eta_i = \frac{h_{in} - h_{out}}{h_{in} - h_{out,s}} = \frac{\sum (\Delta W)}{(\Delta H)_s}, \tag{6.4}$$

where
ΔW = sum of the internal work from all the turbine stages;
ΔH_s = isentropic enthalpy drop for the turbine.

In practice, the internal efficiency of a steam turbine does not include the loss at the turbine exhaust end. **Exhaust end loss** occurs between the last stage of the LP turbine and the condenser inlet. The exhaust end loss represents the amount of steam kinetic energy that is not imparted to the turbine shaft, and is wasted. The exhaust end loss includes the following:

(i) **Actual exit losses** from steam exiting the turbine into the pipe that connects the turbine to the condenser.
(ii) **Gross hood losses**. This represents losses in the exhaust hood that connects the turbine exit to the condenser.
(iii) **Annulus-restriction losses** and **turn-up losses**. The exhaust hood must turn up 90° to allow flow of steam to the condenser. This will result in additional losses.

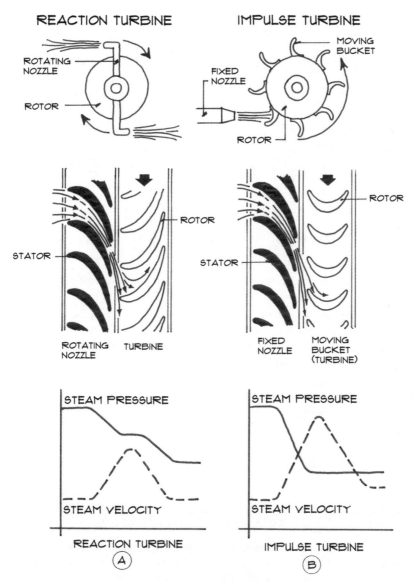

Figure 6.7 Turbine operation

Figure 6.8 shows a schematic of exhaust diffuser of a LP turbine. Note that the steam turns up and is constrained in an annulus before discharge to the condenser.

The exhaust end loss depends heavily on the absolute steam velocity at the last stage of the turbine. The exhaust end loss can be reduced by reducing the absolute steam velocity by (i) increasing the last stage blade length and (ii) increasing the number of steam flows. So, **larger exhaust ends will improve the steam-turbine performance**. As always, the **design engineer** must make a reasonable compromise

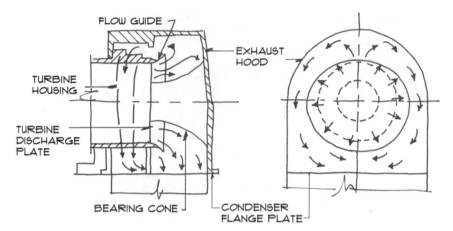

Figure 6.8 Exhaust diffuser of an LP turbine

between increased size for increased performance with the additional costs that will be incurred.

Because of the exhaust end losses, the steam conditions (in particular, the enthalpy) at the exit of the LP turbine and the inlet of the condenser will be different. The steam conditions at the exit of the last stage of the LP turbine are referred to as those at the **expansion-line-end point** (**ELEP**). Steam conditions at the condenser inlet are those at the **used-energy-end point** (**UEEP**). The exhaust end losses (loss in the steam kinetic energy) are such that the enthalpy at the UEEP (condenser inlet) will be larger than the enthalpy at the ELEP (LP turbine outlet).

Therefore,

$$\text{Exhaust end loss} = h_{\text{i,condenser}} - h_{\text{e,turbine}} = \text{UEEP} - \text{ELEP}. \qquad (6.5)$$

Practical Note 6.4   Exhaust End Loss

Typically, the exhaust end loss is approximately 1–3% of the LP turbine work output at **full-load operation**. Full-load operation refers to the maximum amount of power that the turbine is designed to produce. **Part-load operation** may occur during periods of low power demand.

6.5.2 Casing and Shaft Arrangements of Large Steam Turbines

The physical size of steam turbines depends, in part, on the amount of power that they are required to deliver. Their capacities range from a few kilowatts to over 1000000 kW (1000 MW). The inlet pressures to the turbine may be as high as supercritical pressures, and the temperatures may be over 1000°F. Generator drive speeds are typically 3600 and 1800 rpm; geared units can be 10000 rpm or higher. Figure 6.9 shows examples

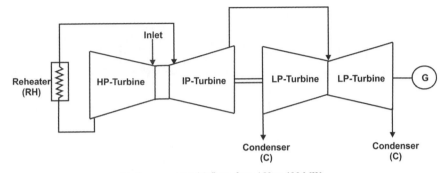

Tandem-compound 2 flows from 150 to 400 MW

(*a*)

Tandem-compound 4 flows from 300 to 800 MW

(*b*)

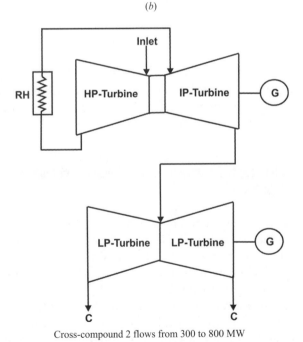

Cross-compound 2 flows from 300 to 800 MW

(*c*)

Figure 6.9 Casing and shaft arrangements for large condensing turbines. (a) Tandem-compound 2 flows from 150 to 400 MW; (b) Tandem-compound 4 flows from 300 to 800 MW; (c) Cross-compound 2 flows from 300 to 800 MW; (d) Cross-compound 4 flows from 800 to 1200 MW

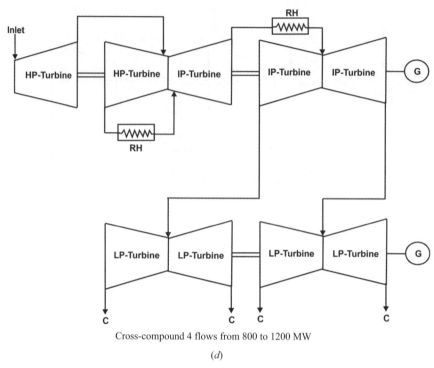

Cross-compound 4 flows from 800 to 1200 MW

(d)

Figure 6.9 (Continued)

of casing and shaft arrangements for large condensing turbines. The figure shows that the cross-compound arrangements produce large amounts of power. They are typically only considered for high-capacity machines.

6.6 Turbine-Cycle Heat Balance and Heat and Mass Balance Diagrams

The law of conservation of energy and mass can be used to show the flow of steam and energy in a steam-turbine power cycle. Heat and mass balances are used to determine the performance/output of the system. A **heat and mass balance diagram** shows a diagrammatic representation of mass flows, pressures, temperatures, enthalpies, and cycle arrangement of the components.

Consider an actual steam-turbine cycle for a conventional fossil-fuel power plant. The plant has one stage of steam reheat between the HP and IP turbines and seven stages of regenerative feedwater heating. The initial pressure to the HP turbine is 3515 psia and the initial temperature is 1000°F. The condenser pressure is set to 2.5 in. Hg. abs. $\approx$ 1.2 psia. After an analysis of the system, a heat and mass balance diagram was generated. Figure 6.10 shows the details of the diagram.

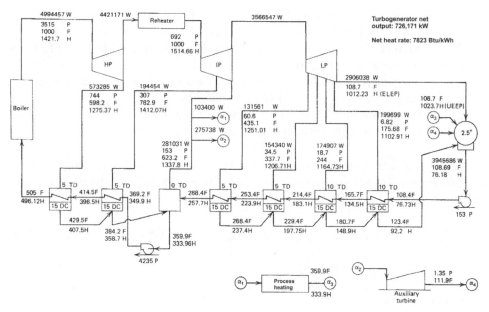

Figure 6.10 Heat-and-mass balance diagram for a fossil-fuel power plant (Li and Priddy (1); reprinted with permission)

The following points are made regarding the heat and mass balance diagram:

(i) There is a single stage of steam reheat between the HP and IP turbines.
(ii) There are seven stages of regenerative feedwater heating. There are six closed feedwater heaters and one open feedwater heater.
(iii) There is a three-stage turbine subsystem—HP, IP, and LP.
(iv) The condenser pressure is set to 2.5 in. Hg. abs. ≈ 1.2 psia.
(v) The following legend may be used:

> **W** = mass flow rate (lb/h)
>
> **P** = pressure (psia)
>
> **F** = temperature (°F)
>
> **H** = enthalpy (Btu/lb)
>
> **TD** = temperature difference in the TTD method
>
> **DC** = temperature difference for the DCA

(vi) An auxiliary steam turbine is used to drive the feedwater heater pump, located at the open feedwater heater. A higher pressure (4235 psia) is required at this point in the cycle.
(vii) Pressure drops occur in the steam piping. At the feedwater heater pump, the discharge pressure is 4235 psia. However, steam exiting the boiler has a pressure

of 3515 psia (as required by the plant). A pressure drop of 720 psia or 17% has occurred in the boiler. A pressure drop also occurred across the reheater.

(viii) The exhaust end loss can be determined. Based on the UEEP and ELEP, the exhaust end loss is 11.5 Btu/lb.

(ix) The **net generator output** or the **turbogenerator net output** supplied by the generator is shown as 726171 kW (power in kW).

(x) The **net heat rate** (NHR) is also shown (7823 Btu/kWh). This is the ratio of the total heat input to the turbogenerator net output.

The **net generator output** includes the network from the cycle, the exhaust end loss, and the efficiencies of the generator and the turbine–generator mechanical coupling. So, for the plant,

$$\dot{W}_{gen} = \eta_{coupling} \eta_{generator} \dot{m}_{steam} \left(w_{cycle} - LOSS_{exhaust} \right). \qquad (6.6)$$

Typically, $\eta_{coupling} = 100\%$ and $\eta_{generator} = 95\%$–98.5%.

The *NHR* is the ratio of the **total heat input** (from all sources: boiler, reheater, etc.) to the net generator output.

For electric motor-driven feedwater pumps,

$$NHR = \frac{\text{total heat input}}{\text{generator output} - \text{electric power for feedwater pumps}}. \qquad (6.7)$$

For shaft-driven or auxiliary turbine-driven feedwater pumps,

$$NHR = \frac{\text{total heat input}}{\text{net generator output}}. \qquad (6.8)$$

It can be seen from Equations (6.7) and (6.8) that lower values of *NHR* indicate improved performance of the power plant since low heat inputs produce high net generator outputs.

Practical Note 6.5   Units of the Net Heat Rate (NHR)

The *NHR* must be reported in Btu/kWh.

6.7 Steam-Turbine Power Plant System Performance Analysis Considerations

Real steam-turbine power plants may become complex with complicated piping arrangements and many different components. The following points may be considered when analyzing the performance of or designing a power plant system:

(i) An appropriate (real) vapor cycle typically consists of a multistage regenerative Rankine cycle with interstage reheating at the turbine.

(ii) Multiple feedwater heaters are usually included to improve the plant efficiency. The total number must be optimized by considering improvements in the plant efficiency and the increased installation and operating costs.

(iii) One stage of reheat is typical. Multiple stages of reheat are not economical.

(iv) Steam is usually expanded to about 1 psia $\approx$ 2.0 in. Hg. abs. in the steam turbine before condensation in the condenser. This will increase the work output from the turbine subsystem.

(v) The net cycle output should be reported in Btu/h or kW.

(vi) During design and determination of the performance of a steam-turbine power plant, the **design engineer** must consider the

(a) main steam pressure and temperature inlet to the steam turbine;

(b) reheat steam temperature and pressure;

(c) boiler, reheater, and pipe pressure drops;

(d) condenser pressure;

(e) number of feedwater heaters (stages of regenerative feedwater heating);

(f) type of feedwater heaters (open or closed feedwater heaters);

(g) heat drain disposal scheme. Will the feedwater heaters drain to other LP feedwater heaters or directly to the condenser? Will drain coolers be used?

(h) DCA;

(i) heater TTD;

(j) steam extraction for auxiliary turbines;

(k) steam extraction for industrial use and heating;

(l) exhaust end loss at the LP turbine.

Practical Note 6.6   How Does One Initiate Operation of a Power Plant System?

Black-start is the process of restoring a power plant station to operation without relying on external energy sources such as from an electric grid. Batteries may be used to start small auxiliary (diesel) generators. These are used to provide power to run the working fluid through bypass lines to ensure flow. For steam power plants, steam may be generated to the required pressure and temperature by burning the fuel. Blowers for forcing combustion gases over the boiler heat exchanger surfaces would receive power from the generators.

The hot, pressurized steam would flow to the steam turbine to initiate power generation. Hydroelectric plants can also be designated as black-start sources.

Example 6.2   Performance Analysis of a Steam-Turbine Power Plant

In a steam-turbine power plant, steam enters a two-stage turbine at 1250 psia and 950°F. There are two extraction points at 100 and 25 psia to two stages of regenerative feedwater heating. The turbine internal efficiency is 90% and the pump efficiency is 85%. The boiler pressure drop is 21%. The steam flow rate at the turbine inlet is 1300000 lb/h. Estimate the performance of this plant.

Further Information: This performance analysis should address and present the following points:

  (i) Net turbogenerator output
 (ii) First-law efficiency of the plant
(iii) NHR
(iv) A heat and mass balance diagram

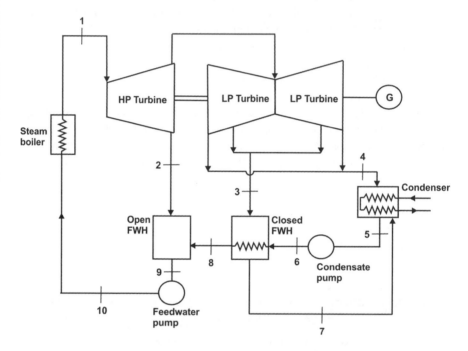

Solution. The performance of this plant will include information on the (i) net turbogenerator output of the plant, (ii) first-law efficiency of the plant, (iii) NHR, and (iv) a heat balance diagram showing the mass flow rates, pressures, temperatures, and enthalpies of the working fluid at different points in the cycle.

Assumptions: (i) The steam finally exhausts to the condenser at 1 psia (typical).
  (ii) The exhaust end loss is 3% of the output of the steam turbine.

Study Each Point in the Cycle

Point 1: Superheated steam at 1250 psia and 950°F. Consult tables.
  With interpolation, $h_1 = 1469$ Btu/lb and $s_1 = 1.60$ Btu/(lb R)
Point 2: Saturated steam at 100 psia after expansion in the HP turbine.
  The entropies at points 1 and 2 are equal for an isentropic expansion: $s_{1s} = s_{2s} = 1.60$ Btu/(lb R).

The quality of the steam at this point is

$$x_2 = \frac{s_{2s} - s_{fs}}{s_{fg,s}} = \frac{(1.60 - 0.474)\ \text{Btu}/(\text{lb R})}{1.129\ \text{Btu}/(\text{lb R})} = 0.997.$$

The enthalpy for the isentropic process is

$$h_{2s} = h_{fs} + x_2 h_{fg,s} = 298.51\ \text{Btu/lb} + (0.997)(888.9)\ \text{Btu/lb} = 1185\ \text{Btu/lb}.$$

The actual enthalpy is calculated by considering the turbine internal efficiency:

$$\eta_t = \frac{h_1 - h_2}{h_1 - h_{2s}}.$$

Therefore,

$$h_2 = h_1 - \eta_t\,(h_1 - h_{2s}) = 1469\ \text{Btu/lb} - 0.90\,(1469 - 1185)\ \text{Btu/lb}$$
$$h_2 = 1213\ \text{Btu/lb}.$$

Point 3: Saturated steam at 25 psia after expansion in the LP turbine.
The entropies at points 1 and 3 are equal for an isentropic expansion: $s_{1s} = s_{3s} = 1.60$ Btu/(lb R).
The quality of the steam at this point is

$$x_3 = \frac{s_{3s} - s_{fs}}{s_{fg,s}} = \frac{(1.60 - 0.354)\ \text{Btu}/(\text{lb R})}{1.361\ \text{Btu}/(\text{lb R})} = 0.916.$$

The enthalpy for the isentropic process is

$$h_{3s} = h_{fs} + x_3 h_{fg,s} = 208.52\ \text{Btu/lb} + (0.916)(952.03)\ \text{Btu/lb} = 1080\ \text{Btu/lb}.$$

The actual enthalpy is calculated by considering the turbine internal efficiency:

$$\eta_t = \frac{h_1 - h_3}{h_1 - h_{3s}}.$$

Therefore,

$$h_3 = h_1 - \eta_t\,(h_1 - h_{3s}) = 1469\ \text{Btu/lb} - 0.90\,(1469 - 1080)\ \text{Btu/lb}$$
$$h_3 = 1119\ \text{Btu/lb}.$$

Point 4: Saturated steam at 1 psia after complete expansion in the LP turbine.
The entropies at points 1 and 4 are equal for an isentropic expansion: $s_{1s} = s_{4s} = 1.60$ Btu/(lb R).

The quality of the steam at this point is

$$x_4 = \frac{s_{4s} - s_{fs}}{s_{fg,s}} = \frac{(1.60 - 0.133) \text{ Btu/(lb R)}}{1.845 \text{ Btu/(lb R)}} = 0.795.$$

The enthalpy for the isentropic process is

$$h_{4s} = h_{fs} + x_4 h_{fg,s} = 69.72 \text{ Btu/lb} + (0.795)(1035.7) \text{ Btu/lb} = 893 \text{ Btu/lb}.$$

The actual enthalpy is calculated by considering the turbine internal efficiency:

$$\eta_t = \frac{h_1 - h_4}{h_1 - h_{4s}}.$$

Therefore,

$$h_4 = h_1 - \eta_t (h_1 - h_{4s}) = 1469 \text{ Btu/lb} - 0.90 (1469 - 893) \text{ Btu/lb}$$
$$h_4 = 951 \text{ Btu/lb}$$

Alternate Approach to Find h_2, h_3, and h_4 Using the Mollier or T-s Diagrams for Water

Use of the Mollier or T-s diagram will provide estimates of the enthalpies for steam extraction from the turbine.

This is mutistage steam turbine with a HP and a LP section. Focus on the HP turbine. Identify the conditions at the inlet of the turbine at 1250 psia and 950°F:

$$h_1 \sim 1470 \text{ Btu/lb and } s_1 \sim 1.61 \text{ Btu/(lb R)}.$$

Identify the conditions at point 2 after isentropic expansion to 100 psia:
On the Mollier diagram, go down to $s_{2s} \sim 1.61$ Btu/(lb R) and 100 psia:

$$h_{2s} \sim 1195 \text{ Btu/lb and } s_{2s} \sim 1.61 \text{ Btu/(lb R)}.$$

Calculate the real enthalpy from the internal efficiency of the turbine:

$$\eta_t = \frac{h_1 - h_2}{h_1 - h_{2s}}.$$

Therefore,

$$h_2 = h_1 - \eta_t \, (h_1 - h_{2s}) = 1470 \text{ Btu/lb} - 0.90 \, (1470 - 1195) \text{ Btu/lb}$$
$$h_2 \sim 1223 \text{ Btu/lb.}$$

Note that this value deviates about 0.8% from that found after calculating the quality. Connect points 1 and 2 on the Mollier diagram with a straight line. The steam at point 2 is superheated (see Mollier diagram). The temperature is $T_2 \sim 390°$F.

Note: The Mollier diagram also shows the amount of superheat of the steam. At point 2, the steam has approximately 65°F superheat. This constant superheat is the temperature difference between the saturation temperature at the operating pressure (100 psia) and the actual temperature. Thus, $T_2 \sim 328°$F $+ \, 65°$F $\sim 393°$F.

For this problem, the inlet conditions to the LP turbine are the same as the exit conditions from the HP turbine, and will be used as the starting point of the expansion line for the LP turbine. Identify the conditions at point 4 after isentropic expansion to 1 psia:

On the Mollier diagram, go down to $s_{4s} \sim 1.61$ Btu/lb-R and 1 psia.

$$h_{4s} \sim 895 \text{ Btu/lb and } s_{4s} \sim 1.61 \text{ Btu/(lb R)}$$

Calculate the real enthalpy from the internal efficiency of the turbine:

$$\eta_t = \frac{h_1 - h_4}{h_1 - h_{4s}}.$$

Therefore,

$$h_4 = h_1 - \eta_t \, (h_1 - h_{4s}) = 1470 \text{ Btu/lb-}0.90 \, (1470 - 895) \text{ Btu/lb}$$
$$h_4 \sim 953 \text{ Btu/lb} = \text{ELEP.}$$

This is at 1 psia. Identify this point on the Mollier diagram. Point 4 lies below the saturation line. Therefore, the saturation temperature at this point is found from the steam tables at 1 psia. Hence, $T_4 = 101.69°$F.

Connect points 2 and 4 on the Mollier diagram with a straight line. Identify the conditions at point 3 (25 psia), along the line.

Therefore,

$$h_3 \sim 1130 \text{ Btu/lb and } T_3 = 240°\text{F (from the steam tables).}$$

This value of the enthalpy at point 3 deviates about 1% from that found after calculating the quality.

Point 5: Saturated water at 1 psia. Consult tables.

$h_5 = 69.72$ Btu/lb and $v_5 = 0.01614$ ft^3/lb.

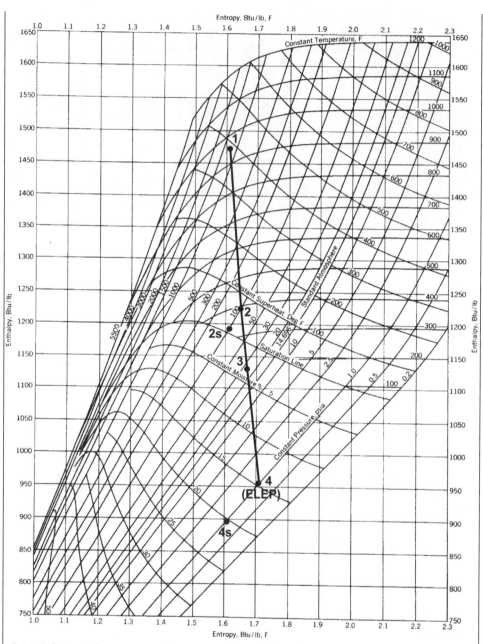

Point 6: Saturated water at condensate pump exit. Consult tables.

The pressure at this point is 100 psia to match the pressure requirement in the open (contact) feedwater heater:

$$h_6 = h_5 - \left[-\frac{v_5(p_6 - p_5)}{\eta_p}\right] = 69.72 \text{ Btu/lb} + 0.01614 \frac{\text{ft}^3}{\text{lb}}(100-1)\frac{\text{lbf}}{\text{in.}^2}$$

$$\times \frac{32.2 \text{ lb/ft}}{\text{s}^2 - \text{lbf}} \times \frac{(12 \text{ in.})^2}{1 \text{ ft}^2} \times \frac{1 \text{ Btu/lb}}{25037 \text{ ft}^2/\text{s}^2} \times \frac{1}{0.85}$$

$$h_6 = 70.07 \text{ Btu/lb}.$$

Point 7: Saturated water in the drain line of the closed feedwater heater
Observation of the schematic presented in the problem preamble suggests that the closed (surface) feedwater heater comes complete with an internal drain and drain cooler. Therefore, the DCA can be used to find the temperature at point 7. This temperature will be used to find the enthalpy:

$$DCA = T_{\text{drain}} - T_x = T_7 - T_6$$

$$T_7 = DCA + T_6.$$

For internal drain coolers, the DCA range is 10–20°F. Assume the DCA to be 15°F. The T_6 temperature is the same as the T_4 temperature.
Therefore,

$$T_7 = (15 + 101.69)°\text{F} = 117°\text{F}.$$

From tables and after interpolation,

$$h_7 = 85 \text{ Btu/lb}.$$

Point 8: Saturated water inlet to the contact (open) feedwater heater
The TTD can be used to find T_8:

$$TTD = T_{\text{sat}} - T_y = T_{3,\text{sat}} - T_8.$$

The saturation temperature of the working fluid in the condensing zone of the feedwater heater at 25 psia is 240°F. Heat exchangers tend to be inefficient due to fouling and real heat losses. Therefore, assume the TTD to be 10°F to be conservative.
Therefore,

$$T_8 = T_{3,\text{sat}} - TTD = (240 - 10)°\text{F} = 230°\text{F}.$$

From tables and after interpolation,

$$h_8 = 198 \text{ Btu/lb}.$$

Point 9: Saturated water at the exit of the contact (open) feedwater heater
The pressure at this point is 100 psia. So,

$h_9 = 298.5$ Btu/lb and $v_9 = 0.01774$ ft^3/lb.

Point 10: Saturated water from the pump

The water in the pipes that pass through the boiler will experience a 21% pressure drop. Therefore, p_{10} must be greater than 1250 psia to compensate for this loss. Hence, $p_{10} = 1583$ psia:

$$h_{10} = h_9 - \left[-\frac{v_9 (p_{10} - p_9)}{\eta_p}\right] = 298.5 \text{ Btu/lb} + 0.01774 \frac{\text{ft}^3}{\text{lb}} (1583 - 100) \frac{\text{lbf}}{\text{in.}^2}$$

$$\times \frac{32.2 \text{ lb/ft}}{\text{s}^2 - \text{lbf}} \times \frac{(12 \text{ in.})^2}{1 \text{ ft}^2} \times \frac{1 \text{ Btu/lb}}{25037 \text{ ft}^2/\text{s}^2} \times \frac{1}{0.85}$$

$$h_{10} = 304.23 \text{ Btu/lb}$$

The total turbine work is the sum of the HP and LP turbine work outputs:

$$\dot{w}_t = \dot{w}_{HP} + \dot{w}_{LP}.$$

The HP and LP turbine work outputs are

$$\dot{w}_{HP} = \dot{m}_1 h_1 - \dot{m}_2 h_2 - \dot{m}_{LP} h_2$$
$$\dot{w}_{HP} = \dot{m}_1 h_1 - \dot{m}_2 h_2 - (\dot{m}_3 + \dot{m}_4) h_2$$
$$\dot{w}_{LP} = (\dot{m}_3 + \dot{m}_4) h_2 - \dot{m}_3 h_3 - \dot{m}_4 h_4 = (h_2 - h_3)\dot{m}_3 + (h_2 - h_4)\dot{m}_4.$$

The mass flow rates at each section of the turbine are needed. Conduct a mass balance:

$$\dot{m}_9 = \dot{m}_2 + \dot{m}_8$$
$$\dot{m}_1 = \dot{m}_2 + \dot{m}_3 + \dot{m}_4$$
$$\dot{m}_5 = \dot{m}_3 + \dot{m}_4$$
$$\dot{m}_1 = \dot{m}_9 = \dot{m}_{10} = 1300000 \text{ lb/h}$$
$$\dot{m}_3 = \dot{m}_7$$
$$\dot{m}_5 = \dot{m}_6 = \dot{m}_8.$$

Energy balance across the open feedwater heater: $\dot{m}_9 h_9 = \dot{m}_2 h_2 + \dot{m}_8 h_8$. Therefore,

$$\dot{m}_2 + \dot{m}_8 = 1300000 \text{ lb/h}$$
$$(1300000 \text{ lb/h})(298.5 \text{ Btu/lb}) = \dot{m}_2(1223 \text{ Btu/lb}) + \dot{m}_8(198 \text{ Btu/lb})$$

These are two equations with two unknowns:

$$\dot{m}_2 = 127384 \text{ lb/h}$$
$$\dot{m}_8 = 1172616 \text{ lb/h}$$

Across the closed (surface) feedwater heater, the energy balance is

$$\dot{m}_3(h_3 - h_7) = \dot{m}_8(h_8 - h_6)$$

$$\dot{m}_3 = \dot{m}_8 \frac{(h_8 - h_6)}{(h_3 - h_7)}$$

$$\dot{m}_3 = (1172616 \text{ lb/h}) \frac{(198 - 70.07) \text{ Btu/lb}}{(1130 - 85) \text{ Btu/lb}} = 143553 \text{ lb/h.}$$

Therefore,

$$\dot{m}_5 = \dot{m}_8 = \dot{m}_3 + \dot{m}_4$$
$$1172616 \text{ lb/h} = 143553 \text{ lb/h} + \dot{m}_4$$
$$\dot{m}_4 = 1029063 \text{ lb/h.}$$

Check the mass flow rates:

$$\dot{m}_1 = \dot{m}_2 + \dot{m}_3 + \dot{m}_4 = (127384 + 143553 + 1029063) \text{ lb/h} = 1300000 \text{ lb/h.}$$

Now that all the mass flow rates are known, the turbine work outputs can be calculated. For the HP turbine,

$$\dot{w}_{HP} = \dot{m}_1 h_1 - \dot{m}_2 h_2 - (\dot{m}_3 + \dot{m}_4) h_2$$
$$\dot{w}_{HP} = (1300000 \text{ lb/h})(1470 \text{ Btu/lb}) - (127384 \text{ lb/h})(1223 \text{ Btu/lb})$$
$$- (143553 \text{ lb/h} + 1029063 \text{ lb/h})(1223 \text{ Btu/lb})$$
$$\dot{w}_{HP} = 321 \times 10^6 \text{ Btu/h.}$$

For the LP turbine,

$$\dot{w}_{LP} = (h_2 - h_3) \dot{m}_3 + (h_2 - h_4) \dot{m}_4$$
$$\dot{w}_{LP} = (1223 \text{ Btu/lb} - 1130 \text{ Btu/lb})(143553 \text{ lb/h})$$
$$+ (1223 \text{ Btu/lb} - 953 \text{ Btu/lb})(1029063 \text{ lb/h})$$
$$\dot{w}_{LP} = 291 \times 10^6 \text{ Btu/h.}$$

The total turbine work output is shown below.

$$\dot{w}_t = \dot{w}_{HP} + \dot{w}_{LP}$$
$$\dot{w}_t = (321 \times 10^6 + 291 \times 10^6) \text{ Btu/h}$$
$$\dot{w}_t = 612 \times 10^6 \text{ Btu/h} = 179447 \text{ kW} = 180 \text{ MW.}$$

Since the turbine *internal* efficiency was given, the exhaust end loss must be considered to find the net turbine work output. Assume that the exhaust end loss is approximately 3% of the LP turbine work output.

Therefore,

$$\dot{w}_{t,loss} = 0.03\,\dot{w}_{LP} = 0.03(291 \times 10^6 \text{ Btu/h}) = 8.7 \times 10^6 \text{ Btu/h}$$

The net turbine work output is

$$\dot{w}_{t,net} = (612 \times 10^6 - 8.7 \times 10^6) \text{ Btu/h} = 603 \times 10^6 \text{ Btu/h} = 176722 \text{ kW} = 177 \text{ MW}.$$

The enthalpy of the steam entering the condenser is larger than that exiting the turbine. The UEEP is

$$\text{UEEP} = h_{4,condenser} = h_{4,turbine} + \frac{\dot{w}_{t,loss}}{\dot{m}_4} = 953 \text{ Btu/lb} + \frac{8.7 \times 10^6 \text{ Btu/h}}{1029063 \text{ lb/h}} = 962 \text{ Btu/lb}.$$

The total pump work is

$$\dot{w}_p = \dot{m}_6 \frac{v_5\,(p_6 - p_5)}{\eta_p} + \dot{m}_{10}\frac{v_9\,(p_{10} - p_9)}{\eta_p}$$

$$\dot{w}_p = (1172616 \text{ lb/h})\left(0.01614\,\frac{\text{ft}^3}{\text{lb}}\right)(100 - 1)\,\frac{\text{lbf}}{\text{in.}^2} \times \frac{32.2 \text{ lb/ft}}{\text{s}^2 - \text{lbf}} \times \frac{(12 \text{ in.})^2}{1 \text{ ft}^2}$$

$$\times \frac{1 \text{ Btu/lb}}{25037 \text{ ft}^2/\text{s}^2} \times \frac{1}{0.85} + (1300000 \text{ lb/h})\,0.01774\,\frac{\text{ft}^3}{\text{lb}}\,(1583 - 100)\,\frac{\text{lbf}}{\text{in.}^2}$$

$$\times \frac{32.2 \text{ lb/ft}}{\text{s}^2 - \text{lbf}} \times \frac{(12 \text{ in.})^2}{1 \text{ ft}^2} \times \frac{1 \text{ Btu/lb}}{25037 \text{ ft}^2/\text{s}^2} \times \frac{1}{0.85}$$

$$\dot{w}_p = 7.85 \times 10^6 \text{ Btu/h} = 2304 \text{ kW} = 2.30 \text{ MW}.$$

The net output of the cycle is

$$\dot{w}_{net} = \dot{w}_{t,net} - \dot{w}_p = 603 \times 10^6 \text{ Btu/h} - 7.85 \times 10^6 \text{ Btu/h}$$

$$\dot{w}_{net} = 595 \times 10^6 \text{ Btu/h} = 175 \times 10^3 \text{ kW} = 175 \text{ MW}.$$

The heat supplied to the working fluid is

$$\dot{q}_h = \dot{m}_1(h_1 - h_{10})$$

$$\dot{q}_h = (1300000 \text{ lb/h})(1470 - 304.23) \text{ Btu/lb}$$

$$\dot{q}_h = 1516 \times 10^6 \text{ Btu/h} = 444 \times 10^3 \text{ kW} = 444 \text{ MW}.$$

The first-law efficiency of the cycle is

$$\eta_{cyc} = \frac{\dot{w}_{net}}{\dot{q}_h} = \frac{595 \times 10^6 \text{ Btu/h}}{1516 \times 10^6 \text{ Btu/h}}$$

$$\eta_{cyc} = \textbf{39.3\%}.$$

The net output of the plant/turbogenerator is (assuming $\eta_{generator} = 95\%$):

$$\dot{W}_{net,plant} = \eta_{coupling}\eta_{generator}\dot{w}_{t,net}$$

$$\dot{W}_{net,plant} = (1.0)(0.95)(603 \times 10^6 \text{ Btu/h})$$

$$\dot{W}_{net,plant} = \textbf{573} \times \textbf{10}^6 \textbf{ Btu/h} = \textbf{167930 kW} = \textbf{168 MW}.$$

The NHR is

$$NHR = \frac{1516 \times 10^6 \text{ Btu/h}}{167930 \text{ kW}}$$

$$NHR = 9028 \text{ Btu/kWh}.$$

The heat and mass balance diagram is shown below.

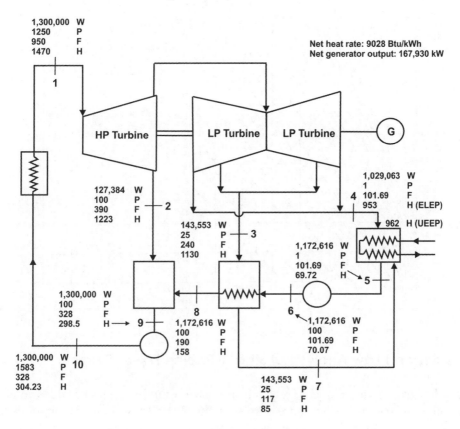

The Mollier diagram showing the uesd-energy end point (UEEP) is presented below:

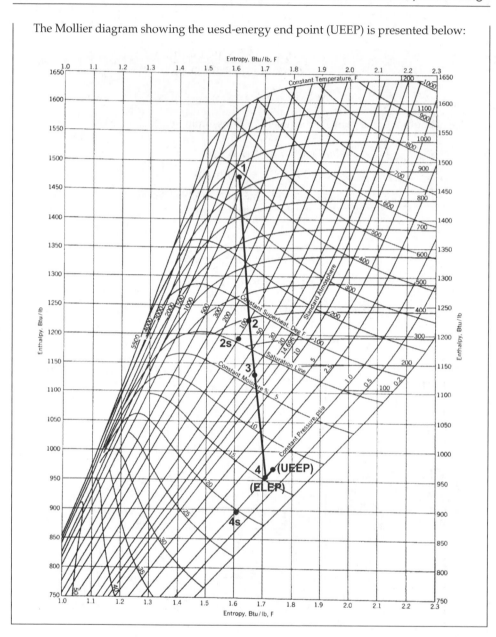

6.8 Second-Law Analysis of Steam-Turbine Power Plants

First-law (conservation of energy) analysis of steam-turbine cycles provides an indication of the overall performance of the system. **Second-law analysis** pinpoints the losses as well as the efficiencies of each component in the system. This type of analysis is useful for identifying components with large losses to target areas that need further improvement.

Second-law analysis is based on the concept of **availability** (exergy, available energy, useful energy). Availability is reversible work produced when the cycle working fluid changes to the state that is in thermal and mechanical equilibrium with the environment or surroundings. In other words, availability can be considered as a property that measures the capacity of the working fluid to do work under environmental conditions.

Therefore,

$$a = \Delta w_{rev}, \tag{6.9}$$

where
a = availability;
w_{rev} = reversible work.

Assuming no kinetic and potential energies, the first law gives

$$\Delta W_{rev} = \Delta m_i h_i - \Delta m_e h_e - \Delta Q_{rev}, \tag{6.10}$$

where
subscript "i" = the initial state;
subscript "e" = the end state;
Q_{rev} = heat transferred.

The second law gives

$$\Delta m_i s_i - \Delta m_e s_e - \frac{\Delta Q_{rev}}{T_0} = 0, \tag{6.11}$$

where T_0 is reference temperature of the surrounding environment.

Therefore,

$$\Delta Q_{rev} = T_0 \left(\Delta m_i s_i - \Delta m_e s_e \right). \tag{6.12}$$

Then,

$$\Delta W_{rev} = \left(\Delta m_i h_i - \Delta m_e h_e \right) - T_0 \left(\Delta m_i s_i - \Delta m_e s_e \right). \tag{6.13}$$

For steady state, steady processes, $\Delta m_i = \Delta m_e = \Delta m$. So,

$$\Delta w_{rev} = \frac{\Delta W_{rev}}{\Delta m} = (h_i - h_e) - T_0 (s_i - s_e). \tag{6.14}$$

With the environment as a reference and the definition of availability,

$$a = (h - h_0) - T_0 (s - s_0). \tag{6.15}$$

Practical Note 6.7   Reference Pressure and Temperature for Availability Analysis

Conditions for the reference environment are typically taken at $P_0 = 14.7$ psia and $T_0 = 77°F$.

The reversible work is the maximum amount of work that can be produced in a process, and is usually greater than the actual work output of the real process for the same end states. So, across a component operating between states 1 and 2,

$$\Delta w_{rev} = a_1 - a_2. \tag{6.16}$$

The difference between the reversible and the actual work is the lost work or **availability loss** due to **irreversibility**.

Therefore,

$$\Delta I = \Delta w_{rev} - \Delta w_{act}, \tag{6.17}$$

or

$$\dot{A}_{loss} = \dot{A}_{input} - \dot{A}_{output}. \tag{6.18}$$

With the input and output availabilities known, the **second-law efficiency** of the steam-turbine plant and **effectiveness** of the components can be determined. The effectiveness is a measure of the quality of component design or performance. The second-law efficiency of a plant or a component is

$$\varepsilon = \frac{\dot{A}_{output}}{\dot{A}_{input}}. \tag{6.19}$$

A second-law analysis contributes significantly to understanding of the performance of a plant. Unlike the first-law efficiency, the second-law efficiency takes into account the availability loss and availability destruction across a component. So, ε is typically lower than η.

Example 6.3   Second-Law Analysis of a Steam-Turbine Power Plant

A simple steam-turbine power plant system operates at steady-state, steady-flow conditions. Steam property data are given. Calculate the second-law efficiency for this power-generating system. Calculate the effectiveness of each component.

State	P (psia)	T (°F)	h (Btu/lb)	s (Btu/lb-R)
1	320	Saturated vapor	1204.3	1.5060
2	1		895.3	1.6031
3	1	Saturated vapor	69.73	0.13266
4	320		70.92	0.13290

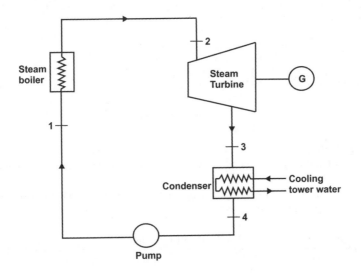

Hard coal is used in the boiler burner. The burner efficiency is approximately 82%.

Solution

System Effectiveness

The second-law efficiency of the system is

$$\varepsilon = \frac{\dot{A}_{\text{output}}}{\dot{A}_{\text{input}}}.$$

For this system, availability output is

$$\dot{A}_{\text{output}} = w_{\text{st}} - w_{\text{p}}.$$

This is the actual work output from the system.
The availability input is

$$\dot{A}_{\text{input}} = q_{\text{h,fuel}}.$$

This is the total heat input to the system. This occurs at the boiler burner, and represents the total heat available from the fuel.
Therefore,

$$\varepsilon = \frac{w_{\text{st}} - w_{\text{p}}}{q_{\text{h,fuel}}}.$$

The steam-turbine work output is

$$w_{\text{st}} = h_1 - h_2 = (1204.3 - 895.3)\ \text{Btu/lb} = 309\ \text{Btu/lb}.$$

The pump work is

$$w_p = h_4 - h_3 = (70.92 - 69.73) \text{ Btu/lb} = 1.19 \text{ Btu/lb}.$$

The total heat available from the fuel is

$$q_{h,fuel} = m_f a_f = m_f(HV_f).$$

The fuel consumption per pound of steam is m_f. Note that the fuel availability is the heating value of the fuel (HV_f). For more conservative analyses, the lower heating value will be used.

Table 5.2 gives the higher heating value of hard coal as $HV_f = 14000$ Btu/lb coal. The fuel consumption per pound of steam is

$$m_f = \frac{h_1 - h_4}{\eta_{burner}(HV_f)} = \frac{(1204.3 - 70.92) \text{ Btu/lb steam}}{0.82\,(14000 \text{ Btu/lb coal})} = 0.0987 \text{ lb coal/lb steam}.$$

Therefore,

$$q_{h,fuel} = (0.0987 \text{ lb coal/lb steam})(14000 \text{ Btu/lb coal})$$

$$q_{h,fuel} = 1382.2 \text{ Btu/lb steam}.$$

The second-law efficiency of the system is

$$\varepsilon = \frac{(309 - 1.19) \text{ Btu/lb steam}}{1382.2 \text{ Btu/lb steam}}$$

$$\varepsilon = 22\%.$$

For comparison, the first-law efficiency is

$$\eta_{cyc} = \frac{w_{st} - w_p}{q_{h,boiler}} = \frac{w_{st} - w_p}{h_1 - h_4}$$

$$\eta_{cyc} = \frac{(309 - 1.19) \text{ Btu/lb steam}}{(1204.3 - 70.92) \text{ Btu/lb steam}} = 27\%.$$

As expected, $\varepsilon > \eta$.

Component Effectiveness—Boiler/Steam Generator

The second-law efficiency of the boiler is

$$\varepsilon_{boiler} = \frac{\dot{A}_{output}}{\dot{A}_{input}}$$

$$\varepsilon_{boiler} = \frac{a_1 - a_4}{m_f(HV_f)}.$$

Find the availabilities at points 1 and 4. Assume that environmental conditions are at $P_0 = 14.7$ psia and $T_0 = 77°F$.
Therefore,

$$a_1 = (h_1 - h_0) - T_0(s_1 - s_0)$$

$$a_1 = (1204.3 - 45.09) \text{ Btu/lb} - (537 \text{ R})(1.5060 - 0.08775) \text{ Btu/(lb R)}$$

$$a_1 = 397.3 \text{ Btu/lb,}$$

and

$$a_4 = (h_4 - h_0) - T_0(s_4 - s_0)$$

$$a_4 = (70.92 - 45.09) \text{ Btu/lb} - (537 \text{ R})(0.13290 - 0.08775) \text{ Btu/lb-R}$$

$$a_4 = 1.585 \text{ Btu/lb.}$$

Hence,

$$\varepsilon_{boiler} = \frac{(397.3 - 1.585) \text{ Btu/lb}}{1382.2 \text{ Btu/lb}}$$

$$\varepsilon_{boiler} = 29\%.$$

The effectiveness of the boiler is very low. In order to improve the overall effectiveness of the system, the performance of the boiler will need to be improved. The availability loss in the boiler is

$$\Delta I = \dot{A}_{loss} = \dot{A}_{input} - \dot{A}_{output} = (1382.2 - 397.3 + 1.585) \text{ Btu/lb}$$

$$\Delta I = \dot{A}_{loss} = 986.5 \text{ Btu/lb.}$$

Of the 1382.2 Btu/lb of energy available from the fuel, 986.5 Btu/lb is lost during operation of the boiler/steam generator.

Component Effectiveness—Steam Turbine

The second-law efficiency of the steam turbine is

$$\varepsilon_{st} = \frac{\dot{A}_{output}}{\dot{A}_{input}}$$

$$\varepsilon_{st} = \frac{w_{st}}{a_1 - a_2} = \frac{h_1 - h_2}{a_1 - a_2}.$$

Find the availability at point 2. Assume that environmental conditions are at $P_0 = 14.7$ psia and $T_0 = 77°F$.
Therefore,

$$a_2 = (h_2 - h_0) - T_0(s_2 - s_0)$$

$$a_2 = (895.3 - 45.09) \text{ Btu/lb} - (537 \text{ R})(1.6031 - 0.08775) \text{ Btu/(lb R)}$$

$$a_2 = 36.47 \text{ Btu/lb.}$$

Hence,

$$\varepsilon_{st} = \frac{309 \text{ Btu/lb}}{(397.3 - 36.47) \text{ Btu/lb}}$$

$$\varepsilon_{st} = 86\%.$$

The effectiveness of the steam turbine is the internal efficiency of the unit. So,

$$\eta_i = \varepsilon_{st} = 86\%.$$

The availability loss in the steam turbine is

$$\Delta I = \dot{A}_{loss} = \dot{A}_{input} - \dot{A}_{output} = (a_1 - a_2) - w_{st} = (360.8 - 309) \text{ Btu/lb}$$

$$\Delta I = \dot{A}_{loss} = 51.8 \text{ Btu/lb}.$$

Component Effectiveness—Condenser

The condenser is a heat exchanger. The effectiveness is determined by considering the availability input from the condensing steam and the availability received by the cooling water in the heat exchanger tubes (availability output).
Therefore,

$$\varepsilon_c = \frac{\dot{A}_{output,cooling \ water}}{\dot{A}_{input,steam}}.$$

For this problem, the inlet and exit conditions of the cooling water are not known, and would be needed to proceed further.

Component Effectiveness—Pump

The second-law efficiency of the pump is

$$\varepsilon_p = \frac{\dot{A}_{output}}{\dot{A}_{input}}$$

$$\varepsilon_{st} = \frac{a_4 - a_3}{w_p} = \frac{a_4 - a_3}{h_4 - h_3}.$$

Find the availability at point 3. Assume that environmental conditions are at $P_0 = 14.7$ psia and $T_0 = 77°F$.
Therefore,

$$a_3 = (h_3 - h_0) - T_0(s_3 - s_0)$$

$$a_3 = (69.73 - 45.09) \text{ Btu/lb} - (537 \text{ R})(0.13266 - 0.08775) \text{ Btu/(lb R)}$$

$$a_3 = 0.5233 \text{ Btu/lb}.$$

Hence,

$$\varepsilon_p = \frac{(1.585 - 0.5233) \text{ Btu/lb}}{1.19 \text{ Btu/lb}}$$

$$\varepsilon_{st} = 89\%.$$

The availability loss in the pump is

$$\Delta I = \dot{A}_{loss} = \dot{A}_{input} - \dot{A}_{output} = (a_3 - a_4) - w_p = -1.06 \text{ Btu/lb} - (-1.19 \text{ Btu/lb})$$

$$\Delta I = \dot{A}_{loss} = 0.13 \text{ Btu/lb}.$$

From this analysis, it is clear that the boiler/steam generator is the most inefficient component of the system.

6.9 Gas-Turbine Power Plant Systems

Power plant systems that operate solely with gases as the working fluid are said to operate on a **gas (power) cycle**. For typical **gas-turbine** power plant systems, air or a combustion gas mixture is the working fluid.

6.9.1 The Ideal Brayton Cycle for Gas-Turbine Power Plant Systems

The simple gas-turbine power plant cycle is based on the **Brayton cycle**. Figure 6.11 shows a schematic of the cycle. Air enters the system and is compressed in a **compressor**, which is then mixed with fuel and burned in a **combustor/burner**. The hot, compressed products of combustion are expanded in a turbine to produce work to drive the compressor and a generator. The combustion gases are exhausted from the system after expansion in the turbine.

The following points are noted regarding Figure 6.11:

(i) Process 1-2 is an isentropic (constant entropy) compression of air in the compressor (see *T-s* diagram of Figure 6.11). The air pressure increases.

(ii) Process 2-3 is a constant pressure heat addition in a combustor or burner. Fuel is mixed with air and burned.

(iii) Process 3-4 is an isentropic expansion of the hot, compressed gas mixture in the gas turbine. The gas pressure decreases and work output occurs. The gas is typically expanded to atmospheric pressure.

This process and working fluid do not follow a thermodynamic cycle like the steam-turbine power plant system. The composition of the working fluid changes due to

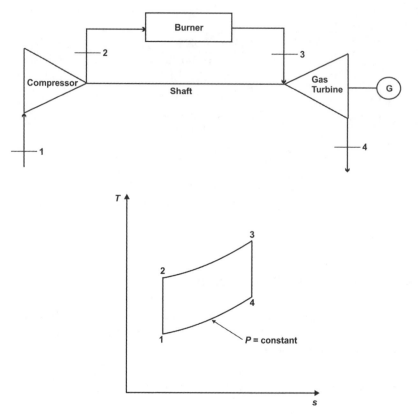

Figure 6.11 Ideal Brayton cycle

fuel mixing and combustion in the burner. In order to analyze a gas-turbine power plant system, several simplifying assumptions are typically made:

  (i) The working fluid is considered as air having a fixed composition. The air is an ideal gas with constant specific heats. This assumption is justified beyond the combustor/burner if the **air/fuel ratio** is high. In this case, a larger amount of air is present during combustion.
 (ii) The combustion process is one that occurs after the mixing of fuel with air. The direct mixing and burning of this air/fuel mixture improves the heat transfer, but the mass of the working fluid will change. The combustion process is assumed to be a heat transfer process from an external source. This is followed by an assumption of constant mass flow rate throughout the system.
(iii) The inlet and exhaust processes can be replaced by constant pressure processes. This will form a complete gas turbine cycle. Typically, air will enter the compressor at 14.7 psia, and will leave the turbine at 14.7 psia.
(iv) To ensure ideality of the cycle, all the processes are internally reversible.

> ## Practical Note 6.8   Combustion Air and Cracking in a Burner
>
> Higher air/fuel ratios in the burner are desired to ensure complete combustion of the fuel (for maximum heat generation) and to reduce cracking of the fuel. Typically, the amount of air will exceed that of the fuel by 15–20% of that required for complete combustion.
>
> Cracking is a process in which hydrocarbon fuels (natural gas/methane) are broken into smaller hydrocarbon chains, and ultimately to carbon. Carbon can deposit on the wall of the burner, blocking the fuel nozzle orifices.

6.9.2 Real Gas-Turbine Power Plant Systems

In real gas-turbine power plant systems, the compressor and turbine will not be 100% efficient. Losses will occur, and the entropies will not be constant in the compression and expansion processes. The T-s diagram shown in Figure 6.12 shows deviations from an isentropic, ideal Brayton cycle for a gas turbine. Note that due to irreversibilities, the entropies of the working fluid that exits the compressor and the turbine are higher than that of the isentropic case.

The first-law efficiency of the real system can be determined by conducting energy balances through the components. Consider a simple gas-turbine system operating on a Brayton cycle. The internal efficiency of the compressor is η_c and the turbine internal efficiency is η_t. The first-law efficiency of the cycle (η_{cycle}) is

$$\eta_{cycle} = \frac{w_t - w_c}{q_h}, \tag{6.20}$$

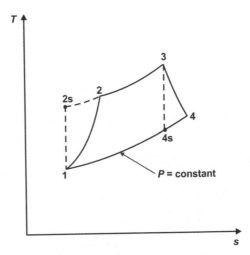

Figure 6.12   Real Brayton cycle

where

w_t = total turbine output;

w_c = energy input required to drive the compressor;

q_h = total heated added to the system (at the burner and/or a reheater).

The total turbine output is

$$w_t = \eta_t c_p \left(T_3 - T_{4s} \right),$$ (6.21)

where c_p = specific heat of the combustion gas mixture and T_{4s} is the temperature after isentropic expansion in the ideal case.

The input required to drive the compressor is

$$w_c = \frac{1}{\eta_c} c_p \left(T_{2s} - T_1 \right),$$ (6.22)

where T_{2s} is the temperature after isentropic compression in the ideal case.

Consider the total turbine output in detail:

$$w_t = \eta_t c_p \left(1 - \frac{T_{4s}}{T_3} \right) T_3.$$ (6.23)

For the ideal isentropic processes between points 1-2 and 3-4,

$$\frac{T_{2s}}{T_1} = \frac{T_3}{T_{4s}} = \left(\frac{P_{2s}}{P_1} \right)^{(k-1)/k} = \left(\frac{P_3}{P_{4s}} \right)^{(k-1)/k} = r_p^{(k-1)/k},$$ (6.24)

where

$k = \frac{c_p}{c_v}$ = specific heat ratio;

r_p = pressure ratio.

Therefore,

$$w_t = \eta_t c_p T_3 \left(1 - \frac{1}{r_p^{(k-1)/k}} \right).$$ (6.25)

A similar analysis can be conducted for the compressor. This analysis gives

$$w_c = \frac{1}{\eta_c} c_p T_1 \left(r_p^{(k-1)/k} - 1 \right).$$ (6.26)

If the total heat input occurs at the burner, the heat input to the system is

$$q_h = c_p \left(T_3 - T_2 \right).$$ (6.27)

$$\text{Let}: q_h = c_p T_1 \left(\frac{T_3}{T_1} - \frac{T_2}{T_1} \right).$$ (6.28)

In practice, the inlet temperatures to the gas turbine (T_3) and the compressor (T_1) will be known. An expression for T_2 will be needed to determine the heat transferred to the working fluid in the burner.

The compressor internal efficiency may be expressed as

$$\eta_c = \frac{h_{2s} - h_1}{h_2 - h_1}. \tag{6.29}$$

If the properties of the working fluid are assumed to be constant so that $c_p = $ constant,

$$\eta_c = \frac{T_{2s} - T_1}{T_2 - T_1}, \tag{6.30}$$

$$T_2 = \frac{T_{2s} - T_1}{\eta_c} + T_1, \tag{6.31}$$

$$\frac{T_2}{T_1} = \frac{1}{\eta_c}\left(\frac{T_{2s}}{T_1} - 1\right) + 1. \tag{6.32}$$

Remember: $\frac{T_{2s}}{T_1} = r_p^{(k-1)/k}$.

Therefore,

$$\frac{T_2}{T_1} = \frac{1}{\eta_c}\left(r_p^{(k-1)/k} - 1\right) + 1, \tag{6.33}$$

and

$$q_h = c_p T_1 \left[\frac{T_3}{T_1} - \frac{r_p^{(k-1)/k} - 1}{\eta_c} - 1\right] \tag{6.34}$$

$$q_h = \frac{c_p T_1}{\eta_c}\left[\eta_c\left(\frac{T_3}{T_1} - 1\right) - r_p^{(k-1)/k} + 1\right]. \tag{6.35}$$

Hence, the first-law efficiency of the cycle becomes

$$\eta_{cycle} = \frac{\eta_t c_p T_3 \left(1 - \frac{1}{r_p^{(k-1)/k}}\right) - \frac{1}{\eta_c} c_p T_1 \left(r_p^{(k-1)/k} - 1\right)}{\frac{c_p T_1}{\eta_c}\left[\eta_c\left(\frac{T_3}{T_1} - 1\right) - r_p^{(k-1)/k} + 1\right]} \tag{6.36}$$

$$\eta_{cycle} = \frac{\eta_t \eta_c T_3 \left(1 - \frac{1}{r_p^{(k-1)/k}}\right) - T_1 \left(r_p^{(k-1)/k} - 1\right)}{T_1 \left[\eta_c\left(\frac{T_3}{T_1} - 1\right) - r_p^{(k-1)/k} + 1\right]} \tag{6.37}$$

$$\eta_{cycle} = \frac{\eta_t \eta_c \frac{T_3}{T_1} \left(1 - \frac{1}{r_P^{(k-1)/k}}\right) - \left(r_P^{(k-1)/k} - 1\right)}{\left[\eta_c \left(\frac{T_3}{T_1} - 1\right) - r_P^{(k-1)/k} + 1\right]}. \tag{6.38}$$

Rearrange

$$\eta_{cycle} = \frac{\eta_t \eta_c \frac{T_3}{T_1} \left(1 - \frac{1}{r_P^{(k-1)/k}}\right) - r_P^{(k-1)/k} \left(1 - \frac{1}{r_P^{(k-1)/k}}\right)}{\left[\eta_c \left(\frac{T_3}{T_1} - 1\right) - r_P^{(k-1)/k} + 1\right]} \tag{6.39}$$

$$\eta_{cycle} = \frac{\left(\eta_t \eta_c \frac{T_3}{T_1} - r_P^{(k-1)/k}\right) \left(1 - \frac{1}{r_P^{(k-1)/k}}\right)}{\left[\eta_c \left(\frac{T_3}{T_1} - 1\right) - r_P^{(k-1)/k} + 1\right]}. \tag{6.40}$$

Analysis of Equation (6.40) will show that for real gas-turbine power plant systems, the first-law cycle efficiency is influenced by the

(i) gas-turbine internal efficiency (η_t);
(ii) compressor internal efficiency (η_c);
(iii) pressure ratio (r_p);
(iv) **maximum temperature ratio,** $\frac{T_3}{T_1}$. It is clear that increasing the turbine inlet temperature (T_3) and decreasing the compressor inlet temperature (T_1) will increase the efficiency of the cycle.

6.9.3 Regenerative Gas-Turbine Power Plant Systems

Regeneration can be introduced into the system to increase the first-law efficiency. A **regenerator** (heat exchanger) can be introduced after the compressor and before the burner. The exhaust duct from the gas turbine can be routed through the regenerator so that heat from the hot exhaust gases can be transferred to the compressed air. This will increase the air temperature at the inlet to the burner, and reduce the fuel requirement, increasing the system efficiency. Figure 6.13 shows a schematic of a gas-turbine system with regeneration.

The internal efficiency (η_{regen}) of the regenerator must be considered when analyzing the performance of gas-turbine power plants.

The first-law efficiency of the plant can be increased further by introducing **intercooling**. Intercoolers (heat exchangers) can be installed between the stages of the compressor. Figure 6.14 shows a regenerative gas-turbine power plant system with

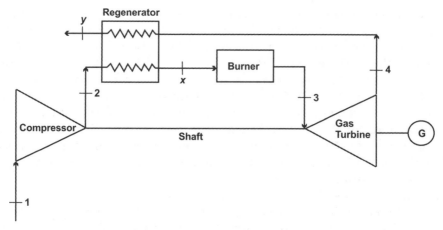

Figure 6.13 Regenerative Brayton cycle

intercooling. Additional reheating via a burner installed in the two-stage gas-turbine assembly is shown.

6.9.4 Operation and Performance of Gas-Turbine Power Plants—Practical Considerations

The first-law efficiency of the plant presents an indication of the overall performance. Gas-turbine first-law efficiencies are typically on the order of 40% (for plants with regeneration). Second-law analysis would be required to determine the efficiencies of the plant components, and to identify components that need improvement.

The gas-turbine unit sizes will range from about 2000 to 100000 kW. Most large units are designed to operate at 3000 or 3600 rpm. Gas turbines have relatively low initial costs. So, they are frequently used for emergency service and to provide extra power during peak (electrical) load periods (high electricity demand periods).

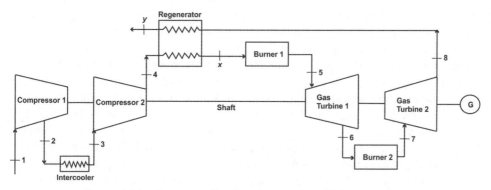

Figure 6.14 Regenerative Brayton cycle with intercooling

Natural gas (methane) is the most suitable fuel for gas turbines because limited fuel preparation is required. In lieu of natural gas, #2 oil may be used. If heavier oils are considered, they must be heated to reduce their viscosities to facilitate piping and **atomization** in the burner. Atomization in the burner results in the formation of droplets of fuel, which are easier to burn since the surface area of droplets are large.

Gas-turbine power plant systems have high **back work ratios**. The back work ratio is the ratio of the compressor work required to the turbine work output. Typically, **about two-thirds of the gas-turbine output is used to drive the compressor**. The remainder is used for power generation. Gas turbines tend to be larger than steam turbines for an equivalent power output to the generator because of the large back work ratios of gas-turbine power plant systems.

Gas-turbine systems can operate at **part-load conditions**. **Full-load operation** is the maximum operation (as specified by the design and the manufacturer) of the gas turbine. As the load decreases to accommodate part-load conditions, the turbine *NHR* increases. For part-load operation, it is typical to keep the gas flow rate constant at the maximum capacity for which the system was designed. Reduction in the fuel flow rate will reduce the turbine inlet temperature (lower energy from combustion), and the work/power output.

In modern gas turbines, the turbine inlet temperatures can be as high as 2600°F. Higher temperatures are known to accelerate **creeping** and **corrosion** of turbine blades. However, the development of protective thermal-sprayed coatings for the blades has enabled the temperature increase. The coatings are usually a bilayer of a Ni/Co alloy and yttria-stabilized zirconia, with the zirconia exposed to the ambient to provide high-temperature and corrosion protection.

Manufactured gas-turbine units will come complete with **design ratings** that are based on standard, outdoor site conditions. The International Standards Organization (ISO) site conditions are most typically used and referenced: $T = 59°F$, $P = 14.7$ psia, and relative humidity of 60%. Some gas-turbine manufacturers have referenced their standard designs to the conditions established by the National Electric Manufacturers Association (NEMA): $T = 80°F$ and $P = 14.17$ psia.

The first-law efficiency of the gas turbine will increase with a decrease in ambient temperature. A 5% increase in the first-law efficiency will occur for a temperature drop from 59°F to 0°F. As the temperature decreases, the air density will also increase. For a constant speed, the air mass flow rate in the compressor will increase by about 12% for a temperature drop from 59°F to 0°F [1].

Pressure losses in the system will affect the power output and efficiency of the gas-turbine power plant. Typically, and based on the arrangement of the gas-turbine power plant system, standard inlet losses are 2 in. of water and exhaust losses are 4 in. of water. These standard losses are due to duct arrangements and standard units installed in the ductwork. Inclusion of other auxiliary units—filters, silencers, dampers, waste heat recovery boilers, regenerators, etc.—will increase the inlet and exhaust end pressure drops. Table 6.1 shows some pressure drop values for some common units that may be installed at the inlet or exhaust of the plant. These pressure drops are in addition to the standard losses.

Table 6.1 Pressure drops at the gas-turbine plant inlet and exhaust [1]

Inlet Losses	Inch of Water
Air filter (protects the air compressor)	0.5
Low-level silencer (for noise reduction)	1.5
High-level silencer (for noise reduction)	3.5
Evaporative air cooler (for water removal)	0.8
Exhaust end losses	**Inch of water**
Waste heat boiler for combined systems	10–14
Bypass duct for combined systems	6
Regenerative air heater (total)	8
Open damper	1

Example 6.4   Real Gas-Turbine Power Plant with Part Loads

A gas-turbine plant is in development for the purpose of power generation. The manufacturer has designed the system as per ISO site conditions for a regenerative gas-turbine model.

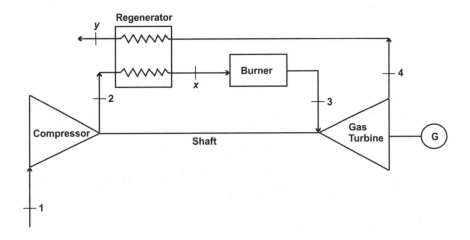

The compressor, turbine, and regenerator efficiencies are noted to be 85%, 88%, and 70%, respectively. Pressure loss in the burner is 6%. The pressure ratio of the compressor will be 4 and the inlet temperature of the turbine cannot exceed 2500°F. The airflow rate will be 1896000 lb/h. Consulting engineers have decided to use an inlet air filter and low-level silencer and an open damper beyond the regenerator in the exhaust duct. Auxiliary power required by the plant for operations will be 7%.

Analyze the performance of the gas-turbine power plant under full-load conditions. Identify the component in the system that requires the most improvement. Determine the part-load fuel consumption percentage, plant efficiency, and *NHR* for a 60% part-load operating condition.

Solution

Full Load Conditions

Assume that the working fluid is air throughout the system. Assume negligible duct pressure losses. Study each point in the plant cycle.

Point 1: Inlet to the compressor

ISO site conditions will govern the conditions at this point. So,

$$P_1 = 14.7 \text{ psia}$$

$$T_1 = 59°\text{F} = 519 \text{ R}.$$

Point 2: Exit of the compressor

The pressure at this point is found by considering the isentropic compression process.

$$\left(\frac{P_{2s}}{P_1}\right)^{(k-1)/k} = r_{\text{p,c}}^{(k-1)/k}$$

$$P_{2s} = r_{\text{p,c}} P_1 = 4\,(14.7 \text{ psia}) = 58.8 \text{ psia}.$$

The pressure loss across the air filter and the silencer will reduce P_{2s}. Assume water properties at 59°F. For the air filter, $\Delta P_a = 0.5$ in. wg. In terms of pounds per square inch (psi), the pressure loss is

$$\Delta P_a = \rho g\,\Delta h = \left(62.36 \text{ lb/ft}^3\right)\left(32.2 \text{ ft/s}^2\right)(0.5 \text{ in.})$$

$$\times \frac{1 \text{ ft}}{12 \text{ in.}} \times \frac{1 \text{ lbf}}{32.2 \text{ lb/ft/s}^2} \times \left(\frac{1 \text{ ft}}{12 \text{ in.}}\right)^2 = 0.018 \text{ psi.}$$

For the low-level silencer, $\Delta P_s = 1.5$ in. wg. In terms of pounds psi, the pressure loss is

$$\Delta P_s = \rho g\,\Delta h = \left(62.36 \text{ lb/ft}^3\right)\left(32.2 \text{ ft/s}^2\right)(1.5 \text{ in.})$$

$$\times \frac{1 \text{ ft}}{12 \text{ in.}} \times \frac{1 \text{ lbf}}{32.2 \text{ lb/ft/s}^2} \times \left(\frac{1 \text{ ft}}{12 \text{ in.}}\right)^2 = 0.054 \text{ psi.}$$

Therefore,

$$P_2 = (58.8 - 0.018 - 0.054) \text{ psia} = 58.7 \text{ psia.}$$

The temperature at this point is found by considering the isentropic compression and the compressor efficiency:

$$\frac{T_{2s}}{T_1} = r_{\text{p,c}}^{(k-1)/k}$$

Let $k \approx 1.4$ for air at 59°F. So,

$$T_{2s} = T_1 r_{\text{p,c}}^{(k-1)/k} = (519 \text{ R})\,(4)^{(1.4-1)/1.4} = 771 \text{ R} = 311°\text{F.}$$

The real temperature, T_2, is found from the definition of the compressor efficiency. Assume constant specific heats of air:

$$\eta_c = \frac{T_{2s} - T_1}{T_2 - T_1}$$

$$T_2 = T_1 + \frac{T_{2s} - T_1}{\eta_c} = 519\,R + \frac{(771 - 519)\,R}{0.85}$$

$T_2 = 816\,R = 356°F.$

Point 3: Inlet to the gas turbine

There is a 6% pressure drop across the burner. All the losses in the regenerator will be considered in the exhaust end pressure loss.

Therefore,

$$P_3 = (1 - 0.06)P_2 = 0.94(58.7\text{ psia})$$

$$P_3 = 55.2\text{ psia}$$

$$T_3 = 2500°F = 2960\,R.$$

Point 4: Exit of the gas turbine

Due to losses in the regenerator (ΔP_{reg}) and the open damper (ΔP_d), the pressure of the gas mixture leaving the turbine must be greater than the ambient pressure to prevent backflow into the gas turbine.

The total exhaust end pressure loss is

$$\Delta P_T = \Delta P_{reg} + \Delta P_d = (8 + 1)\text{ in. wg.} = 9\text{ in. wg.}$$

In terms of psi:

$$\Delta P_T = \left(62.36\text{ lb/ft}^3\right)\left(32.2\text{ ft/s}^2\right)(9.0\text{ in.}) \times \frac{1\text{ ft}}{12\text{ in.}} \times \frac{1\text{ lbf}}{32.2\text{ lb/ft/s}^2} \times \left(\frac{1\text{ ft}}{12\text{ in.}}\right)^2 = 0.33\text{ psi}$$

Therefore,

$$P_4 = 14.7\text{ psia} + 0.33\text{ psia} = 15.0\text{ psia.}$$

The pressure ratio of the turbine is

$$r_{p,t} = \frac{55.2\text{ psia}}{15.0\text{ psia}} = 3.67.$$

The temperature of the exhaust gas is found by considering the isentropic expansion of the gas mixture in the turbine and the turbine efficiency:

$$\frac{T_3}{T_{4s}} = r_{p,t}^{(k-1)/k}$$

$$T_{4s} = \frac{T_3}{r_{p,t}^{(k-1)/k}}.$$

Assume that the gas mixture has properties of air at 2500°F. So, $k \approx 1.3$.

$$T_{4s} = \frac{2960 \text{ R}}{(3.67)^{(1.3-1)/1.3}} = 2193 \text{ R} = 1733°\text{F}.$$

The real exhaust gas temperature is found by using the definition of the turbine efficiency:

$$\eta_t = \frac{T_3 - T_4}{T_3 - T_{4s}}$$

$$T_4 = T_3 - \eta_t (T_3 - T_{4s}) = 2960 \text{ R} - 0.88 (2960 - 2193) \text{ R}$$

$T_4 = 2285 \text{ R} = 1825°\text{F}.$
Point x: Exit of the regenerator (compressed air to burner)
Assume no pressure drop in the regenerator between points 2 and *x*. So,
$P_2 = P_x = 58.7$ psia.
The temperature is found by considering the efficiency of the regenerator:

$$\eta_{reg} = \frac{h_x - h_2}{h_4 - h_y} \approx \frac{h_x - h_2}{h_4 - h_2} \approx \frac{T_x - T_2}{T_4 - T_2}.$$

Note that T_4–T_2 is the temperature difference that will give the maximum heat transfer in the regenerator. As T_y becomes equal to T_2, the performance of the regenerator (a heat exchanger) improves. T_x–T_2 is the temperature difference that will give the actual heat transfer in the regenerator.
Therefore,

$$T_x = T_2 + \eta_{reg} (T_4 - T_2) = 816 \text{ R} + (0.70) (2285 - 816) \text{ R}$$
$$T_x = 1844 \text{ R} = 1384°\text{F}.$$

Point y: Exit of the regenerator (exhaust gas from the turbine)
This point is after the regenerator. So, the turbine exhaust gas pressure will be reduced by the losses in the regenerator. So,
$P_y = P_4 - \Delta P_{gen}$

$$P_y = 15.0 \text{ psia} - \left(62.36 \text{ lb/ft}^3\right) \left(32.2 \text{ ft/s}^2\right) (8 \text{ in.}) \times \frac{1 \text{ ft}}{12 \text{ in.}} \times \frac{1 \text{ lbf}}{32.2 \text{ lb/ft/s}^2} \times \left(\frac{1 \text{ ft}}{12 \text{ in.}}\right)^2$$

$P_y = 14.71$ psia.
The temperature is found by conducting an energy balance across the regenerator. The subscript "a" represents the compressed air and subscript "g" represents exhaust gas mixture.

$$\dot{m}_a c_p (T_x - T_2) = \dot{m}_g c_p (T_4 - T_y)$$

For constant gas flow rates: $\dot{m}_a c_p = \dot{m}_g c_p$. Therefore,

$$T_x - T_2 = T_4 - T_y$$

$$T_y = T_4 - T_x + T_2 = (2285 - 1844 + 816) \text{ R}$$

$$Ty = 1257 \text{ R} = 797°\text{F}.$$

It is important to check the exhaust gas temperature to ensure that T_y is greater than the dew point of water at the exhaust gas conditions. The dew point temperature is the temperature at which water will condense from the exhaust gases at constant vapor pressure. The liquid condensate can mix with other chemicals in the exhaust gas to cause corrosion and material degradation of the exhaust stack. The relative humidity is needed to calculate the dew point temperature. However, in this problem and at the exhaust temperature, the water vapor will be completely superheated.

The work output from the turbine is

$$w_t = \eta_t c_p (T_3 - T_{4s}).$$

For air properties at $T_{ave} = 2117°\text{F}$,

$$w_t = 0.88(0.28 \text{ Btu/lb-R})(2960 - 2193) \text{ R}$$

$$w_t = 189 \text{ Btu/lb}.$$

The auxiliary power required for plant operations is

$$w_a = 0.07 w_t$$

$$w_a = 0.07(189 \text{ Btu/lb}) = 13.2 \text{ Btu/lb}.$$

The work required by the compressor is

$$w_c = \frac{1}{\eta_c} c_p (T_{2s} - T_1).$$

For air properties at $T_{ave} = 185°\text{F}$,

$$w_c = \frac{1}{0.85} (0.24 \text{ Btu/(lb R)}) (771 - 519) \text{ R}$$

$$w_c = 71.2 \text{ Btu/lb}.$$

The net generator output is

$$\dot{W}_{net} = \eta_{gen} \eta_{coupling} \dot{m}_a (w_t - w_c - w_a).$$

Assume: $\eta_{gen} = 95\%$ and $\eta_{coupling} = 100\%$. So,

$$\dot{W}_{net} = (0.95)\,(1.00)\,(1896000\ \text{lb/h})\,(189 - 71.2 - 13.2)\ \text{Btu/lb}$$
$$\dot{W}_{net} = 1.89 \times 10^8\ \text{Btu/h} = 552\,41\ \text{kW} = 55.2\ \text{MW}.$$

The back work ratio is

$$\text{Back work ratio} = \frac{w_c}{w_t} = \frac{71.2\ \text{Btu/lb}}{189\ \text{Btu/lb}}$$

Back work ratio $= 0.38$
The heat added at the burner is

$$q_h = c_p(T_3 - T_x).$$

For air properties at $T_{ave} = 1942°\text{F}$,

$$q_h = (0.286\ \text{Btu/(lb R)})(2960 - 1844)\ \text{R}$$
$$q_h = 319.2\ \text{Btu/lb}.$$

In kilowatts,

$$\dot{q}_h = \dot{m}_a q_h$$
$$\dot{q}_h = (1896000\ \text{lb/h})(319.2\ \text{Btu/lb})$$
$$\dot{q}_h = 6.05 \times 10^8\ \text{Btu/h} = 177354\ \text{kW} = 177.4\ \text{MW}.$$

The first-law efficiency of the plant is

$$\eta_{cyc} = \frac{\dot{W}_{net}}{\dot{q}_h} = \frac{55241\ \text{kW}}{177354\ \text{kW}}$$
$$\eta_{cyc} = 31.2\%.$$

The plant *NHR* is

$$NHR = \frac{\text{heat input}}{\text{generator output}}$$
$$NHR = \frac{6.05 \times 10^8\ \text{Btu/h}}{55241\ \text{kW}}$$
$$\textbf{\textit{NHR} = 10952 Btu/kWh}.$$

Note that the low cycle efficiency results in a high *NHR*.
Below is a heat and mass balance diagram for this problem.

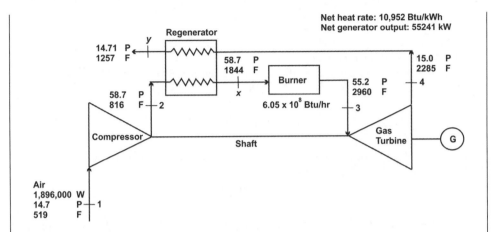

A second-law analysis must be conducted to identify the component of the plant that needs the most improvement. Some of the efficiencies are already known.

$$
\begin{aligned}
&\text{For the compressor:} && \varepsilon_c = 85\% \\
&\text{For the turbine:} && \varepsilon_t = 88\% \\
&\text{For the regenerator:} && \varepsilon_{reg} = 70\%.
\end{aligned}
$$

For the burner (a heat exchanger), the effectiveness is the ratio of the availability increase in the working fluid (gas mixture) to the availability supplied to the burner. The availability supplied to the burner is the heat added (q_h). Therefore,

$$
\varepsilon_b = \frac{a_3 - a_x}{q_h} = \frac{(h_3 - h_x) - T_0\,(s_3 - s_x)}{h_3 - h_x} = 1 - \frac{T_0\,(s_3 - s_x)}{h_3 - h_x}.
$$

Remember that for constant c_p, $s_3 - s_x \approx c_p \ln\frac{T_3}{T_x} - R\ln\frac{P_3}{P_x}$. Thus,

$$
\varepsilon_b = 1 - \frac{T_0\left[c_p \ln\dfrac{T_3}{T_x} - R\ln\dfrac{P_3}{P_x}\right]}{h_3 - h_x}.
$$

Let $T_0 = 77°F = 537\ R$, and

$$
\varepsilon_b = 1 - \frac{(537\ R)\left[(0.286\ \text{Btu/(lb R)})\ln\dfrac{2960\ R}{1844\ R} - (0.0686\ \text{Btu/lb-R})\ln\dfrac{55.2\ \text{psia}}{58.7\ \text{psia}}\right]}{319.2\ \text{Btu/lb}}
$$

$$
\varepsilon_b = 76.5\%.
$$

Based on the analysis, the **burner** and the **regenerator** require the most improvement. In practice, regenerator efficiencies are usually lower than 85%.

Part-Load Operation—60%

At full-load operation, $w_t = 189$ Btu/lb.

At 60% part-load operation, $w_t = 113$ Btu/lb.

For a constant air mass flow rate, the compressor work will remain unchanged. Therefore,

$$w_c = 71.2 \text{ Btu/lb.}$$

Assume that the auxiliary power required for plant operations remains the same. Hence,

$$w_a = 13.2 \text{ Btu/lb.}$$

At part-load operation, the turbine inlet temperature will decrease. The expression for the turbine work output can be used to find the turbine inlet temperature.

$$w_t = \eta_t c_p \left(T_3 - T_{4s}\right) = \eta_t c_p \left(T_3 - \frac{T_3}{r_{p,t}^{(k-1)/k}}\right) = \eta_t c_p T_3 \left(1 - \frac{1}{r_{p,t}^{(k-1)/k}}\right)$$

Therefore,

$$T_3 = \frac{w_t}{\eta_t c_p \left(1 - \dfrac{1}{r_{p,t}^{(k-1)/k}}\right)}$$

$$T_3 = \frac{113.4 \text{ Btu/lb}}{(0.88)(0.28 \text{ Btu/lb-R})\left(1 - \dfrac{1}{3.67^{(1.3-1)/1.3}}\right)}$$

$$T_3 = 1775.5 \text{ R} = 1315.5°\text{F}$$

The exhaust gas temperature (from the turbine) is

$$T_{4s} = \frac{T_3}{r_{p,t}^{(k-1)/k}} = \frac{1775.5 \text{R}}{(3.67)^{(1.3-1)/1.3}} = 1315.3 \text{ R} = 855.3°\text{F}.$$

Thus, the real exhaust gas temperature is

$$T_4 = T_3 - \eta_t \left(T_3 - T_{4s}\right) = 1775.5 \text{ R} - 0.88 \left(1775.5 - 1315.3\right) \text{ R}$$

$$T_4 = 1370.5 \text{ R} = 910.5°\text{F}.$$

The temperature at point x is

$$T_x = T_2 + \eta_{reg} \left(T_4 - T_2\right) = 816 \text{ R} + (0.70)(1370.5 - 816) \text{ R}$$

$$T_x = 1204.2 \text{ R} = 744.2°\text{F}.$$

The heat added at the burner for a 60% part-load operation is

$$q_h = c_p(T_3 - T_x)$$

$$q_h = (0.26 \text{ Btu}/(\text{lb} - \text{R}))(1775.5 - 1204.2) \text{ R}$$

$$q_h = 148.5 \text{ Btu}/\text{lb}.$$

The first-law efficiency of the plant at 60% part-load operation is

$$\eta_{cyc} = \frac{w_{net}}{q_h} = \frac{w_t - w_c - w_a}{q_h} = \frac{(113.4 - 71.2 - 13.2) \text{ Btu}/\text{lb}}{148.5 \text{ Btu}/\text{lb}}$$

$$\eta_{cyc} = \textbf{19.5\%}.$$

The plant *NHR* is

$$NHR = \frac{\text{heat input}}{\text{generator output}}.$$

The heat input in Btu/h is

$$\dot{q}_h = \dot{m}_a q_h$$

$$\dot{q}_h = (1896000 \text{ lb}/\text{h})(148.5 \text{ Btu}/\text{lb})$$

$$\dot{q}_h = 2.82 \times 10^8 \text{ Btu}/\text{h}.$$

The net generator output for a 60% part-load situation is

$$\dot{W}_{net} = \eta_{gen} \eta_{coupling} \dot{m}_a (w_t - w_c - w_a).$$

Assume: $\eta_{gen} = 95\%$ and $\eta_{coupling} = 100\%$. So,

$$\dot{W}_{net} = (0.95)(1.00)(1896000 \text{ lb}/\text{h})(113.4 - 71.2 - 13.2) \text{ Btu}/\text{lb}$$

$$\dot{W}_{net} = 5.22 \times 10^7 \text{ Btu}/\text{h} = 15309 \text{ kW} = 15.3 \text{ MW}.$$

Thus,

$$NHR = \frac{2.82 \times 10^8 \text{ Btu}/\text{h}}{15309 \text{ kW}}$$

$$NHR = \textbf{18421 Btu/kWh}.$$

As the efficiency decreases for part-load operation, the *NHR* increases.

Fuel Requirements

Assume that #2 oil is used for the power plant burner.
   At full-load operation,

$$\dot{V}_{fuel} = \frac{\dot{q}_h}{HV_f} = \frac{6.05 \times 10^8 \text{ Btu}/\text{h}}{140000 \text{ Btu}/\text{gal}}$$

$$\dot{V}_{fuel} = 4321 \text{ gph}.$$

At 60% part-load operation:

$$\dot{V}_{fuel} = \frac{\dot{q}_h}{HV_f} = \frac{2.82 \times 10^8 \text{ Btu/h}}{140000 \text{ Btu/gal}}$$

$$\dot{V}_{fuel} = 2014 \text{ gph}.$$

The part-load fuel consumption percentage is

$$\frac{2014 \text{ gph}}{4321 \text{ gph}} = 47\%.$$

6.10 Combined-Cycle Power Plant Systems

The efficiency and output of a power plant can be increased by combining different types of cycles for power generation. The combined-cycle plant may consist of one or more complete gas-turbine units and a steam-turbine unit. The efficiency of this type of **combined-cycle power plant system** may exceed 50%.

Figure 6.15 shows a schematic of a combined-cycle power plant with one gas-turbine unit and one steam-turbine unit (Brayton–Rankine combined-cycle power plant).

In this power plant, electrical power is generated by the gas turbine. The hot exhaust gas mixture is discharged to a **waste heat recovery boiler**, where steam is generated for the steam-turbine unit. The cooled exhaust gas is discharged from the **boiler stack**. In this way, a greater amount of the fuel supplied at the gas-turbine burner is used for electrical power generation.

The power plant shown in Figure 6.15 is very simple. Typically, regeneration may be included in the gas turbine (with regenerators) and steam turbine (with feedwater heaters). There may also be intercooling in multistage compressors of the gas-turbine

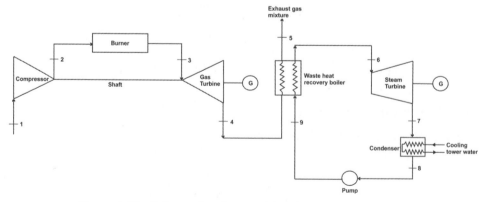

Figure 6.15   Schematic of a combined-cycle power plant

unit. The waste heat recovery boiler may also receive additional heating through the supplementary burner. In addition to all the losses previously discussed, an additional loss of 4% may be included to account for losses in the piping and ductwork of the systems.

Determination of the efficiency and *NHR* of the combined-cycle plant considers the total net output of all the turbine units and the total heat added to the plant station through all sources. For the combined-cycle plant,

$$\eta_{cyc} = \frac{w_{net,gt} + w_{net,st}}{Q_T},$$
(6.41)

where the subscript "gt" is gas turbine and the subscript "st" is steam turbine.

6.10.1 The Waste Heat Recovery Boiler

The **waste heat recovery boiler** is a key component in a combined-cycle power plant system. The boiler is arranged in three main sections: an **economizer**, an **evaporator**, and a **superheater**. Figure 6.16 shows a schematic of a single-pressure waste heat recovery boiler. The waste heat recovery boiler is a heat exchanger, and the sections may consist of rows of bare (typical) or finned tubes over which the hot exhaust gases from the gas turbine will flow to exchange heat. Compressed liquid will enter the boiler at the economizer. As heat is exchanged with the exhaust gases, the liquid becomes saturated, evaporated to vapor in the evaporator, and superheated further in the superheater. Figure 6.16 also shows the waste heat recovery boiler with a **drum**. The drum serves to mix the fluids from the economizer and the evaporator to increase the temperature of the fluid that enters the superheater.

Figure 6.17 shows a schematic of a typical temperature profile in a single-pressure waste heat recovery boiler. As the temperature of the exhaust gas decreases, the temperature of the fluid will increase, eventually becoming superheated. Not shown

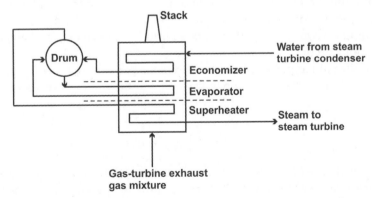

Figure 6.16   Piping schematic of a single-pressure waste heat recovery boiler

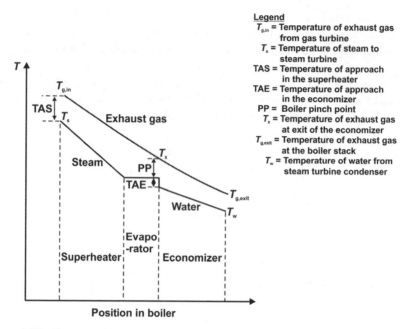

Figure 6.17 Temperature profile in a single-pressure waste heat recovery boiler

is the increase in the temperature of the superheated fluid if supplementary firing of the burner was used.

The selection of a waste heat recovery boiler and its operating parameters are based on experience and economic considerations. The following are several important points that the design engineer should bear in mind when selecting a waste heat boiler for a combined-cycle gas-turbine steam-tubine power plant station:

(i) The economizer is the section of the boiler in which compressed water is heated, but not vaporized. This will avoid the occurrence of **water hammer** and **steam blanketing**. Water hammer is a large pressure surge that shakes and breaks pipes. Steam blanketing occurs when steam and liquid water are not mixed. The fluids become stratified (separated into distinct layers), and erosion of the pipes may occur.

(ii) The **boiler pinch point** (**PP**) is the difference between the exhaust gas temperature and the water saturation temperature at the exit of the economizer.
    So,

$$PP = T_x - T_{\text{sat,water}}. \tag{6.42}$$

Lower PPs will result in greater heat transfer from the exhaust gas. This occurs because as the boiler PP decreases, the temperature difference between the exhaust gas and the water at the inlet of the economizer will increase, promoting heat transfer. A reduction in the boiler PP will result in greater steam production. However, the size of the boiler economizer will increase (to promote heat transfer between the hotter exhaust gas and the water at the inlet), which will increase space and financial requirements. In practice, **the boiler PP ranges from 50°F to 80°F.**

(iii) The **temperature of approach in the economizer (TAE)** is the difference between the steam saturation temperature and the temperature of the water leaving the economizer. If the water leaving the economizer is saturated, TAE = 0. In practice, this may be difficult to attain. So, for nonsteaming economizers, **the typical TAE is about 40°F.** Similar to the PP, lower TAE values will result in greater heat transfer and steam production. However, the size of the economizer and overall cost of the boiler will increase.

(iv) The evaporator is the section of the boiler that serves to reduce the exhaust gas temperature and bring the water to the boiling point. Addition of heat beyond this point will superheat the vapor.

(v) In the superheater section, the steam is superheated for the steam turbine. At this point, the exhaust gas temperature is high.

(vi) The **temperature of approach in the superheater (TAS)** is the difference between the inlet exhaust gas temperature and the temperature of the superheated steam leaving the superheater. **For an unfired boiler, TAS should be greater than 50°F.** Lower values would result in larger boiler sizes and higher costs that would need to be justified. Supplementary firing of the boiler would remove this restriction. In smaller boilers, the heat difference would be provided by the burners during supplementary firing.

(vii) The steam pressure and temperature required by the steam turbine will influence the size and cost of the waste heat recovery boiler. In addition, it will also be used to determine if supplementary firing of the burners will be needed. Table 6.2 shows some common combinations of steam pressure and temperature for the steam turbine. The pressure of the steam will depend on the pressure of

Table 6.2 Common steam conditions for waste heat recovery boilers [1]

Steam Pressure (psig)	Steam Temperature (°F)
150	450
250	550
400	650
600	750
850	825
1000	900
1250	950

the compressed liquid from the pump (condensate or feedwater heater pump). Losses or leakage will reduce this pressure. The temperature of the superheated steam will develop in the waste heat recovery boiler.

(viii) The amount of steam produced by the waste heat recovery boiler will depend on the mass flow rate of exhaust gas and conditions of both the exhaust gas and steam. Conservation of energy across the superheater and evaporator could be used to find the mass flow rate ratio. Alternatively, conservation of energy across the entire boiler (superheater to economizer) could be used to find the mass flow ratio. In fact, the energy balance could be conducted across any section of the boiler where the conditions are known or can be easily determined. So, across the superheater and evaporator,

$$\dot{m}_{gas} c_{p,gas} \left(T_{g,in} - T_x \right) = \dot{m}_{steam} \left(h_{steam,out} - h_{water,in} \right) \quad (6.43)$$

$$\frac{\dot{m}_{steam}}{\dot{m}_{gas}} = \frac{c_{p,gas} \left(T_{g,in} - T_x \right)}{\left(h_{steam,out} - h_{water,in} \right)}. \quad (6.44)$$

(ix) The **allowable back pressure** is an important parameter that must be considered. This is the pressure of the gas mixture at the exhaust nozzles of the gas turbine. This pressure controls the size of the free flow area of the hot gas between the boiler tubes. High back pressures from the gas turbine will reduce the size of the boiler. However, the work output of the gas turbine will decrease with increasing back pressures since the exhaust pressure from the gas turbine is high. The output of the gas turbine will decrease at a rate of 0.3% for every inch of water increase in back pressure. Typically, the back pressure ranges from 10 to 15 in. of water gage [1].

(x) The **stack** allows for final expulsion of the exhaust gases to the open atmosphere. Low temperatures of the exhaust gas from the stack indicate high heat transfer rates in the boiler. However, that may also result in a large and expensive boiler. As a minimum, the temperature of the exhaust gas mixture should be higher than the dew point temperature. This will prevent condensation in the stack and the economizer, which will result in corrosion.

Example 6.5   Performance of a Waste Heat Recovery Boiler

An applications engineer has proposed the following flow diagram and determined the temperature profiles for a single-pressure waste heat recovery boiler subsystem in a combined-cycle power plant system:

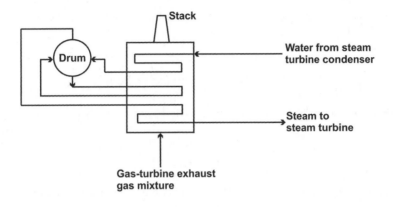

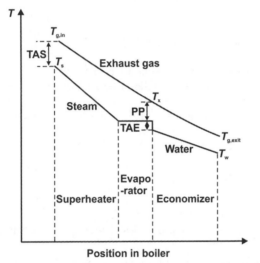

Position in boiler

To simplify the plant load control system, the engineer has decided to use an unfired waste heat recovery boiler. Estimate the steam production rate and the stack gas exit temperature.

Data

Parameter	Value
Exhaust gas flow rate	2255670 lb/h
Exhaust gas temperature	914.3°F
Steam pressure	1250 psia
Boiler water inlet temperature	326.4°F

Solution. Let the subscript "g" refer to the exhaust gas and "s" refer to steam. Refer to the temperature profile for additional terms. In this problem, some of the conditions of the steam are given. To estimate the steam production rate ($\dot{m}_s$), conduct an energy (heat balance) balance across the superheater and the evaporator, where steam will be produced in the pipes. Heat transferred by the cooling exhaust gas will be directly transferred to the saturated liquid water to convert it to steam.

Therefore,

$$\dot{m}_g c_{p,g} \left(T_{g,in} - T_x \right) = \dot{m}_s \left(h_s - h_{water} \right).$$

The definition of the boiler PP can be used to find T_x. Assume that $PP = 60°F$.

$$PP = T_x - T_{sat,water}$$
$$T_x = PP + T_{sat,water}.$$

The saturated water temperature at 1250 psia is found from the steam tables and interpolation: $T_{sat,water} = 572.2°F$.

Therefore,

$$T_x = 60°F + 572.2°F = 632.2°F.$$

For steam at 1250 psia, the common temperature is 950°F (see Table 6.2). The enthalpy for steam at 1250 psia and 950°F is found from the steam tables and interpolation: $h_s = 1469$ Btu/lb.

The enthalpy for water at 1250 psia is found from the steam tables and interpolation: $h_{water} = 579$ Btu/lb.

The steam production rate is

$$\dot{m}_s = \frac{\dot{m}_g c_{p,g} \left(T_{g,in} - T_x \right)}{\left(h_s - h_{water} \right)}.$$

Assuming that the exhaust gas has the properties of air: $c_{p,g} = 0.240$ Btu/(lb R),

$$\dot{m}_s = \frac{(2255670 \text{ lb/h}) (0.240 \text{ Btu/(lb R)}) (914.3 + 460 - 632.2 - 460) \text{ R}}{(1469 \text{ Btu/lb} - 579 \text{ Btu/lb})}$$

$\dot{m}_s = \textbf{171593 lb/h}.$

The stack gas exit temperature is found by conducting an energy (heat) balance across the economizer section of the boiler. Note that the energy balance could have been conducted across the superheater to the economizer section. Try it!

$$\dot{m}_g c_{p,g} \left(T_x - T_{g,exit} \right) = \dot{m}_s \left(h_{water,1250} - h_{water,inlet} \right),$$

where $h_{water,1250}$ is the enthalpy of the saturated water at 1250 psia in the economizer pipe section of the boiler and $h_{water,inlet}$ is the enthalpy of the saturated water at 326.4°F, inlet to the economizer.

Therefore,

$$T_{g,exit} = T_x - \frac{\dot{m}_s \left(h_{water,1250} - h_{water,inlet} \right)}{\dot{m}_g c_{p,g}}.$$

From the steam tables and after interpolation: $h_{water,inlet} = 297.1$ Btu/lb.

$$T_{g,exit} = (632.2 + 460)\ R - \frac{(171593\ lb/h)\,(579 - 297.1)\ Btu/lb}{(2255670\ lb/h)\,(0.24\ Btu/(lbR))}$$

$T_{g,exit} = \mathbf{1002.8\ R = 542.9°F}.$

In this case, the water vapor in the exhaust gas is not superheated. The saturation pressure of water at $T_{g,exit} = 542.9°F$ is approximately $P_{sat} = 963$ psia. Assume that the relative humidity of the exhaust gas is 30%. Then, the vapor pressure of the water is

$$P_v = \phi P_{sat}$$
$$P_v = (0.30)\,(963\ psia) = 289\ psia.$$

The dew point temperature is

$$T_{dp} = T_{sat,P_v} \approx 415°F.$$

Since $T_{g,exit} > T_{dp}$, no condensation of water vapor will occur, and corrosion of the boiler stack will be avoided.

Problems

6.1. Epcor has contracted General Electric Energy to build and install a combined gas-turbine–steam-turbine power plant system. A senior engineer would like to reuse an existing recovery waste heat boiler as the heat exchanger interface between the two turbine subsystems. Due to age, the efficiency of this boiler is 45%, and will produce steam with a quality of 0.77 for the steam turbine. Two junior design engineers object, stating that this existing recovery waste heat boiler could significantly undermine the performance of the steam-turbine subsystem. Are their concerns justified?

6.2. An engineer at Westinghouse has been charged with the responsibility of designing a gas-turbine system to deliver 40 MW of power to a small rural community. The load requirements vary from 5 to 40 MW over the course of a given day. To meet the daily load variations, the engineer has decided to vary the gas mass flow rate. The client objects to this plan, stating that this would undermine the performance of the compressor and turbine components of the system, given that these components are fluid machines like a pump or a fan. Are their concerns justified? Comment.

6.3. The demand to utility networks is not usually constant and steam-turbine generating units do not always operate at full-load capacity to produce maximum power. To achieve part-load operation, a throttle governing method, complete with a throttle valve may be used to control steam consumption by the turbine. Other appropriate methods may involve bypass governing with the use of bypass lines. A single-stage steam-turbine power plant operates with steam generator (steam boiler) exit conditions of 1000 psia, 900°F, and 850000 lb/h of steam. The condenser working pressure is restricted to 3 in. Hg. abs., and the line that connects the steam turbine exit to the condenser is relatively short. The turbine internal efficiency is 0.89 and the pump efficiency is 92%. It is expected that a 14% pressure drop will occur in the steam boiler, and due to the design of the plant and its auxiliaries, a pressure drop of 48 psi gage will occur in the line that connects the steam boiler and steam turbine. These pressure drops are independent of steam flow rate. The lines are wrapped in high-quality fiberglass, which has a high-temperature range of 1000°F. Determine the plant net output needed to satisfy 40% part-load operation and the first-law efficiency under these same operating conditions.

   Note: Do not present a heat and mass balance diagram

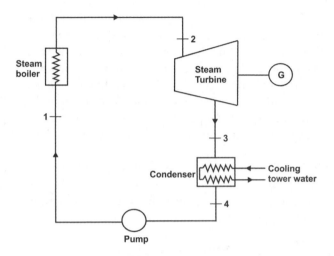

6.4. A two-stage steam-turbine power plant with two-stages of regenerative feed-water heating is being proposed for development. The design specifies one contact heater that receives extracted steam at a pressure of 150 psia from a HP turbine and a surface-type heater, complete with a drain cooler, that uses extracted steam at 50 psia. The turbine inlet conditions are 1250 psia and 850°F. The condenser working pressure is restricted to 3.5 in. Hg. abs. The turbine internal efficiency is 0.88 and the pump efficiency is 0.85. Analyze the performance of the proposed plant design and estimate the steam flow rate for a plant net output of 200 MW.

6.5. The Con Edison Company of New York provides electric service to New York City (except for a small area in Queens), and most of Westchester County. A senior plant engineer has presented the following diagram for one of their steam-turbine power plants:

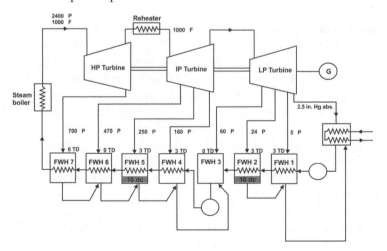

Since this is an existing system, it is known that the HP turbine, IP turbine, and LP turbine internal efficiencies are 90%, 92%, and 89%, respectively. The pump efficiency is 90% and the generator efficiency is 98%. The feedwater pump drive is motor driven. There is a 7% reheater pressure loss and a 16% boiler pressure loss. Only FWH 2 and FWH 5 have drain cooling. The steam flow rate is 4×10^6 lb/h. In an effort to improve company-wide operations, Management has requested a performance analysis of all existing plants. Prepare an analysis of the aforementioned plant for submission to Management.

6.6. The Con Edison Company of New York provides electric service to New York City (except for a small area in Queens), and most of Westchester County. A senior plant engineer has presented the following diagram for one of their steam-turbine power plants:

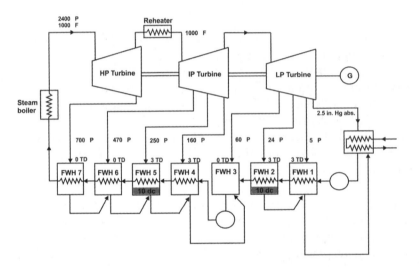

Since this is an existing system, it is known that the HP turbine, IP turbine, and LP turbine internal efficiencies are 90%, 92%, and 89%, respectively. The pump efficiency is 90%, the generator efficiency is 98%, and the burner efficiency is 87%. Anthracite coal is used in the boiler burner and natural gas is used in the reheater burner. The feedwater pump drive is motor driven. There is a 7% reheater pressure loss and a 16% boiler pressure loss. Only FWH 2 and FWH 5 have drain cooling. The steam flow rate is 4×10^6 lb/h. In an effort to improve company-wide operations, Management has requested an analysis of all existing plants to identify components in each plant that may need improvement. Prepare the analysis of the aforementioned plant for submission to Management.

6.7. A gas-turbine power plant operates on a regenerative Brayton cycle with air as the working fluid into the compressor.

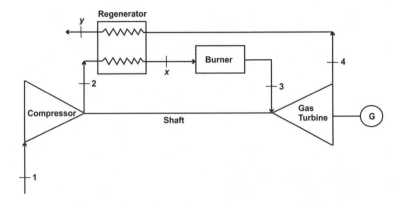

The following conditions were provided:

Compressor inlet conditions	$P_1 = 14.7$ psia
	$T_1 = 60°F$
Turbine inlet temperature	$T_3 = 2000°F$
Pressure ratio	4
Compressor efficiency	0.85
Turbine efficiency	0.88
Regenerator efficiency	0.70
Gas flow rate	360000 lb/h

Calculate the net cycle output and the thermal efficiency (first-law efficiency). Assume that no pressure drops occur in the system.

6.8. The schematic diagram of a closed-cycle gas-turbine system is shown below. The system is directly coupled with a high-temperature gas-cooled reactor. The working fluid is helium, and since no combustion is involved, the helium has a constant composition. Since helium is expensive, the flow rate is limited to 100000 lb/h. Estimate the performance of this plant.

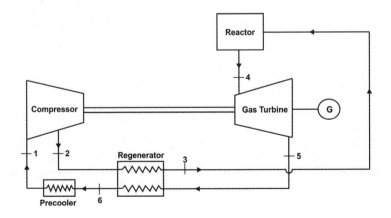

Data

State	Pressure (psia)	Temperature (°F)
1	445	105
2	1000	
3	980	
4	960	1500
5	460	
6	450	

Turbine efficiency, 0.90
Regenerator efficiency, 0.82
Compressor efficiency, 0.90

6.9. The following binary cycle has been proposed by a client. Dry saturated mercury vapor enters a mercury turbine at 225 psia and exhausts at 4 psia. In the steam side, steam enters the turbine at 680 psia and 900°F and exhausts to the condenser at 1 psia. Both turbine processes can be approximated as isentropic, and all pump work is negligible. Estimate the thermal efficiency of the proposed binary cycle.

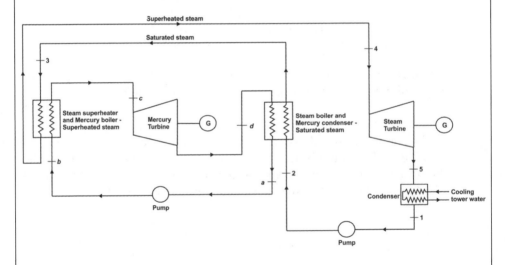

Properties of mercury

P (psia)	T (°F)	h_f (Btu/lb)	h_v (Btu/lb)	s_f Btu/(lb R)	s_v Btu/(lb R)
225	1038	32.20	156.32	0.03565	0.11852
4	557.8	17.16	143.44	0.02373	0.14787

6.10. A combined-cycle system consists mainly of a gas turbine, a waste heat recovery boiler, and a steam turbine that will be operated in the summer. A schematic and some system conditions are given below:

Gas Turbine

Gas flow rate	400000 lb/h
Pressure ratio	9
Turbine inlet temperature	2100°F
Compressor efficiency	85%
Turbine efficiency	90%
Burner efficiency	95%

Boiler and Steam Turbine

Steam pressure and temperature	600 psia, 700°F
Turbine internal efficiency	80%
Pump efficiency	78%

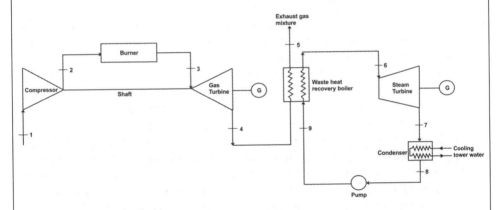

Considering all component inefficiencies, estimate the first-law efficiency, plant output, and *NHR* of this combined-cycle system. A heat and mass balance diagram is not required.

6.11. A single-pressure waste heat recovery boiler is being considered in a combined-cycle system design. The boiler is expected to provide steam at 1250 psia and 950°F at an inlet water temperature of 274°F. Calculate the steam flow rate and stack gas temperature for the following conditions:

Exhaust gas flow rate	1.68×10^6 lb/h
Exhaust gas temperature	960°F
Supplementary heat addition rate	26.57×10^6 Btu/h at superheater

Conduct an analysis to comment on the acceptability of the value of the stack gas temperature with respect to longevity of the exhaust stack.

6.12. Baltimore Gas and Electric (BGE) has recently installed a reheater for use in a plant in Prince Georges County, MD. The reheater will receive steam at 850 psia and 700°F and release it at 1200°F. Because of losses in the tubing, there is a 12% pressure drop in the steam. The steam flow rate is 2000000 lb/h. Hot flue gases from a burner enter the reheater at 1900°F and 3500000 lb/h to flow over the steam tubes that are arranged in-line. The overall heat transfer coefficient is approximately 8.5 Btu/(h ft² °F). Estimate the effectiveness of the reheater by using the ε-NTU and by conducting a second-law analysis. Specify all assumptions and comment on the results of the calculations.

6.13. It is well known that exhaust end losses between the LP steam turbine exit and the condenser inlet will result in different values of enthalpy at the two points. In other words, the ELEP will differ from the UEEP. Conduct an analysis to show that the exhaust end loss represents the amount of steam kinetic energy that is lost between the turbine exit and the condenser inlet. Present all applicable assumptions and explain the results of the analysis.

6.14. Syncrude Canada Ltd. has contracted the services of a consulting firm to assess the performance of a small steam-turbine plant. This existing plant has a single-stage steam turbine and one stage of closed feedwater heating, which comes complete with drain cooling. It has been specified that the full-load steam mass flow rate is 255000 lb/h. The turbine inlet conditions are 1000 psia and 900°F. Steam is extracted from the turbine at 200 psia for feedwater heating. Losses in the plant system result in a turbine internal efficiency of 86%, a pump efficiency of 70%, and a total boiler pressure drop of 60 psia. An auxiliary power requirement of 8% will be derived from the steam turbine to maintain plant operations. Using practical knowledge of steam-turbine plants (where applicable), conduct an appropriate analysis to prepare a heat-and-mass balance diagram for full-load operation of the plant. Was a desuperheating zone of tubes installed in the feedwater heat exchanger?

References and Further Reading

[1] Li, K. and Priddy, A. (1985) *Power Plant System Design*, John Wiley & Sons, Ltd, New York.
[2] Mollier, R. (1906) *Neue Tabellen und Diagramme für Wasserdampf*, Springer-Verlag, Berlin.
[3] Mollier, R. (1927) *The Mollier Steam Tables and Diagrams, Extended to the Critical Pressure* (English edition adapted and extended from the third German edition by H. Moss), Pitman and Sons, London.

Appendix A

Pipe and Duct Systems

Table A.1 Average roughness of commercial pipes

New Material	ε ft	mm
Riveted steel	0.003–0.03	0.9–9.0
Concrete	0.001–0.01	0.3–3.0
Wood stave	0.0006–0.003	0.18–0.9
Cast iron	0.00085	0.26
Galvanized iron	0.0005	0.15
Asphalted cast iron	0.0004	0.12
Commercial steel	0.00015	0.046
Drawn tubing	0.000005	0.0015
Glass	"Smooth"	"Smooth"

Source: Fox, R. and McDonald, A. (1998) *Introduction to Fluid Mechanics*, 5th edn, John Wiley & Sons, Inc., New York

Introduction to Thermo-Fluids Systems Design, First Edition. André G. McDonald and Hugh L. Magande.
© 2012 André G. McDonald and Hugh L. Magande. Published 2012 by John Wiley & Sons, Ltd.

Table A.2   Correlation equations for friction factors

Swamee-Jain formulae	$f_o = 0.25 \left[\log \left(\dfrac{\varepsilon/D}{3.7} + \dfrac{5.74}{Re_D^{0.9}} \right) \right]^{-2}$
Colebrook equation	$\dfrac{1}{\sqrt{f}} = -2.0 \log \left(\dfrac{\varepsilon/D}{3.7} + \dfrac{2.51}{Re_D\sqrt{f}} \right)$
Haaland's approximation	$\dfrac{1}{\sqrt{f}} = -1.8 \log \left[\dfrac{6.9}{Re_D} + \left(\dfrac{\varepsilon/D}{3.7} \right)^{1.11} \right]$
Blasius correlation for smooth pipes	$f = \dfrac{0.316}{Re_D^{0.25}}$, for $Re_D \leq 10^5$
Churchill's equation	$f = 8 \left[\left(\dfrac{8}{Re_D} \right)^{12} + (A+B)^{-3/2} \right]^{1/12}$, $A = \left[2.457 \ln \left(\dfrac{1}{C} \right) \right]^{16}$, $B = \left(\dfrac{37530}{Re_D} \right)^{16}$, $C = \left(\dfrac{7}{Re_D} \right)^{0.9} + 0.27 \left(\dfrac{\varepsilon}{D} \right)$
First Petukhov equation for turbulent flow in smooth tubes	$f = [0.79 \ln Re - 1.64]^{-2}$, for $3000 < Re < 5 \times 10^6$

Table A.3 Circular equivalents of rectangular ducts for equal friction and capacity

	Diameter of Circular Duct																
y / x	6	7	8	9	10	11	12	13	14	15	16	17	18	19	20	22	24
6	6.6																
7	7.1	7.7															
8	7.5	8.2	8.8														
9	8.0	8.6	9.3	9.9													
10	8.4	9.1	9.8	10.4	10.9												
11	8.8	9.5	10.2	10.8	11.4	12.0											
12	9.1	9.9	10.7	11.3	11.9	12.5	13.1										
13	9.5	10.3	11.1	11.8	12.4	13.0	13.6	14.2									
14	9.8	10.7	11.5	12.2	12.9	13.5	14.2	14.7	15.3								
15	10.1	11.0	11.8	12.6	13.3	14.0	14.6	15.3	15.8	16.4							
16	10.4	11.4	12.2	13.0	13.7	14.4	15.1	15.7	16.3	16.9	17.5						
17	10.7	11.7	12.5	13.4	14.1	14.9	15.5	16.1	16.8	17.4	18.0	18.6					
18	11.0	11.9	12.9	13.7	14.5	15.3	16.0	16.6	17.3	17.9	18.5	19.1	19.7				
19	11.2	12.2	13.2	14.1	14.9	15.6	16.4	17.1	17.8	18.4	19.0	19.6	20.2	20.8			
20	11.5	12.5	13.5	14.4	15.2	15.9	16.8	17.5	18.2	18.8	19.5	20.1	20.7	21.3	21.9		
22	12.0	13.1	14.1	15.0	15.9	16.7	17.6	18.3	19.1	19.7	20.4	21.0	21.7	22.3	22.9	24.1	
24	12.4	13.6	14.6	15.6	16.6	17.5	18.3	19.1	19.8	20.6	21.3	21.9	22.6	23.2	23.9	25.1	26.2
26	12.8	14.1	15.2	16.2	17.2	18.1	19.0	19.8	20.6	21.4	22.1	22.8	23.5	24.1	24.8	26.1	27.2
28	13.2	14.5	15.6	16.7	17.7	18.7	19.6	20.5	21.3	22.1	22.9	23.6	24.4	25.0	25.7	27.1	28.2
30	13.6	14.9	16.1	17.2	18.3	19.3	20.2	21.1	22.0	22.9	23.7	24.4	25.2	25.9	26.7	28.0	29.3
32	14.0	15.3	16.5	17.7	18.8	19.8	20.8	21.8	22.7	23.6	24.4	25.2	26.0	26.7	27.5	28.9	30.1
34	14.4	15.7	17.0	18.2	19.3	20.4	21.4	22.4	23.3	24.2	25.1	25.9	26.7	27.5	28.3	29.7	31.0
36	14.7	16.1	17.4	18.6	19.8	20.9	21.9	23.0	23.9	24.8	25.8	26.6	27.4	28.3	29.0	30.5	32.0
38	15.0	16.4	17.8	19.0	20.3	21.4	22.5	23.5	24.5	25.4	26.4	27.3	28.1	29.0	29.8	31.4	32.8
40	15.3	16.8	18.2	19.4	20.7	21.9	23.0	24.0	25.1	26.0	27.0	27.9	28.8	29.7	30.5	32.1	33.6

Dimensions in inches, feet, or meters.

Table A.4 Approximate equivalent lengths for selected fittings in circular ducts

Fitting	Diameter (in.)	6	8	10	12	L_e/D
Elbows						
Pleated, 90°		8	10	13	15	15
Pleated, 45°		5	6	8	9	9
Mitered, 90°		30	40	50	60	60
Mitered with vanes		5	7	8	10	10
Transitions						
Converging, 20°		2	3	3	4	4
Diverging, 120°		20	27	33	40	40
Abrupt expansion		30	40	50	60	60
Round to rectangular boot, 90°		25	33	40	50	50
Round to rectangular boot, straight		5	7	8	10	10
Entrances						
Abrupt, 90°		15	20	25	30	30
Bellmouth		6	8	10	12	12
Branch fittings, diverging						
Wye, 45°, branch		10	13	17	20	20
Wye, 45°, through		4	5	7	8	8
Tee, branch		20	27	33	40	40
Tee, through		4	5	7	8	8
Branch fittings, converging						
Wye, 45°, branch		10	13	17	20	20
Wye, 45°, through		5	7	8	10	10
Tee, branch		20	27	33	40	40
Tee, through		6	8	10	12	12

Equivalent Length (ft)

Source: McQuiston, F., Parker, J., and Spitler, J. (2000) *Heating, Ventilating, and Air Conditioning: Analysis and Design*, 5th edn, John Wiley & Sons, Inc., New York, p. 433.

Table A.5 Approximate equivalent lengths for elbows in ducts

Fitting	Diameter (in.)	6	8	10	12	L_e/D
Elbows $R/D = 1.5$						
Smooth, 90°		4.5	6	–	–	9
5-piece, 90°		6	8	10	12	12
3-piece, 90°		12	16	20	24	24
Smooth, 45°		2.3	3	–	–	4.5
3-piece, 45°		3	4	5	6	6

Equivalent Length (ft)

Source: *System Design Manual, Part 2*: Air Distribution, Carrier Air Conditioning Co., Syracuse, NY, 1974.

Table A.6 Data for copper pipes

Material	Diameter (in.)			Weight per Linear Foot of Pipe and Water (lb)	Gallons of Water per Linear Foot
	Nominal	Inner	Outer		
Copper					
Type L	1/4	0.315	0.375	0.16	0.004
Type L	3/8	0.430	0.500	0.26	0.008
Type L	1/2	0.545	0.625	0.39	0.012
Type L	5/8	0.666	0.750	0.512	0.018
Type L	3/4	0.785	0.875	0.67	0.025
Type L	1	1.025	1.125	1.01	0.043
Type L	11/4	1.265	1.375	1.43	0.065
Type L	11/2	1.505	1.625	1.91	0.093
Type L	2	1.985	2.125	3.09	0.161
Type L	21/2	2.465	2.625	4.55	0.248
Type L	3	2.945	3.125	6.29	0.354
Type L	31/2	3.425	3.625	8.29	0.479
Type L	4	3.905	4.125	10.58	0.622
Type L	5	4.875	5.125	15.70	0.970
Type L	6	5.845	6.125	21.81	1.394
Type L	8	7.725	8.125	39.58	2.435
Type L	10	9.625	10.125	61.61	3.780
Type L	12	11.565	12.125	85.89	5.457
Type K	1/4	0.305	0.375	0.177	0.004
Type K	3/8	0.402	0.500	0.275	0.007
Type K	1/2	0.527	0.625	0.438	0.011
Type K	5/8	0.652	0.750	0.563	0.017
Type K	3/4	0.745	0.875	0.829	0.023
Type K	1	0.995	1.125	1.176	0.040
Type K	11/4	1.245	1.375	1.570	0.063
Type K	11/2	1.481	1.625	2.109	0.089
Type K	2	1.959	2.125	3.364	0.157
Type K	21/2	2.435	2.625	4.927	0.242
Type K	3	2.907	3.125	6.870	0.345
Type K	31/2	3.385	3.625	9.051	0.467
Type K	4	3.857	4.125	11.564	0.607
Type K	5	4.805	5.125	17.532	0.942
Type K	6	5.741	6.125	25.132	1.345
Type K	8	7.583	8.125	45.494	2.346
Type K	10	9.449	10.125	70.689	3.643
Type K	12	11.315	12.125	101.355	5.224

Table A.7 Data for schedule 40 steel pipes

Material	Diameter (in.)			Weight per Linear Foot of Pipe and Water (lb)	Gallons of Water per Linear Foot
	Nominal	Inner	Outer		
Steel					
Schedule 40	1/4	0.364	0.540	0.475	0.005
Schedule 40	3/8	0.493	0.675	0.647	0.010
Schedule 40	1/2	0.622	0.840	0.992	0.016
Schedule 40	3/4	0.824	1.050	1.372	0.028
Schedule 40	1	1.049	1.315	2.055	0.045
Schedule 40	1 1/4	1.380	1.660	2.929	0.077
Schedule 40	1 1/2	1.610	1.900	3.602	0.106
Schedule 40	2	2.067	2.375	5.114	0.174
Schedule 40	2 1/2	2.469	2.875	7.873	0.248
Schedule 40	3	3.068	3.500	10.781	0.383
Schedule 40	3 1/2	3.548	4.000	13.397	0.513
Schedule 40	4	4.026	4.500	16.316	0.660
Schedule 40	5	5.047	5.563	23.280	1.039
Schedule 40	6	6.065	6.625	31.490	1.501
Schedule 40	8	7.981	8.625	47.150	2.599
Schedule 40	10	10.020	10.750	74.600	4.096
Schedule 40	12	11.938	12.750	102.100	5.815
Schedule 40	14	13.126	14.000	121.870	7.029
Schedule 40	16	15.000	16.000	159.500	9.180
Schedule 40	18	16.874	18.000	202.200	11.617
Schedule 40	20	18.814	20.000	243.400	14.442
Schedule 40	24	22.626	24.000	345.200	20.887

Table A.8 Data for schedule 80 steel pipes

Material	Diameter (in.)			Weight per Linear Foot of Pipe and Water (lb)	Gallons of Water per Linear Foot
	Nominal	Inner	Outer		
Steel					
Schedule 80	3/8	0.423	0.675	0.798	0.007
Schedule 80	1/2	0.546	0.840	1.189	0.012
Schedule 80	3/4	0.742	1.050	1.686	0.026
Schedule 80	1	0.957	1.315	2.483	0.037
Schedule 80	1 1/4	1.278	1.660	3.551	0.067
Schedule 80	1 1/2	1.500	1.900	4.396	0.092
Schedule 80	2	1.939	2.375	6.302	0.154
Schedule 80	2 1/2	2.323	2.875	9.491	0.220
Schedule 80	3	2.900	3.500	13.122	0.344
Schedule 80	3 1/2	3.364	4.000	16.225	0.458
Schedule 80	4	3.826	4.500	19.953	0.597
Schedule 80	5	4.813	5.563	28.650	0.945
Schedule 80	6	5.761	6.625	39.860	1.354
Schedule 80	8	7.625	8.625	63.200	2.372
Schedule 80	10	9.564	10.750	95.840	3.732
Schedule 80	12	11.376	12.750	132.600	5.280
Schedule 80	14	12.500	14.000	158.200	6.375
Schedule 80	16	14.314	16.000	206.700	8.360
Schedule 80	18	16.126	18.000	259.500	10.610
Schedule 80	20	17.938	20.000	318.400	13.128
Schedule 80	24	21.564	24.000	455.200	18.972

Table A.9   Data for class 150 cast iron pipes

Material	Diameter (in.) Nominal	Inner	Outer	Weight per Linear Foot of Pipe and Water (lb)	Gallons of Water per Linear Foot
Cast Iron					
Class 150	3	3.32	3.96	15.92	0.450
Class 150	4	4.10	4.80	21.97	0.686
Class 150	6	6.14	6.90	38.43	1.538
Class 150	8	8.23	9.05	59.66	2.763
Class 150	10	10.22	11.10	73.94	4.261
Class 150	12	12.24	13.20	113.82	6.113
Class 150	14	14.28	15.30	148.05	8.320
Class 150	16	16.32	17.40	185.30	10.867
Class 150	18	18.34	19.50	228.69	13.723
Class 150	20	20.36	21.60	277.44	16.913
Class 150	24	24.34	25.80	391.31	24.171
Class 150	30	30.30	32.00	589.19	37.458
Class 150	36	36.42	38.30	814.90	54.118
Class 150	42	42.40	44.50	1087.30	73.348
Class 150	48	48.52	50.80	1392.60	96.051

Table A.10   Data for glass pipes

Material	Diameter (in.) Nominal	Inner	Outer	Weight per Linear Foot of Pipe and Water (lb)	Gallons of Water per Linear Foot
Glass					
Regular Schedule	$1\frac{1}{2}$	1.60	1.84	0.89	0.104
Regular Schedule	2	2.06	2.34	1.45	0.173
Regular Schedule	3	3.07	3.41	3.19	0.385
Regular Schedule	4	4.13	4.53	5.79	0.696
Regular Schedule	6	6.18	6.66	12.78	1.558
Heavy Schedule	1	1.00	1.31	0.95	0.041
Heavy Schedule	$1\frac{1}{2}$	1.50	1.84	1.63	0.092
Heavy Schedule	2	2.00	2.34	2.46	0.163
Heavy Schedule	3	3.00	3.41	5.06	0.367
Heavy Schedule	4	4.00	4.53		
Heavy Schedule	6	6.00	6.66		

Table A.11 Data for PVC plastic pipes

Material	Diameter (in.)			Weight per Linear Foot of Pipe and Water (lb)	Gallons of Water per Linear Foot
	Nominal	Inner	Outer		
PVC Plastic					
Schedule 40	1/8	0.269	0.405	0.068	0.003
Schedule 40	1/4	0.364	0.540	0.119	0.005
Schedule 40	3/8	0.493	0.675	0.183	0.010
Schedule 40	1/2	0.622	0.840	0.282	0.016
Schedule 40	3/4	0.784	1.050	0.429	0.025
Schedule 40	1	1.049	1.315	0.669	0.045
Schedule 40	1 1/4	1.380	1.660	1.047	0.078
Schedule 40	1 1/2	1.610	1.900	1.360	0.106
Schedule 40	2	2.067	2.375	2.095	0.174
Schedule 40	2 1/2	2.469	2.875	3.092	0.249
Schedule 40	3	3.068	3.500	4.533	0.384
Schedule 40	3 1/2	3.548	4.000	5.878	0.514
Schedule 40	4	4.026	4.500	7.409	0.661
Schedule 40	5	5.047	5.563	11.430	1.039
Schedule 40	6	6.065	6.625	15.489	1.501
Schedule 40	8	7.981	8.625	26.880	2.599
Schedule 40	10	10.018	10.750	41.605	4.095
Schedule 40	12	11.938	12.750	58.523	5.815
Schedule 80	1/8	0.215	0.405	0.071	0.002
Schedule 80	1/4	0.302	0.540	0.125	0.004
Schedule 80	3/8	0.423	0.675	0.190	0.007
Schedule 80	1/2	0.546	0.840	0.251	0.012
Schedule 80	3/4	0.742	1.050	0.481	0.022
Schedule 80	1	0.957	1.315	0.693	0.037
Schedule 80	1 1/4	1.278	1.660	1.082	0.067
Schedule 80	1 1/2	1.500	1.900	1.404	0.092
Schedule 80	2	1.939	2.375	2.163	0.153
Schedule 80	2 1/2	2.323	2.875	3.184	0.220
Schedule 80	3	2.900	3.500	4.664	0.343
Schedule 80	3 1/2	3.364	4.000	6.045	0.462
Schedule 80	4	3.826	4.500	7.616	0.597
Schedule 80	5	4.780	5.563	11.996	0.932
Schedule 80	6	5.761	6.625	16.318	1.354
Schedule 80	8	7.625	8.625	27.823	2.372
Schedule 80	10	9.564	10.750	42.994	3.732
Schedule 80	12	11.376	12.750	60.365	5.280

Source: Some data from Erico International Corp., *Pipe Hanger and Support Recommended Specifications*, Erico Corp., Solon, OH, 2010, pp. 114–117.

Table A.12 Typical average velocities for selected pipe flows[a]

Fluid	Application	Velocity (fps)	Velocity (m/s)
Steam	Superheated process steam	148–328	45–100
	Auxiliary heat steam	98–246	30–75
	Saturated and low-pressure steam	98–164	30–50
Water	Centrifugal pump suction lines	3–4.9 (must be <4.9 fps)[b]	0.9–1.5 (must be < 1.5 m/s)[b]
	Power plant feedwater	7.9–15	2.4–4.6
	General building service	3.9–10.2	1.2–3.1
	Potable water	Up to 6.9 (must be <9.8 fps)[b]	Up to 2.1 (must be <3.0 m/s)[b]

[a] Adapted from the US Department of the Army, TM 5-810-15, Central Boiler Plants, August 1995.
[b] Adapted from *2005 Fundamentals* American Society of Heating, Refrigerating, and Air-Conditioning Engineers, Atlanta, GA, 2005, pp. 36–11.

Table A.13 Erosion limits: maximum design fluid velocities for water flow in small tubes

Low carbon steel	10 ft/s
Stainless steel	15 ft/s
Aluminum	6 ft/s
Copper	6 ft/s
90–10 Cupronickel	10 ft/s
70–30 Cupronickel	15 ft/s
Titanium	50 ft/s
For other liquids	$V_{liq,max} = V_{water,max}\left[\dfrac{\rho_{water}}{\rho_{liq}}\right]$
For gases and dry vapors (ft/s), where M = molecular weight	$V_{gas,max} = \sqrt{\dfrac{1800}{PM}}$

Source: Adapted from Wolverine Tube Inc., *Wolverine Tube Heat Transfer Data Book*, Wolverine Tube Inc., Huntsville, AL, 2009, p. 48.

Table A.14   Loss coefficients for pipe fittings

Nominal Diameter (in.)	Screwed				Flanged				
	1/2	1	2	4	1	2	4	8	20
Valves (FO)									
Globe	14	8.2	6.9	5.7	13	8.5	6.0	5.8	5.5
Gate	0.30	0.24	0.16	0.11	0.80	0.35	0.16	0.07	0.03
Swing check	5.1	2.9	2.1	2.0	2.0	2.0	2.0	2.0	2.0
Angle	9.0	4.7	2.0	1.0	4.5	2.4	2.0	2.0	2.0
Ball valve[a]	0.05	0.05	0.05	0.05	0.05	0.05	0.05	0.05	0.05
Gate valve[a]	1/4C	1/2 C	3/4 C						
	0.3	2.1	17						
Foot valve with strainer[a,b]	Poppet disk	Hinged disk							
	7	1.25							
Elbows									
45° regular	0.39	0.32	0.30	0.29					
45° long radius					0.21	0.20	0.19	0.16	0.14
90° regular	2.0	1.5	0.95	0.64	0.50	0.39	0.30	0.26	0.21
90° long radius	1.0	0.72	0.41	0.23	0.40	0.30	0.19	0.15	0.10
180° regular	2.0	1.5	0.95	0.64	0.41	0.35	0.30	0.25	0.20
180° long radius					0.40	0.30	0.21	0.15	0.10
Tees									
Line flow	0.90	0.90	0.90	0.90	0.24	0.19	0.14	0.10	0.07
Branch flow	2.4	1.8	1.4	1.1	1.0	0.80	0.64	0.58	0.41
Expansion[c]	d/D	d/D	d/D	d/D					
	0.2	0.4	0.6	0.8					
$K_{expansion}$	0.30	0.25	0.15	0.10					
Contraction[c]:	60° contraction angle								
	0.07								

[c]*Source*: Çengel, Y. and Cimbala, J. (2009) *Fluid Mechanics: Fundamentals and Applications*, 2nd edn., New York.

FO, fully open; C, closed.

[a]These are representative loss coefficient values. Consult manufacturer's data for final design values.

[b]Values estimated with data from: Flow of fluids through valves, fittings, and pipe, Technical Paper No. 410, Crane Company, New York, 1982.

Table A.15 Typical pipe data format

Pipe Data							
Pipe Section No.	Pipe Length (ft)	Flow Rate (gpm)	Lost Head (ft/100 ft)	Fluid Velocity (ft/s)	Nominal Size (in.)	Minor Losses (ft)	Total Head Loss (ft)

Table A.16 Typical pump schedule format

Pump Schedule									
				Fluid			Electrical		
Tag	Manufacturer and Model Number	Type	Construction	Flow Rate (gpm)	Working Fluid	Head Loss (ft)	Motor Size (hp)	Motor Speed (rpm)	Volt/ pH/ Hz

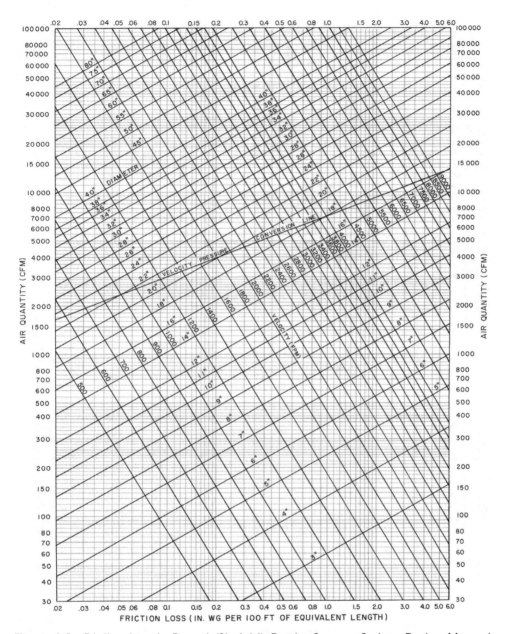

Figure A.1 Friction Loss in Round (Straight) Ducts. *Source: System Design Manual, Part 2: Air Distribution*, Carrier Air Conditioning Co., Syracuse, NY, 1974 (Reprinted with permission)

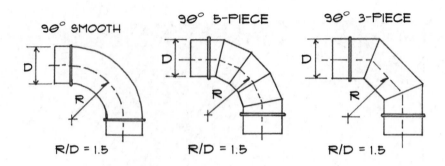

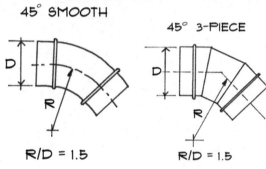

Figure A.2   Schematics elbows in ducts

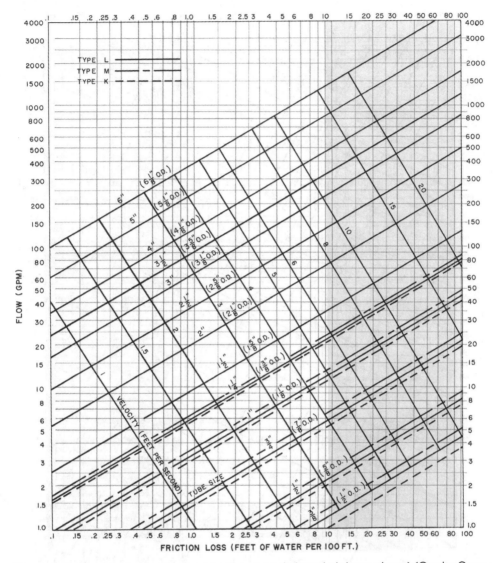

Figure A.3 Copper tubing friction loss (open and closed piping systems) (Carrier Corp.; reprinted with permission)

(*a*) Open Piping Systems

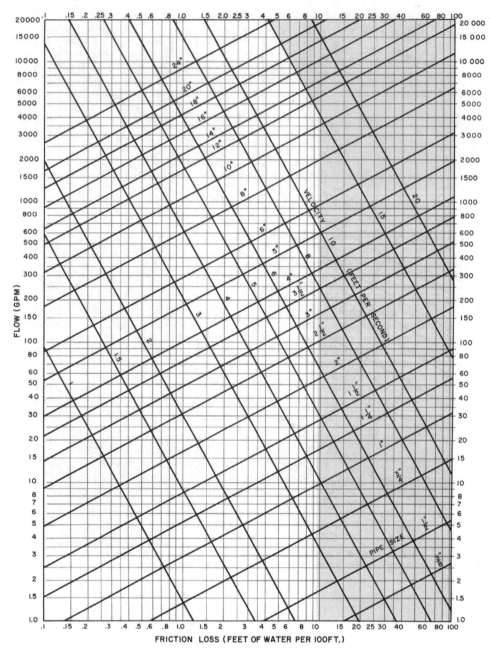

Figure A.4 Commercial steel pipe (Schedule 40) friction loss. (a) *Open piping systems* (Carrier Corp.; reprinted with permission); (b) *closed piping systems* (Carrier Corp.; reprinted with permission)

(b) Closed Piping Systems

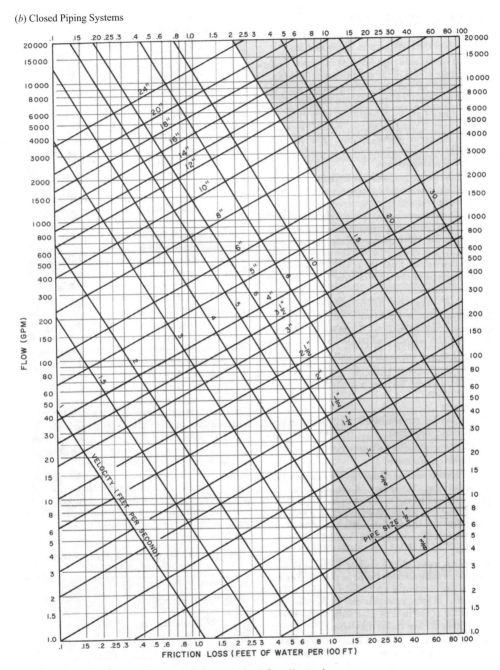

Figure A.4 (*Continued*)

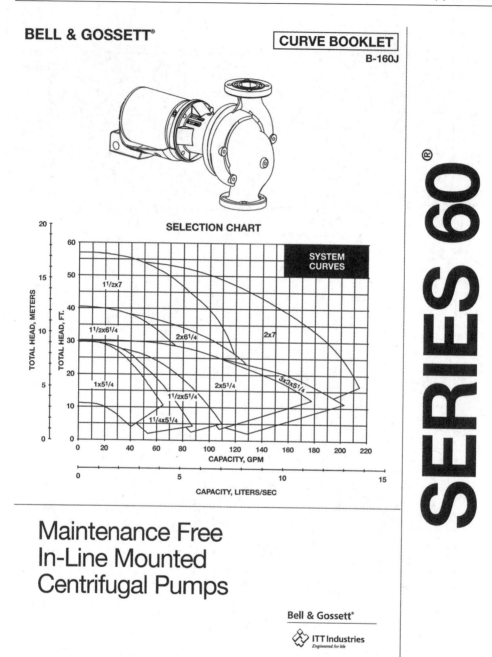

Figure A.5   Bell & Gosset pump catalog (ITT Bell & Gossett; reprinted with permission)

SERIES 60 STOCK PUMPS

BRONZE FITTED CONSTRUCTION

SINGLE PHASE UNITS

PART #	MODEL	PUMP SIZE	MOTOR HP	IMPELLER DIA.
172701	601S	1x1x5^1/$_4$	^{1}/$_4$	4.38"
172702	602S	1x1x5^1/$_4$	^{1}/$_3$	4.84"
172703	603S	1x1x5^1/$_4$	^{1}/$_2$	5.25"
172707	604S	1^1/$_4$x1^1/$_4$x5^1/$_4$	^{1}/$_4$	4.38"
172708	605S	1^1/$_4$x1^1/$_4$x5^1/$_4$	^{1}/$_3$	4.84"
172667	606S	1^1/$_4$x1^1/$_4$x5^1/$_4$	^{1}/$_2$	5.25"
172712	607S	1^1/$_2$x1^1/$_2$x5^1/$_4$	^{1}/$_3$	4.38"
172713	608S	1^1/$_2$x1^1/$_2$x5^1/$_4$	^{1}/$_2$	4.94"
172668	609S	1^1/$_2$x1^1/$_2$x5^1/$_4$	^{3}/$_4$	5.25"
172717	610S	2x2x5^1/$_4$	^{1}/$_2$	4.12"
172718	611S	2x2x5^1/$_4$	^{3}/$_4$	4.75"
172669	612S	2x2x5^1/$_4$	1	5.25"
172755	621S	1^1/$_2$x1^1/$_2$x6^1/$_4$	^{1}/$_2$	4.88"
172722	613S	1^1/$_2$x1^1/$_2$x6^1/$_4$	^{3}/$_4$	5.75"
172670	614S	1^1/$_2$x1^1/$_2$x6^1/$_4$	1	6.25"
172723	615S	2x2x6^1/$_4$	^{3}/$_4$	5.62"
172671	616S	2x2x6^1/$_4$	1	6.22"
172757	622S	1^1/$_2$x1^1/$_2$x7	^{3}/$_4$	5.50"
172758	623S	1^1/$_2$x1^1/$_2$x7	1	6.00"
172724	617S	1^1/$_2$x1^1/$_2$x7	1^1/$_2$	6.50"
172761	624S	2x2x7	1	5.69"
172762	625S	2x2x7	1^1/$_2$	6.12"

THREE PHASE UNITS

PART #	MODEL	PUMP SIZE	MOTOR HP	IMPELLER DIA.
172725	601T	1x1x5^1/$_4$	^{1}/$_4$	4.38"
172726	602T	1x1x5^1/$_4$	^{1}/$_3$	4.84"
172727	603T	1x1x5^1/$_4$	^{1}/$_2$	5.25"
172731	604T	1^1/$_4$x1^1/$_4$x5^1/$_4$	^{1}/$_4$	4.38"
172732	605T	1^1/$_4$x1^1/$_4$x5^1/$_4$	^{1}/$_3$	4.84"
172733	606T	1^1/$_4$x1^1/$_4$x5^1/$_4$	^{1}/$_2$	5.25"
172737	607T	1^1/$_2$x1^1/$_2$x5^1/$_4$	^{1}/$_3$	4.38"
172738	608T	1^1/$_2$x1^1/$_2$x5^1/$_4$	^{1}/$_2$	4.94"
172739	609T	1^1/$_2$x1^1/$_2$x5^1/$_4$	^{3}/$_4$	5.25"
172743	610T	2x2x5^1/$_4$	^{1}/$_2$	4.12"
172744	611T	2x2x5^1/$_4$	^{3}/$_4$	4.75"
172745	612T	2x2x5^1/$_4$	1	5.25"
172756	621T	1^1/$_2$x1^1/$_2$x6^1/$_4$	^{1}/$_2$	4.88"
172749	613T	1^1/$_2$x1^1/$_2$x6^1/$_4$	^{3}/$_4$	5.75"
172750	614T	1^1/$_2$x1^1/$_2$x6^1/$_4$	1	6.25"
172751	615T	2x2x6^1/$_4$	^{3}/$_4$	5.62"
172752	616T	2x2x6^1/$_4$	1	6.22"
172759	622T	1^1/$_2$x1^1/$_2$x7	^{3}/$_4$	5.50"
172760	623T	1^1/$_2$x1^1/$_2$x7	1	6.00"
172753	617T	1^1/$_2$x1^1/$_2$x7	1^1/$_2$	6.50"
172672	618T	1^1/$_2$x1^1/$_2$x7	2	7.00"
172763	624T	2x2x7	1	5.69"
172764	625T	2x2x7	1^1/$_2$	6.12"
172754	619T	2x2x7	2	6.50"
172673	620T	2x2x7	3	7.00"

Pump Construction: Standard Buna/Carbon-Ceramic Seal, maximum 175 psi working pressure, motors are 1750 RPM ODP. Three Phase are 208-230/460 volts, Single Phase 115/230 volts.

Figure A.5 *(Continued)*

SERIES 60 STOCK PUMPS

ALL BRONZE CONSTRUCTION

SINGLE PHASE UNITS

PART #	MODEL	PUMP SIZE	MOTOR HP	IMPELLER DIA.
172704	B601S	1x1x5$\frac{1}{4}$	$\frac{1}{4}$	4.38"
172705	B602S	1x1x5$\frac{1}{4}$	$\frac{1}{3}$	4.84"
172706	B603S	1x1x5$\frac{1}{4}$	$\frac{1}{2}$	5.25"
172709	B604S	1$\frac{1}{4}$x1$\frac{1}{4}$x5$\frac{1}{4}$	$\frac{1}{4}$	4.38"
172710	B605S	1$\frac{1}{4}$x1$\frac{1}{4}$x5$\frac{1}{4}$	$\frac{1}{3}$	4.84"
172711	B606S	1$\frac{1}{4}$x1$\frac{1}{4}$x5$\frac{1}{4}$	$\frac{1}{2}$	5.25"
172714	B607S	1$\frac{1}{2}$x1$\frac{1}{2}$x5$\frac{1}{4}$	$\frac{1}{3}$	4.38"
172715	B608S	1$\frac{1}{2}$x1$\frac{1}{2}$x5$\frac{1}{4}$	$\frac{1}{2}$	4.94"
172716	B609S	1$\frac{1}{2}$x1$\frac{1}{2}$x5$\frac{1}{4}$	$\frac{3}{4}$	5.25"
172719	B610S	2x2x5$\frac{1}{4}$	$\frac{1}{2}$	4.12"
172720	B611S	2x2x5$\frac{1}{4}$	$\frac{3}{4}$	4.75"
172721	B612S	2x2x5$\frac{1}{4}$	1	5.25"

THREE PHASE UNITS

PART #	MODEL	PUMP SIZE	MOTOR HP	IMPELLER DIA.
172728	B601T	1x1x5$\frac{1}{4}$	$\frac{1}{4}$	4.38"
172729	B602T	1x1x5$\frac{1}{4}$	$\frac{1}{3}$	4.84"
172730	B603T	1x1x5$\frac{1}{4}$	$\frac{1}{2}$	5.25"
172734	B604T	1$\frac{1}{4}$x1$\frac{1}{4}$x5$\frac{1}{4}$	$\frac{1}{4}$	4.38"
172735	B605T	1$\frac{1}{4}$x1$\frac{1}{4}$x5$\frac{1}{4}$	$\frac{1}{3}$	4.84"
172736	B606T	1$\frac{1}{4}$x1$\frac{1}{4}$x5$\frac{1}{4}$	$\frac{1}{2}$	5.25"
172740	B607T	1$\frac{1}{2}$x1$\frac{1}{2}$x5$\frac{1}{4}$	$\frac{1}{3}$	4.38"
172741	B608T	1$\frac{1}{2}$x1$\frac{1}{2}$x5$\frac{1}{4}$	$\frac{1}{2}$	4.94"
172742	B609T	1$\frac{1}{2}$x1$\frac{1}{2}$x5$\frac{1}{4}$	$\frac{3}{4}$	5.25"
172746	B610T	2x2x5$\frac{1}{4}$	$\frac{1}{2}$	4.00"
172747	B611T	2x2x5$\frac{1}{4}$	$\frac{3}{4}$	4.75"
172748	B612T	2x2x5$\frac{1}{4}$	1	5.25"

Pump Construction: Standard Buna/Carbon-Ceramic Seal,
maximum 175 psi working pressure,
motors are 1750 RPM ODP.
Three Phase are 208-230/460 volts,
Single Phase 115/230 volts.

Figure A.5 *(Continued)*

SERIES 60 BUILT-TO-ORDER PUMP PERFORMANCE CURVES

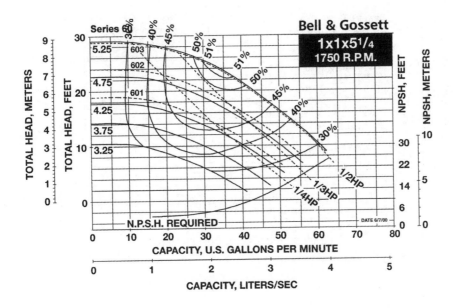

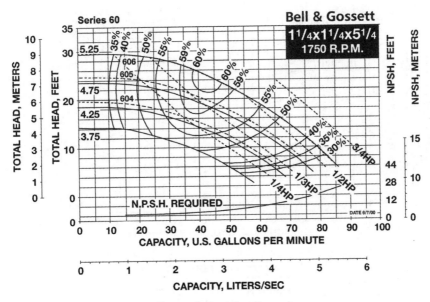

Figure A.5 *(Continued)*

SERIES 60 BUILT-TO-ORDER PUMP PERFORMANCE CURVES

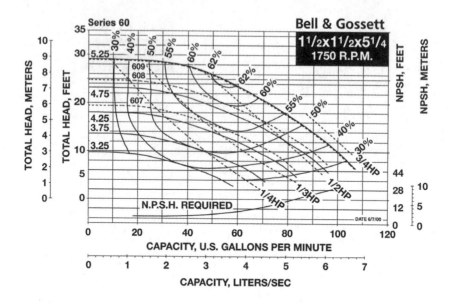

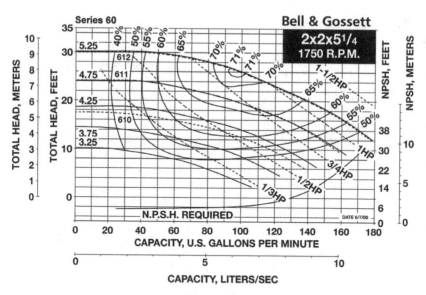

Figure A.5 (*Continued*)

SERIES 60 BUILT-TO-ORDER PUMP PERFORMANCE CURVES

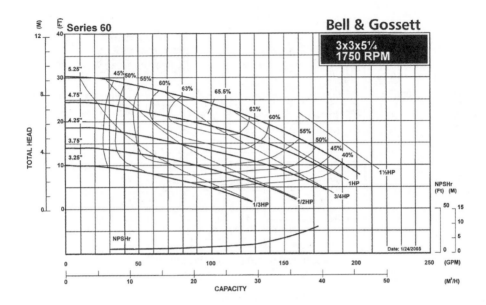

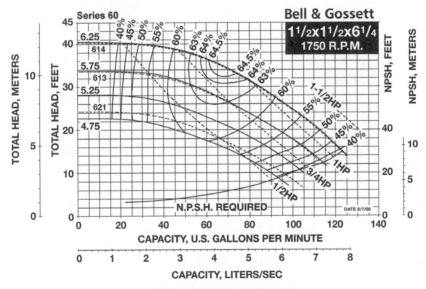

Figure A.5 (*Continued*)

SERIES 60 BUILT-TO-ORDER PUMP PERFORMANCE CURVES

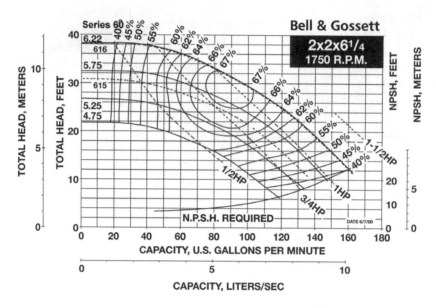

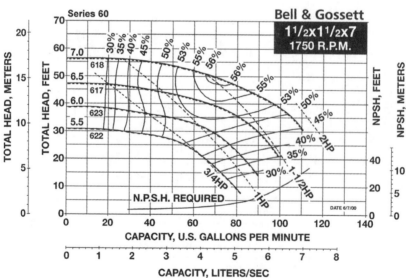

Figure A.5 *(Continued)*

B-160J

SERIES 60 BUILT-TO-ORDER PUMP PERFORMANCE CURVES

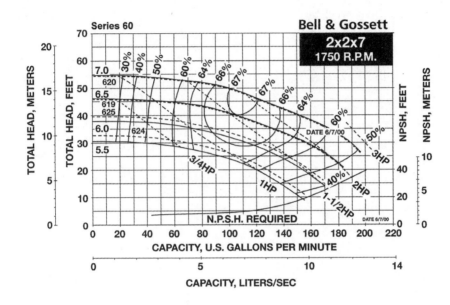

Bell & Gossett

© COPYRIGHT 1967, 2005 BY ITT INDUSTRIES, INC.

PRINTED IN U.S.A. 5-05

USA
Bell & Gossett
8200 N. Austin Avenue
Morton Grove, IL 60053
Phone: (847) 966-3700
Facsimile: (847) 966-9052
http://www.bellgossett.com

**ISO 9001
Certified**

INTL..
Bell & Gossett / Export Dept.
8200 N. Austin Avenue
Morton Grove, IL 60053
Phone: (847) 966-3700
Facsimile: (847) 966-8366
http://www.bellgossett.com

CANADA
Fluid Products Canada
55 Royal Road
Guelph, Ontario,
N1H 1T1, Canada
Phone: (519) 821-1900

Figure A.5   (*Continued*)

Appendix B

Symbols for Drawings

Table B.1 Air moving devices and ductwork symbols

Air Movement	
Fans	
Axial flow fan	
Centrifugal fan	
Propeller fan	
Intake roof ventilator (supply)	
Exhaust roof ventilator	
Ductwork	
Direction of flow	
Change of elevation [rise (R), drop (D)]	
Vertical or horizontal access doors	
Acoustical lining (sound insulation)	
Cowl, gooseneck, and flashing	

(continued)

Introduction to Thermo-Fluids Systems Design, First Edition. André G. McDonald and Hugh L. Magande.
© 2012 André G. McDonald and Hugh L. Magande. Published 2012 by John Wiley & Sons, Ltd.

Table B.1   *(Continued)*

Ductwork *(cont.)*	
Flexible connection	
Flexible duct	
Sound attenuator	SA
Mixing terminal unit	H C → TU M-1
Reheat terminal unit	TU RH-1
Variable air volume terminal unit	TU VAV-1
Turning vanes	
Fire and/or smoke detectors	1 2

Source: HVAC Duct Construction Standards: Metal and Flexible, Sheet Metal and Air Conditioning Contractors National Association, Inc., Vienna, VA, 1985.

Table B.2 Piping symbols

Piping

Heating

High-pressure steam	———— HPS ————
Medium-pressure steam	———— MPS ————
Low-pressure steam	———— LPS ————
High-pressure condensate	———— HPC ————
Medium-pressure condensate	———— MPC ————
Low-pressure condensate	———— LPC ————
Boiler blowdown	———— BBD ————
Pumped condensate	———— PC ————
Vacuum pump discharge	———— VPD ————
Makeup water	———— MU ————
Atmospheric vent	———— ATV ————
Fuel oil discharge	———— FOD ————
Fuel oil gage	———— FOG ————
Fuel oil suction (supply)	———— FOS ————
Fuel oil return	———— FOR ————
Low-temperature hot water supply	———— HWS ————
Medium-temperature hot water supply	———— MTWS ————
High-temperature hot water supply	———— HTWS ————
Low-temperature hot water return	———— HWR ————
Medium-temperature hot water return	———— MTWR ————
High-temperature hot water return	———— HTWR ————
Compressed air	———— A ————
Vacuum (air)	———— VAC ————
Existing piping	———— (NAME) ————
Piping to be removed	XX——(NAME) — XX

Air Conditioning and Refrigeration

Refrigerant discharge	———— RD ————
Refrigerant suction (supply)	———— RS ————
Brine supply	———— B ————
Brine return	———— BR ————
Condenser water supply	———— C ————
Condenser water return	———— CR ————
Chilled water supply	———— CWS ————
Chilled water return	———— CWR ————
Fill line	———— FILL ————
Humidification line	———— H ————
Drain	———— D ————
Hot/chilled water supply	———— HCS ————
Hot/chilled water return	———— HCR ————
Refrigerant liquid	———— RL ————
Heat pump water supply	———— HPWS ————
Heat pump water return	———— HPWR ————

Table B.3 Symbols for piping specialities

Piping Specialities	
Automatic air vent (AV)	AV
Manual air vent (MV)	MV
Air separator	S
Alignment guide	
Intermediate anchor	
Main anchor	
Ball joint	
Expansion joint (with drawing number)	EJ-1
Flexible connector	
Orifice flowmeter	
Venturi flowmeter	
Flow switch	FS
Hanger rod	H
Hanger spring	H
Liquid–liquid heat exchanger	
Pipe pitch [rise (R), drop (D)]	R
Pressure gage and cock	
Pressure switch	PS
Pump (indicate use and drawing number)	CW-1
Pump suction diffuser	PSD
Strainer	
Blow off strainer	
Duplex strainer	
Tank (indicate use)	FO
Thermometer	

Table B.4 Additional/alternate valve symbols

Valves	
Air line valve	
Ball valve	
Butterfly valve	
Diaphragm valve	
Gate valve	
Gate (angle) valve	
Globe valve	
Globe (angle) valve	
Plug valve	
Three way valve	

Special Duty Valves

Swing gate check valve	
Spring check valve	
Electric–pneumatic control valve	EP
Pneumatic–electric control valve	PE
Hose end drain	
Lock shield valve	
Needle valve	
Pressure reducing valve	
Quick opening valve	
Quick closing (fusible link) valve	
Safety (S) or relief (R) valve	
Solenoid valve	S

Table B.5 Fittings

Fittings	
Connections (e.g., 90° elbow)	
Flanged connection	
Threaded connection	
Bell and spigot connection	
Welded connection	
Soldered connection	
Solvent cement	
*Fittings**	
Bushing	
Cap	
Connection (from the bottom)	
Connection (from the top)	
Coupling (joint)	
Cross	
90° elbow	
45° elbow	
Elbow (turned up)	
Elbow (turned down)	
Reducing elbow (sizes in inches)	
Elbow (base)	
Long radius elbow	

(*continued*)

Table B.5 *(Continued)*

Fittings* (cont.)

Double branch elbow	
Elbow (side outlet, outlet up)	
Elbow (side outlet, outlet down)	

Fittings*

Lateral	
Concentric reducer	
Eccentric (straight invert) reducer	
Eccentric (straight crown) reducer	
Tee	
Tee (outlet up)	
Tee (outlet down)	
Reducing tee (sizes in inches)	
Tee (side outlet, outlet up)	
Tee (side outlet, outlet down)	
Single sweep tee	
Screwed union	
Flanged union	

[a]Where applicable, the fittings are shown for threaded screw connections.
Source: *ASHRAE Handbook*, Fundamentals Volume, 2005.

Table B.6   Radiant Panel Symbols

Radiant Panel Symbols	
Hydronic heating element	○
Electric heating element	●

Radiant Ceiling Panels

Embedded	
Above ceiling	
Surface mounted	
Suspended	

Radiant Floor Panels

Slab on grade	
Above subfloor	
Below subfloor	
Slab above subfloor	

Radiant Wall Panels

Embedded	
Surface mounted	

Source: *ASHRAE Handbook*, Fundamentals Volume, 2005.

Appendix C

Heat Exchanger Design

Table C.1 Representative values of the overall heat transfer coefficients (US)

Type of Heat Exchanger	U (Btu/(h ft^2 °F))
Water-to-water	150–300
Water-to-oil	18–60
Water-to-gasoline or kerosene	55–180
Feedwater heaters	180–1500
Steam-to-light fuel oil	35–70
Steam-to-heavy fuel oil	10–35
Steam condenser	180–1060
Freon condenser (water cooled)	55–180
Ammonia condenser (water cooled)	140–250
Alcohol condenser (water cooled)	45–125
Gas-to-gas	2–7
Water-to-air in finned tubes (water in tubes)	5–10 (air); 70–150 (water)
Steam-to-air in finned tubes (steam in tubes)	5–50 (air); 70–705 (water)

Introduction to Thermo-Fluids Systems Design, First Edition. André G. McDonald and Hugh L. Magande.
© 2012 André G. McDonald and Hugh L. Magande. Published 2012 by John Wiley & Sons, Ltd.

Table C.2 Representative values of the overall heat transfer coefficients (SI)

Type of Heat Exchanger	U (W/(m^2 °C))
Water-to-water	850–1700
Water-to-oil	100–350
Water-to-gasoline or kerosene	300–1000
Feedwater heater	1000–8500
Steam-to-light fuel oil	200–400
Steam-to-heavy fuel oil	50–200
Steam condenser	1000–6000
Freon condenser (water cooled)	300–1000
Ammonia condenser (water cooled)	800–1400
Alcohol condenser (water cooled)	250–700
Gas-to-gas	10–40
Water-to-air in finned tubes (water in tubes)	30–60 (air); 400–850 (water)
Steam-to-air in finned tubes (steam in tubes)	30–300 (air); 400–4000 (water)

Source: Çengel, Y.A. (2007) *Heat and Mass Transfer: A Practical Approach*, 3rd edn, McGraw-Hill, Inc., New York.

Table C.3 Representative fouling factors in heat exchangers

Fluid	R_f ((ft^2 h °F)/Btu)
Gas oil	0.00051
Transformer oil	0.00102
Lubrication oil	0.00102
Heat transfer oil	0.00102
Hydraulic oil	0.00102
Fuel oil	0.0051
Hydrogen	0.00999
Engine exhaust	0.00999
Steam (oil-free)	0.00051
Steam with oil traces	0.0010
Cooling fluid vapors with oil traces	0.00199
Organic solvent vapors	0.0010
Alcohol vapors	0.00057
Refrigerants (vapor)	0.0023
Compressed air	0.00199
Natural gas	0.0010
Distilled water, seawater, river water, boiler feedwater: below 122°F	0.00057
Distilled water, seawater, river water, boiler feedwater: above 122°F	0.0011
Refrigerants (liquid)	0.0011
Cooling fluid	0.0010
Organic heat transfer fluids	0.0010
Salts	0.00051
Liquefied petroleum gas (LPG), liquefied natural gas (LNG)	0.0010
MEA and DEA (amines) solutions	0.00199
DEG and TEG (glycols) solutions	0.00199
Vegetable oils	0.0030

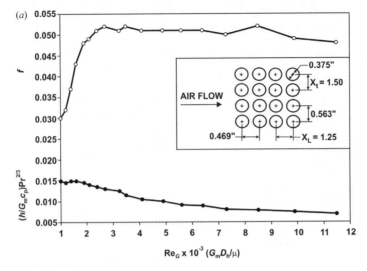

DATA

Tube outside diameter: 0.375 in.

Hydraulic diameter, D_h: 0.0248 ft

Free-flow area/Frontal area, σ: 0.333

Heat transfer area/Total volume, α: 53.6 ft^2/ft^3

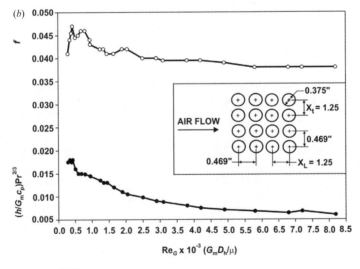

DATA

Tube outside diameter: 0.375 in.

Hydraulic diameter, D_h: 0.01237 ft

Free-flow area/Frontal area, σ: 0.200

Heat transfer area/Total volume, α: 64.4 ft^2/ft^3

Figure C.1 j-factor versus Re_G charts for in-line tube banks. Transient tests (2 charts): (a) For $X_t = 1.50$ and $X_L = 1.25$; (b) For $X_t = 1.25$ and $X_L = 1.25$. (Kays, W. and London, A. (1964) *Compact Heat Exchangers*, 2nd edn, McGraw-Hill, Inc., New York)

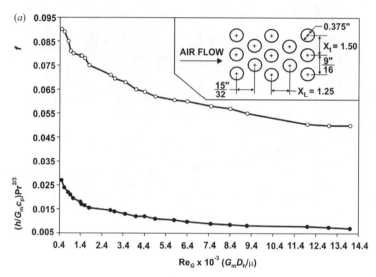

DATA
Tube outside diameter: 0.375 in.

Hydraulic diameter, D_h: 0.0249 ft

Free-flow area/Frontal area, σ: 0.333

Heat transfer area/Total volume, α: 53.6 ft^2/ft^3

Note: Minimum free-flow area is in the spaces
transverse to the flow

Figure C.2 *j*-factor versus Re_G charts for staggered tube banks. Transient tests (6 charts): (a) For $X_t = 1.50$ and $X_L = 1.25$; (b) For $X_t = 1.25$ and $X_L = 1.25$; (c) For $X_t = 1.50$ and $X_L = 1.0$; (d) For $X_t = 1.5$ and $X_L = 1.5$; (e) For $X_t = 2$ and $X_L = 1$; (f) For $X_t = 2.5$ and $X_L = 0.75$.
(Kays, W. and London, A. (1964) *Compact Heat Exchangers*, 2nd edn, McGraw-Hill, Inc., New York)

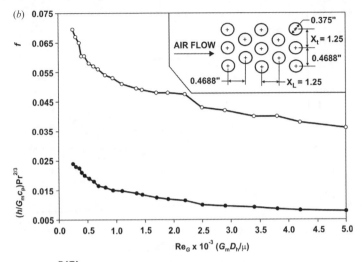

(b)

DATA

Tube outside diameter: 0.375 in.

Hydraulic diameter, D_h: 0.0125 ft

Free-flow area/Frontal area, σ: 0.200

Heat transfer area/Total volume, α: 64.4 ft^2/ft^3

Note: Minimum free-flow area is in the spaces
transverse to the flow

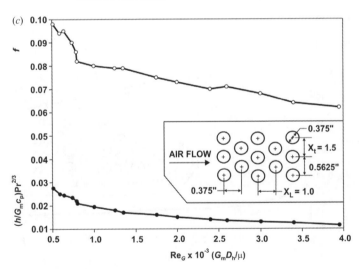

(c)

DATA

Tube outside diameter: 0.375 in.

Hydraulic diameter, D_h: 0.0196 ft

Free-flow area/Frontal area, σ: 0.333

Heat transfer area/Total volume, α: 67.1 ft^2/ft^3

Note: Minimum free-flow area is in the spaces
transverse to the flow

Figure C.2 (*Continued*)

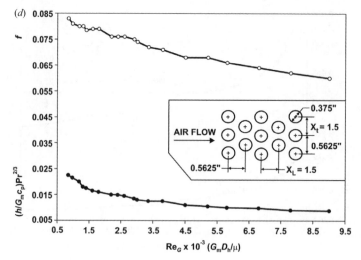

(d)

DATA

Tube outside diameter: 0.375 in.

Hydraulic diameter, D_h: 0.0298 ft

Free-flow area/Frontal area, σ: 0.333

Heat transfer area/Total volume, α: 44.8 ft^2/ft^3

Note: Minimum free-flow area is in the spaces
transverse to the flow

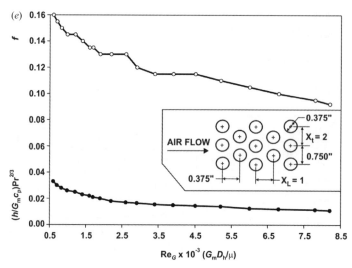

(e)

DATA

Tube outside diameter: 0.375 in.

Hydraulic diameter, D_h: 0.0327 ft

Free-flow area/Frontal area, σ: 0.414

Heat transfer area/Total volume, α: 50.3 ft^2/ft^3

Note: Minimum free-flow area is in the spaces
transverse to the flow

Figure C.2 *(Continued)*

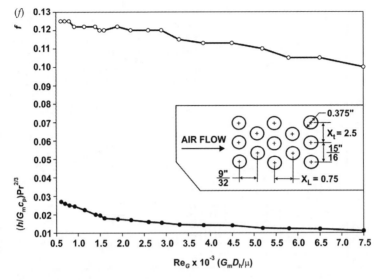

DATA

Tube outside diameter: 0.375 in.

Hydraulic diameter, D_h: 0.0271 ft

Free-flow area/Frontal area, σ: 0.366

Heat transfer area/Total volume, α: 53.6 ft^2/ft^3

Note: Minimum free-flow area is in the spaces
        transverse to the flow

Figure C.2   *(Continued)*

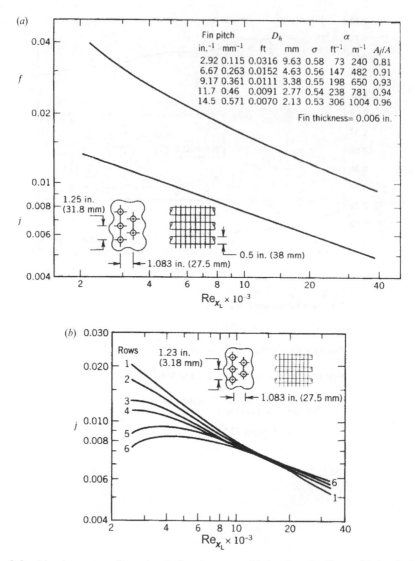

Figure C.3 *j*-factor versus Re$_{x_L}$ charts for staggered tube banks (finned tubes): (a) five rows of tubes (*ASHRAE Transactions*, vol. 79, Part II, 1973; reprinted with permission); (b) multiple rows of tubes (*ASHRAE Transactions*, vol. 81, Part I, 1975; reprinted with permission)

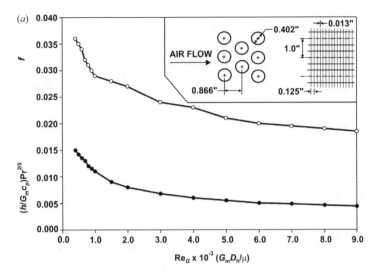

DATA

Tube outside diameter: 0.402 in.

Fin pitch: 8 fins per in.

Fin thickness: 0.013 in.

Hydraulic diameter, D_h: 0.01192 ft

Free-flow area/Frontal area, σ: 0.534

Heat transfer area/Total volume, α: 179 ft^2/ft^3

Fin area/Total area: 0.913

**Note: Minimum free-flow area is in the spaces
transverse to the flow**

Figure C.4 j-factor versus Re_G charts for staggered tube banks (finned tubes). (a) Tube outer diameter = 0.402 in.; (b) tube outer diameter = 0.676 in.
(Kays, W. and London, A. (1964) *Compact Heat Exchangers*, 2nd edn, McGraw-Hill, Inc., New York)

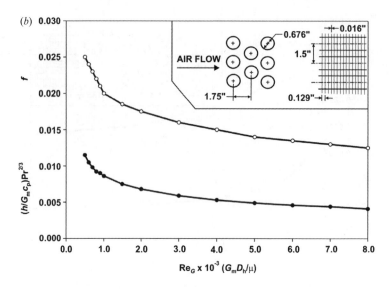

DATA

Tube outside diameter: 0.676 in.

Fin pitch: 7.75 fins per in.

Fin thickness: 0.016 in.

Hydraulic diameter, D_h: 0.0114 ft

Free-flow area/Frontal area, σ: 0.481

Heat transfer area/Total volume, α: 169 ft^2/ft^3

Fin area/Total area: 0.950

Note: Minimum free-flow area is in the spaces
        transverse to the flow

Figure C.4 *(Continued)*

Appendix D

Design Project—
Possible Solution

D.1 Fuel Oil Piping System Design

A major part of electric power supply is from large steam power plants with steam produced by the combustion of fossil fuels or from nuclear reactors. However, for the supply of power to small communities, it may be economical to use engines to provide electricity and heat. The expected load and daily power demand for a small community is provided by the utility company.

Time	0000–0500 h	0500–0700 h	0700–1800 h	1800–2200 h	2200–0000 h
Load (kW)	1650	3100	15200	3100	1650

Engines that operate on #2 oil or other light oil could be purchased. The combined alternator-transmission efficiency for these engines is typically 90–98%. *Motor Ship* (October 2000) has shown that specific fuel consumption of slow-speed engines at peak load varies from approximately 0.30–0.35 lb/(bhp h).

Design the complete fuel oil piping system to supply the engines that are required by this community.

Possible Solution

Definition

Design a complete fuel oil piping system for engines used to produce power for a small community.

Introduction to Thermo-Fluids Systems Design, First Edition. André G. McDonald and Hugh L. Magande.
© 2012 André G. McDonald and Hugh L. Magande. Published 2012 by John Wiley & Sons, Ltd.

Preliminary Specifications and Constraints

(i) Engines will be used to provide power to a small community.
(ii) The daily power demand is shown below.

Time	0000–0500 h	0500–0700 h	0700–1800 h	1800–2200 h	2200–0000 h
Load (kW)	1650	3100	15200	3100	1650

(iii) The working fluid could be #2 fuel oil.
(iv) The alternator-transmission efficiency of the engine is typically 90%.
(v) Fuel consumption could be on the order of 0.30–0.35 lb/(bhp h) for slow-speed engines.

Detailed Design

Objective

To size the pipes in the system and to size and select appropriate pumps. Piping accessories will also be specified.

Data Given or Known

(i) The daily power load requirements are known.
(ii) The alternator-transmission efficiency of the engine is typically 90%.
(iii) Fuel consumption can range from 0.30 to 0.35 lb/(bhp h).

Assumptions/Limitations/Constraints

(i) Let the piping material be Schedule 40 steel. Steel is an acceptable material as per the National Fire Protection Association (NFPA) standard 30, section 27.4 and NFPA 31, section 8.2.
(ii) Let the flow velocity be about 5 fps. The maximum erosion limit for #2 fuel oil is about 12 fps. It was assumed that the specific gravity of #2 oil was approximately 0.83 (Material Data Safety Sheet (MSDS) supplied by NOVA Chemicals, Calgary, AB).
(iii) Limit pipe frictional losses (major losses) to 5 ft of water per 100 ft of pipe.
(iv) Pipe changes should be gradual to reduce losses.
(v) All bends will be 90° threaded (screwed) regular bends.
(vi) All fittings and pipe accessories will be threaded to reduce losses.
(vii) Negligible elevation head in the piping system between the engines. Assume that all components are on the same level. Some fluid lifting may be needed on the suction line between the pump and the fuel storage tank.
(viii) Assume that the viscosity (at 104°F) and specific gravity (at 59°F) of #2 fuel oil is 3.4 cSt and 0.83, respectively. Note that 3.4 cSt = 3.66×10^{-5} ft^2/s (Material Data Safety Sheet (MSDS) supplied by NOVA Chemicals, Calgary, AB).

(ix) Assume the vapor pressure of the #2 fuel oil to be 1.5 mm Hg = 0.059 in. Hg = 0.80 in. wg = 0.0290 psi at 68°F (Material Data Safety Sheet (MSDS) supplied by NOVA Chemicals, Calgary, AB).

(x) Assume the power requirement of the plant that will house the engines is about 500 kW. According to ABB Automation (*Smart Generation*: Energy Efficient Design of Auxiliary Systems in Fossil-fuel Power Plants), oil-fired power plants will require 3.5–7.5% of the plant output at full load to run the plant auxiliaries.

(xi) Assume that the maximum efficiency of an engine will occur when it operates at about 80% of its rated output.

(xii) The fuel storage tank(s) will be filled once every week. This can be modified based on the desires of the client.

Sketch

There are many additional parameters that will be needed before a drawing or sketch can be provided. A full drawing will be provided toward the end of the design. A sketch may be provided in Section "Analysis."

Analysis

Bear the following points in mind for this design problem:

(i) Pipe sizing for this system can be done quickly by using the appropriate friction loss charts for Schedule 40 pipe. However, if the required sizes, flow rates, velocities, etc., are not included on the published charts or if the design is based on a specialized pipe material, then the designer should use Colebrook's equation (or other correlation equation) to find the friction factor (f) and iterate to find the pipe diameters.

(ii) Lower fuel velocities will be used to keep head loss low, reduce pipe erosion, and provide longer, trouble-free service.

(iii) It is possible that the pump will need to lift oil from an underground oil storage tank. Underground storage will provide protection against accidental damage to the tanks. On the other hand, above ground tanks will be easier to maintain and problems will be easier to identify.

(iv) All attempts will be made to install the pump as close as possible to the fuel oil storage tank.

(v) Shorter, larger suction pipes are usually recommended.

(vi) Fuel oils will produce vapor that are flammable, posing a safety problem. In addition, vapor formation in the pump and suction line can result in cavitation, noise, and reduced performance of the pump. The net positive suction heads (NPSH) should be calculated.

(vii) A strainer will be installed in the pump suction line to prevent or reduce the amount of solid particulates that enter the pump and engines.

Number and Type of Engines

The lowest load will be 1650 kW. Considering the 500 kW plant load, the minimum load of this community will be 2150 kW. Operating at 80% of the rated output of the engine, and with a 90% alternator-transmission efficiency, the smallest engine size should be at least 2986 kW ≈ 3000 kW. The following table summarizes the total power required from the engines over the course of a day:

Time	0–5 h	5–7 h	7–18 h	18–22 h	22–0 h
Load (kW)	1650	3100	15200	3100	1650
Total load (kW)	2150	3600	15700	3600	2150
Minimum engine requirement (kW)	**2986**	**5000**	**21806**	**5000**	**2986**

Based on the load requirements for this community, a group of medium-speed stationary engines would probably be sufficient to provide the power needed. Two companies that design and manufacture engines are Wärtsilä (www.wartsila.com) and MAN Diesel (www.mandiesel.com). To determine a suitable number of engines, the maximum load is divided by the minimum load. Seven (7) engines at 3115 kW/engine would be selected. A review of the MAN Diesel catalog for stationary engines shows that seven engines would require a significant amount of space and a large plant footprint. To aid the client to reduce the initial construction and future operation costs of the project, **five engines** will be selected. So, to produce the maximum engine power requirement of 21806 kW, each engine will need to provide at least 4361 kW of power.

A MAN Diesel engine will be selected. From the stationary engines catalog, the **MAN-9L32/44CR 9-cylinder, 440 mm stroke engine** will be chosen (see the appendix to this project report for manufacturer's catalog sheets). Under maximum operation at 60 Hz, **this engine will produce 4889 kW of electrical power**. Note that the manufacturer, in this case, provides both the mechanical and electrical power output of the engines.

At the peak demand period (0700–1800 h), the five engines will operate close to 80% of the rated output (89%). However, at off-peak periods, the engines will operate at less than 80% of the rated output. For the 0000–0500 h and the 2200–0000 h periods, one engine will be needed, and the percentage of the rated output will be

$$x\,(4889\ \text{kW}) = 2150\ \text{kW}$$

$$x = 44\%.$$

So, for 7 h, one engine will be working at 44% of its maximum rated capacity. The efficiency of the engine operating under this condition will be lower than the maximum possible efficiency.

For the 0500–0700 h and the 1800–2200 h periods, two engines will be needed. Both these engines will operate at the same percentage of the rated output. The percentage of the rated output of these engines will be

$$2\,(x)\,(4889\text{ kW}) = 7200\text{ kW}$$

$$x = 74\%.$$

So, for 4 h, two engines will be working at 74% of their maximum rated capacity. The efficiency of the engine operating under this condition will be slightly lower than the maximum possible efficiency.

The following table shows the number of engines that will be operating during the day and their percentages of the rated output.

Time	0000–0500 h	0500–0700 h	0700–1800 h	1800–2200 h	2200–0000 h
Load (kW)	1650	3100	15200	3100	1650
Total load (kW)	2150	3600	15700	3600	2150
Minimum engine requirement (kW)	2986	5000	21806	5000	2986
No. of engines	1	2	5	2	1
Output %	44	74	89	74	44

One additional engine will be purchased for use when each engine is undergoing maintenance and another engine will be purchased as a spare. Therefore, a total of **seven engines** will be required.

The specific fuel oil consumption (SFOC) at the maximum engine rated output is not directly provided by MAN Diesel in the engine catalog. However, with the heat rate (HR) known, the SFOC can be estimated. The HR for the selected engine (MAN-9L32/44CR) is 7879 kJ/kWh. The HR is the amount of fuel required to produce 1 kWh of power. The value specified above is based on electrical power and the 2007/2008 World Bank emission control guidelines. The SFOC is

$$SFOC = \frac{HR}{HV_{fuel}},$$

where HV_{fuel} is the heating value of #2 fuel oil shown in Table 5.2 (140000 Btu/gal). It should be noted that the lower heating value (LHV) of the fuel should be used. So,

$$SFOC = \frac{7879\text{ kJ/kWh}}{140000\text{ Btu/gal}} \times 0.83 \times 62\text{ lb/ft}^3 \times \frac{1\text{ ft}^3}{7.48\text{ gal}} \times \frac{1\text{ Btu}}{1.06\text{ kJ}} \times \frac{1\text{ kWh}}{1.34\text{ bhp h}}$$

$$SFOC = 0.27\text{ lb/(bhp h)}.$$

Size of Fuel Oil Storage Tank

Now that the number of engines and their capacities are known (to provide the maximum demand), the total volume of fluid, the size of the tank, and the specifications of the tank can be determined. The volume flow rate of fuel required will be

$$\dot{\Omega} = \left(\frac{0.35 \text{ lb}}{\text{bhp h}}\right) (5 \text{ engines}) \left(\frac{4440 \text{ kW}}{\text{engine}}\right) \left(\frac{1.341 \text{ bhp}}{1 \text{ kW}}\right) \left(\frac{1}{(0.83)\left(62.36 \text{ lb/ft}^3\right)}\right)$$

$$\left(\frac{264.17 \text{ gal}}{35.315 \text{ ft}^3}\right) \left(\frac{1 \text{ h}}{60 \text{ min}}\right)$$

$$\dot{\Omega} = 25 \text{ gpm}.$$

The volume of fluid required every week is

$$\Omega = \left(\frac{0.35 \text{ lb}}{\text{bhp h}}\right) (5 \text{ engines}) \left(\frac{4440 \text{ kW}}{\text{engine}}\right) \left(\frac{1.341 \text{ bhp}}{1 \text{ kW}}\right) (1 \text{ week}) \left(\frac{168 \text{ h}}{1 \text{ week}}\right)$$

$$\left(\frac{1}{(0.83)\left(62.36 \text{ lb/ft}^3\right)}\right)$$

$$\Omega = 33820 \text{ ft}^3 = 252988 \text{ US gal}.$$

For the five engines that will operate at 79% of their maximum rated output, only 79% of the fuel will be used over the 1-week period (177092 gal). However, the final size of the storage tank will be selected based on this maximum volume. In this case, there will be enough fuel for approximately 2 days beyond the 1-week period (36141 gal per day).

Five above ground fuel storage tanks will be selected. Clemmer Steelcraft Technologies (www.clemmertech.com) Model AGDW1581 tanks will be chosen. See the equipment schedule for additional information. The size of each tank is 52835 US gal.

It is expected that the total length of piping between the fuel storage tanks and the pump will be large. So, the pump size will be large. Consistently operating a large pump to move fluid to the engines may be costly and frequent maintenance may be needed. Therefore, **two fuel oil day tanks will be installed inside the plant**. They will be filled to provide 12 h of oil, and a smaller pump will move fuel from the day tanks to the engines. Clemmer Steelcraft Technologies Model AGDW1573 tanks will be chosen. See the equipment schedule for additional information. The size of each tank is 9246 US gal. About 9035 US gal are needed per tank per day.

Pipe Sizes

Charts can be used to determine the pipe sizes that will be needed for the fuel oil system. The total volume flow rate required for the five engines operating during the maximum load period is 25 gpm. Each engine will require 5 gpm.

At this point, a preliminary, tentative layout of the system would be useful.

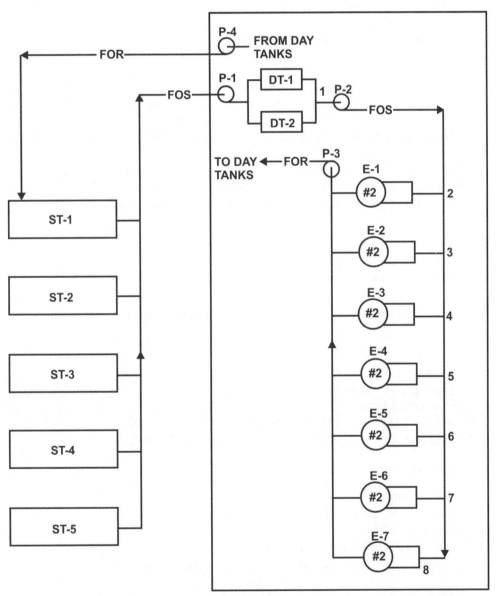

Since day tanks (DT) are present in the plant, fuel from the outdoor storage tanks (ST) will be pumped at a rate of 50 gpm from the pipe manifold. This rate is double that required by the five engines (E) to ensure that fuel will always be in the day tank. In addition, it should take about 6 h to fill the two day tanks.

With the Schedule 40 friction loss chart in Figure A.4, the fuel oil supply (FOS) piping between the storage tanks and the day tanks will have the following approximate parameters:

Pipe size: 2 in. (nominal)
Fuel velocity: 4.8 fps
Major loss: 4.9 ft of water per 100 ft of pipe
Reynolds number: $\text{Re}_D = \frac{VD}{\nu} = \frac{(4.8 \text{ ft/s})(0.172 \text{ ft})}{3.66 \times 10^{-5} \text{ ft}^2/\text{s}} = 22590 > 4000$

The fuel velocity and major head loss are lower than the constraints established in Section "Assumptions/Limitations/Constraints.". The fuel oil return (FOR) piping from the day tanks to storage tank 1 will have the same pipe size and flow parameters. Note that it will probably not be used often, except for complete draining for cleaning and routine maintenance.

Each engine will require 5 gpm of fuel. Note that Engines 1 and 2 are spares, and will normally be closed. Engines 3 to 7 will operate on a regular basis. The FOS piping sections for the engines will have the following approximate parameters:
Section 1–4:

Flow rate: 25 gpm
Pipe size: $1\frac{1}{2}$ in. (nominal)
Fuel velocity: **3.9 fps**
Major loss: 4.5 ft of water per 100 ft of pipe
Reynolds number: $\text{Re}_D = \frac{VD}{\nu} = \frac{(3.9 \text{ ft/s})(0.134 \text{ ft})}{3.66 \times 10^{-5} \text{ ft}^2/\text{s}} = 14297 > 4000$

Section 4–5:

Flow rate: 20 gpm
Pipe size: $1\frac{1}{2}$ in. (nominal)
Fuel velocity: **3.1 fps**
Major loss: 3.0 ft of water per 100 ft of pipe
Reynolds number: $\text{Re}_D = \frac{VD}{\nu} = \frac{(3.1 \text{ ft/s})(0.134 \text{ ft})}{3.66 \times 10^{-5} \text{ ft}^2/\text{s}} = 11350 > 4000$

Section 5–6:

Flow rate: 15 gpm
Pipe size: $1\frac{1}{4}$ in. (nominal)
Fuel velocity: **3.2 fps**
Major loss: 4.0 ft of water per 100 ft of pipe
Reynolds number: $\text{Re}_D = \frac{VD}{\nu} = \frac{(3.2 \text{ ft/s})(0.115 \text{ ft})}{3.66 \times 10^{-5} \text{ ft}^2/\text{s}} = 10055 > 4000$

Section 6–7:

Flow rate: 10 gpm
Pipe size: $1\frac{1}{4}$ in. (nominal)

Fuel velocity: **2.1 fps**
Major loss: 1.8 ft of water per 100 ft of pipe
Reynolds number: $\text{Re}_D = \frac{VD}{v} = \frac{(2.1 \text{ ft/s})(0.115 \text{ ft})}{3.66 \times 10^{-5} \text{ ft}^2/\text{s}} = 6598 > 4000$

Section 7–8:

Flow rate: 5 gpm
Pipe size: 1 in. (nominal)
Fuel velocity: **1.9 fps**
Major loss: 2.0 ft of water per 100 ft of pipe
Reynolds number: $\text{Re}_D = \frac{VD}{v} = \frac{(1.9 \text{ ft/s})(0.0874 \text{ ft})}{3.66 \times 10^{-5} \text{ ft}^2/\text{s}} = 4538 > 4000$

For sections 6–7 and 7–8 where the major head loss is low, balancing valves may need to be installed to increase the head loss in that section of the FOS piping to the engines. In that case, the loss coefficients of the valves will be calculated to facilitate selection by the contractor. All the pipe flows are fully turbulent.

The adjacent FOR piping sections from the engines will have the same pipe sizes and flow parameters.

Pipe Lengths and Major Head Losses

The lengths of the pipe sections for the supply and return lines will depend on the installation distance between the tanks and engines. Of interest will also be the distance of the tanks and engines from the walls of the building.

The National Fire Protection Association standards will be used to specify the distances. NFPA 30 (Flammable and Combustible Liquids Code), section 22.4 require that outdoor aboveground storage tanks with capacities between 50001 and 100000 gal be located **at least 15 ft from the plant building** (see the appendix to this project report). The same section of the code requires that each tank shall **not be required to be separated by more than 3 ft**. This part of the code refers to crude petroleum. However, it will be applied to #2 light fuel oil. Note that no attempt was made to classify #2 fuel oil as a stable liquid. If that is the case, an 8-ft distance between the tanks and the building would satisfy the requirements of the code.

For consistency, a 3 ft distance will be placed between the engines, around the pumps, and between the day tanks. Note the following dimensions:

Storage tank (AGDW1581)	Diameter = 14 ft
	Length = 53 ft
Day tank (AGDW1573)	Diameter = 9 ft
	Length = 24 ft
Engine (MAN-9L32/44CR 9-cyl.)	Length = 40 ft
	Width = 9 ft

With these distances and equipment dimensions, the preliminary layout of the system, complete with lengths, becomes

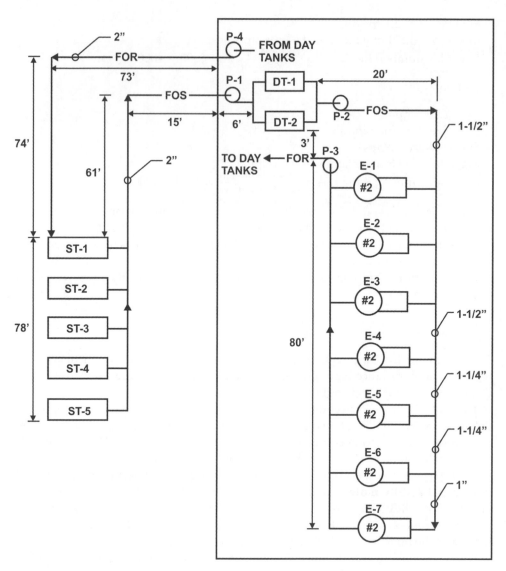

The longest run of fuel oil supply piping between the storage tanks (ST-5) and pump, P-1 is approximately,

$$L_{ST\text{-}S} = (78 + 61 + 15 + 6 + 3 + 4.5 + 4.5) \text{ ft} = 172 \text{ ft}.$$

The last 3 ft, and two 4.5 ft lengths were added based on the distance between the day tanks (3 ft) and half the diameter of the tank (4.5 ft).

The total major head loss for this section of supply piping is

$$H_{ST\text{-}S} = \frac{4.9 \text{ ft}}{100 \text{ ft}} \times 172 \text{ ft} = 8.43 \text{ ft}.$$

For the return line between ST-1 and P-4,

$$L_{ST\text{-}S} = (6 + 73 + 74) \text{ ft} = 153 \text{ ft.}$$

The total major head loss for this section of return piping is

$$H_{ST\text{-}S} = \frac{4.9 \text{ ft}}{100 \text{ ft}} \times 153 \text{ ft} = 7.50 \text{ ft.}$$

The total major head loss that will be experienced by pump, P-2.
FOS section between the day tanks (DT-1 and DT-2) and engine 3 (E-3):

$$\frac{4.5 \text{ ft}}{100 \text{ ft}} \times (1.5 + 4.5 + 20 + 1.5 + 9 + 3 + 3 + 9 + 3 + 9 + 3 + 4.5) \text{ ft} = 3.06 \text{ ft.}$$

FOS section between E-3 and E-4,

$$\frac{3.0 \text{ ft}}{100 \text{ ft}} \times (4.5 + 3 + 4.5) \text{ ft} = 0.36 \text{ ft.}$$

FOS section between E-4 and E-5,

$$\frac{4.0 \text{ ft}}{100 \text{ ft}} \times (4.5 + 3 + 4.5) \text{ ft} = 0.48 \text{ ft.}$$

FOS section between E-5 and E-6,

$$\frac{1.8 \text{ ft}}{100 \text{ ft}} \times (4.5 + 3 + 4.5) \text{ ft} = 0.22 \text{ ft.}$$

FOS section between E-6 and E-7,

$$\frac{2.0 \text{ ft}}{100 \text{ ft}} \times (4.5 + 3 + 4.5) \text{ ft} = 0.24 \text{ ft.}$$

The total major head loss that will be experienced by pump, P-2 will be

$$H_{FOS\text{-}E} = (3.06 + 0.36 + 0.48 + 0.22 + 0.24) \text{ ft} = 4.36 \text{ ft.}$$

The total major head loss that will be experienced by pump, P-3 will be approximately

$$H_{FOR\text{-}E} = 4.36 \text{ ft.}$$

Pump Sizing

Four pumps will need to be sized. To size the pumps, the total head loss will be needed to determine the pump head required. At this point, all the pipe specialties will be selected and shown on the sketch. Note the following points:

 (i) Gate valves will be installed on the suction and/or discharge lines of the outdoor storage tanks, the day tanks, and the engines.
 (ii) The supply and discharge lines will be attached to the bottom of the day tanks. This will be easier to install and will be more in-line with the pumps.
 (iii) A check valve will be installed between P-1 and the day tanks and P-3 and the day tanks.
 (iv) A strainer will be installed in the FOS line between P-1 and the outdoor storage tanks.

The sketch showing all the piping specialties is shown below. Note that screwed unions were included in the FOS and FOR lines between the day tanks and the storage tanks since it may not be possible to purchase 2-in. diameter pipes in single runs of 60 ft or greater.

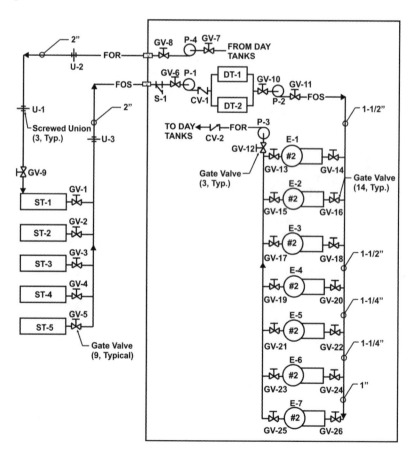

Now that the pipe diameters are known, the minor losses for each section can be estimated. The loss coefficients (K) for the piping specialties are given below (consult Table A.14). Note that the pipe diameters were considered when reporting the loss coefficients.

Fuel oil supply piping between the storage tanks and pump, P-1:

Pipe Specialty	Number	Pipe Diameter (in.)	Pipe Velocity (fps)	K	Minor Loss (ft)
GV-5, GV-6	2	2	4.8	0.16	0.114
U-3	1	2	4.8	0.08	0.029
S-1	1	2	4.8	1.5	0.537
CV-1	1	2	4.8	2.1	0.751
Screwed 90° bend	3	2	4.8	0.95	1.02
Line flow tee	4	2	4.8	0.90	1.29
Branch flow tee	1	2	4.8	1.4	0.501
Sharp-edged entrance	1	2	4.8	0.5	0.179
Rounded exit	1	2	4.8	1.05	0.376
Total					**4.80**

The total head loss for the fuel oil supply piping between the storage tanks and pump, P-1 is

$$H_{T,\text{ST-S}} = (8.43 + 4.80) \text{ ft} = 13.23 \text{ ft}.$$

The pump head (in feet) is

$$H_{P\text{-}1} = \left(\frac{p_2}{\rho g} + \frac{V_2^2}{2g} + z_2 \right) - \left(\frac{p_1}{\rho g} + \frac{V_1^2}{2g} + z_1 \right) + H_{T,\text{ST}-S}.$$

Let point 1 be at the bottom of storage tank 5 and point 2 be at the bottom of the day tanks. So, $V_1 = V_2$ and $z_1 = z_2$.

$$H_{P\text{-}1} = \frac{p_2 - p_1}{\rho g} + H_{T,\text{ST-S}}$$

To ensure that the pump will be sized adequately for all operation scenarios, it will be assumed that the storage tank is empty and the day tank is filled to maximum capacity. Assume that the tanks are vented to the atmosphere.
So,

$$p_1 = p_{\text{atm}}$$
$$p_2 = p_{\text{atm}} + \rho g h = p_{\text{atm}} + \rho g D_{\text{tank}}.$$

Then,

$$H_{P\text{-}1} = D_{\text{tank}} + H_{T,\text{ST-S}}$$
$$H_{P\text{-}1} = (9 + 13.23) \text{ ft} = 22.23 \text{ ft} \approx 23 \text{ ft}.$$

The total flow rate through P-1 will be 50 gpm.

Fuel oil supply piping between the day tanks and the engines through pump, P-2:

Pipe Specialty	Number	Pipe Diameter (in.)	Pipe Velocity (fps)	K	Minor Loss (ft)
GV-10, GV-11	2	$1\frac{1}{2}$	3.9	0.24	0.113
Sharp-edged entrance	1	$1\frac{1}{2}$	3.9	0.50	0.118
Branch flow tee	1	$1\frac{1}{2}$	3.9	1.8	0.425
Screwed 90° bend	2	$1\frac{1}{2}$	3.9	1.5	0.709
Line flow tee	3	$1\frac{1}{2}$	3.9	0.90	0.638
Line flow tee	1	$1\frac{1}{4}$	3.2	0.90	0.143
Gradual Contraction	1	$1\frac{1}{2}$ to $1\frac{1}{4}$	3.1	0.07	0.011
Gradual Contraction	1	$1\frac{1}{4}$ to 1	2.1	0.07	0.005
Line flow tee	1	$1\frac{1}{4}$	3.2	0.90	0.143
Line flow tee	1	$1\frac{1}{4}$	3.2	0.90	0.143
90° bend	1	1	1.9	1.5	0.084
GV-26	1	1	1.9	0.24	0.014
Rounded exit	1	1	1.9	1.05	0.059
Total					**2.61**

As mentioned earlier, the head loss may become very small as one gets farther away from P-2. This will result in unbalanced flow through the pipeline. Check sections 1–8 in this line and determine if balancing valves will be needed. Note that section 1–4 is ignored because there is no other path for fluid flow between points 1 and 4 if engines 1 and 2 are not operational.

Section	Major Head Loss (ft)	Minor Head Loss (ft)	Total Head Loss (ft)
Section 1–4	3.06	2.00	5.06
Section 4–5	0.36	0.154	0.514
Section 5–6	0.48	0.143	0.623
Section 6–7	0.22	0.148	0.368
Section 7–8	0.24	0.157	0.397

The total head losses in sections 4–8 are close. So, no balancing valves will be recommended.

The total head loss for the fuel oil supply piping between the day tanks and pump, P-2 is

$$H_{T,FOS-E} = (4.36 + 2.61) \text{ ft} = 6.97 \text{ ft.}$$

The pump head (in ft) is

$$H_{P-2} = \left(\frac{p_4}{\rho g} + \frac{V_4^2}{2g} + z_4 \right) - \left(\frac{p_3}{\rho g} + \frac{V_3^2}{2g} + z_3 \right) + H_{T,FOS-E}.$$

Let point 3 be at the bottom of the day tank and point 4 be at the bottom of the small engine reservoir. So, $V_1 = V_2$ and $z_1 = z_2$.

$$H_{P\text{-}2} = \frac{p_4 - p_3}{\rho g} + H_{T,FOS\text{-}E}$$

To ensure that the pump will be sized adequately for all operation scenarios, it will be assumed that the engine fuel reservoir is small and at atmospheric pressure and the day tank is empty. Assume that the day tanks are vented to the atmosphere.

So,

$$p_4 = p_{atm}$$
$$p_3 = p_{atm}.$$

Then,

$$H_{P\text{-}2} = H_{T,FOS\text{-}E} = 6.97 \text{ ft} \approx 7 \text{ ft}.$$

The flow rate through this pump will be 25 gpm.

FOR piping between the engines and the day tanks through pump, P-3, is

Pipe Specialty	Number	Pipe Diameter (in.)	Pipe Velocity (fps)	K	Minor Loss (ft)
GV-25	1	1	1.9	0.24	0.014
Sharp-edged entrance	1	1	1.9	0.50	0.028
Screwed 90° bend	1	1	1.9	1.5	0.084
Line flow tee	1	$1\frac{1}{4}$	2.1	0.90	0.062
Line flow tee	1	$1\frac{1}{4}$	3.2	0.90	0.143
Line flow tee	1	$1\frac{1}{2}$	3.2	0.90	0.143
Gradual expansion	1	1 to $1\frac{1}{4}$ ($d/D = 0.8$)	2.1	0.10	0.007
Gradual expansion	1	$1\frac{1}{4}$ to $1\frac{1}{2}$ ($d/D = 0.83$)	3.2	0.10	0.016
Line flow tee	3	$1\frac{1}{2}$	3.9	0.90	0.638
GV-12	1	$1\frac{1}{2}$	3.9	0.24	0.057
90° bend	3	$1\frac{1}{2}$	3.9	1.5	1.06
CV-2	1	$1\frac{1}{2}$	3.9	2.9	0.685
Branch flow tee	1	$1\frac{1}{2}$	3.9	1.8	0.425
Rounded exit	1	$1\frac{1}{2}$	3.9	1.05	0.248
Total					**3.62**

The total head loss for the FOR piping between the engines and pump, P-3, is

$$H_{T,FOR\text{-}E} = (4.36 + 3.62) \text{ ft} = 7.98 \text{ ft}.$$

The pump head (in feet) is

$$H_{P\text{-}3} = \left(\frac{p_6}{\rho g} + \frac{V_6^2}{2g} + z_6 \right) - \left(\frac{p_5}{\rho g} + \frac{V_5^2}{2g} + z_5 \right) + H_{T,FOR\text{-}E}.$$

Let point 5 be at the bottom of the E-7 and point 6 be at the bottom of the day tank. So, $V_1 = V_2$ and $z_1 = z_2$.

$$H_{P\text{-}3} = \frac{p_6 - p_5}{\rho g} + H_{T,FOR\text{-}E}$$

To ensure that the pump will be sized adequately for all operation scenarios, it will be assumed that the engine fuel reservoir is small and at atmospheric pressure and the day tank is full to the maximum capacity. Assume that the day tanks are vented to the atmosphere.
So,

$$p_5 = p_{atm}$$
$$p_6 = p_{atm} + \rho g h = p_{atm} + \rho g D_{tank}.$$

Then,

$$H_{P\text{-}3} = D_{tank} + H_{T,FOR\text{-}E} = (9 + 7.98)\ \text{ft} = 16.98\ \text{ft} \approx 17\ \text{ft}.$$

The flow rate through this pump will be 25 gpm.

FOR piping between the day tanks and storage tank, ST-1 through P-4:

Pipe Specialty	Number	Pipe Diameter (in.)	Pipe Velocity (fps)	K	Minor Loss (ft)
GV-7, GV-8, GV-9	3	2	4.8	0.16	0.172
Sharp-edged entrance	1	2	4.8	0.50	0.179
Branch flow tee	1	2	4.8	1.4	0.500
Screwed 90° bend	3	2	4.8	0.95	1.02
U-2	2	2	4.8	0.08	0.057
Rounded exit	1	2	4.8	1.05	0.376
Total					**2.30**

The total head loss for the FOR piping between the day tanks and pump, P-4 is

$$H_{T,ST\text{-}R} = (7.50 + 2.30)\ \text{ft} = 9.80\ \text{ft}.$$

The pump head (in feet) is

$$H_{P\text{-}4} = \left(\frac{p_8}{\rho g} + \frac{V_8^2}{2g} + z_8 \right) - \left(\frac{p_7}{\rho g} + \frac{V_7^2}{2g} + z_7 \right) + H_{T,ST\text{-}R}.$$

Let point 7 be at the bottom of the day tank and point 8 be at the bottom of the ST-1 So, $V_1 = V_2$ and $z_1 = z_2$.

$$H_{P-4} = \frac{p_8 - p_7}{\rho g} + H_{T,ST-R}$$

To ensure that the pump will be sized adequately for all operation scenarios, it will be assumed that the day tank is empty and at atmospheric pressure and ST-1 is full to the maximum capacity. Assume that the tanks are vented to the atmosphere.
So,

$$p_7 = p_{atm}$$
$$p_8 = p_{atm} + \rho g h = p_{atm} = \rho g D_{ST-1}.$$

Then,

$$H_{P-4} = D_{ST-1} + H_{T,ST-R} = (14 + 9.80)\ \text{ft} = 23.8\ \text{ft} \approx 24\ \text{ft}.$$

The flow rate through this pump will be 50 gpm.

Pump Selection

For this fuel oil piping system, pumps rated with the following capacities are required:

P-1: 23 ft of head and 50 gpm;
P-2: 7 ft of head and 25 gpm;
P-3: 17 ft of head and 25 gpm;
P-4: 24 ft of head and 50 gpm.

Use manufacturer's charts to select an appropriate pump. A review of the various Bell & Gossett pumps shows that Series 60 in-line mounted pumps would be suitable. The pump performance plots are shown below.
From the Master Selection chart, the 2 in. $\times$ 2 in. $\times$ $5\frac{1}{4}$ in. casing pump will be chosen for P-1 and P-4. The 1 in. $\times$ 1 in. $\times$ $5\frac{1}{4}$ in. casing pump will be chosen for P-2 and the $1\frac{1}{4}$ in. $\times$ $1\frac{1}{4}$ in. $\times$ $5\frac{1}{4}$ in. casing pump will be chosen for P-3.
From the performance plots (see below), the final choices are as follows.

P-1 and P-4:	2 in. $\times$ 2 in. $\times$ $5\frac{1}{4}$ in. casing
	5.25 in. impeller diameter
	$\frac{3}{4}$ hp motor
	1750 rpm speed

The manufacturer's cut sheet will show that a Series 60, 612T is the best option in stock.

P-2:	1 in. × 1 in. × $5\frac{1}{4}$ in. casing
	3.75 in. impeller diameter
	$\frac{1}{4}$ hp motor
	1750 rpm speed

The manufacturer's cut sheet will show that a Series 60, 601T is the best option in stock.

P-3:	$1\frac{1}{4}$ in. × $1\frac{1}{4}$ in. × $5\frac{1}{4}$ in. casing
	4.75 in. impeller diameter
	$\frac{1}{4}$ hp motor
	1750 rpm speed

The manufacturer's cut sheet will show that a Series 60, 605T is the best option in stock (see Figure A.5).

Note that "to-the-point" design was avoided, and slightly larger sizes and powers were chosen.

SELECTION CHART

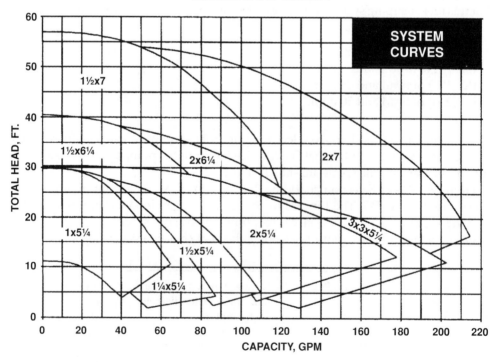

B-160J

SERIES 60 BUILT-TO-ORDER PUMP PERFORMANCE CURVES

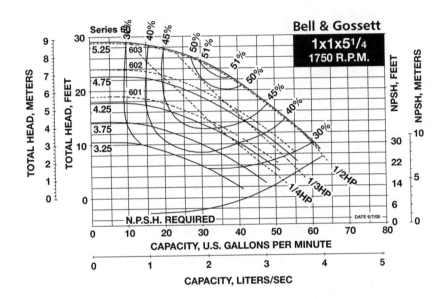

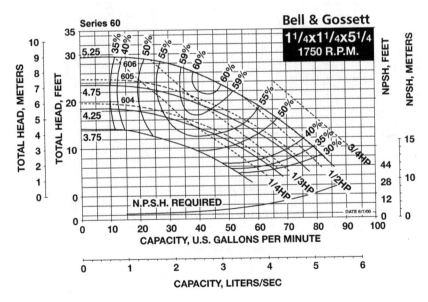

B-160J

SERIES 60 BUILT-TO-ORDER PUMP PERFORMANCE CURVES

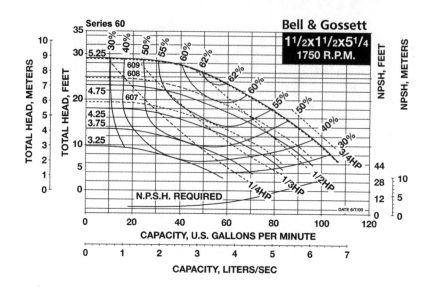

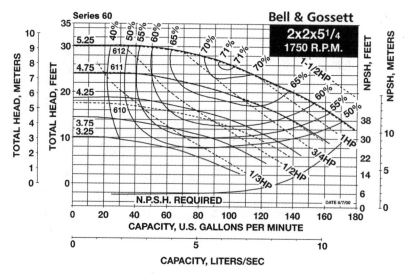

Cavitation and NPSH

Pump P-1 will run the risk of experiencing cavitation because the maximum suction line length from storage tank ST-5 is very long. The NPSHA will be determined for comparison with the NPSHR obtained from the manufacturer's pump performance plot. So,

$$\text{NPSHA} = \left(\frac{p_s}{\rho g} + \frac{V_s^2}{2g} \right)_{\text{pump,inlet}} - \frac{P_{\text{vapor}}}{\rho g}.$$

The energy equation will be used to find the pump inlet pressure, p_s:

$$\frac{P_1}{\rho g} + \alpha_1 \frac{V_1^2}{2g} + z_1 = \frac{p_s}{\rho g} + \alpha_s \frac{V_s^2}{2g} + z_s + H_{T,\text{ST-S}}$$

Point 1 is the free surface of the oil in storage tank ST-5. To determine the most conservative value of NPSHA, it will be assumed that the fluid height in the tank is at the same elevation as pump suction. So, $V_1 \approx 0$, $P_1 = P_{\text{atm}}$, and $z_1 = z_s$.
So,

$$\frac{p_s}{\rho g} = \frac{P_{\text{atm}}}{\rho g} - \alpha_s \frac{V_s^2}{2g} - H_{T,\text{ST-S}}.$$

The flow is turbulent in the suction pipe of pump P-1 ($\text{Re}_D = 22590$). So, $\alpha_s \approx 1$. So,

$$\frac{p_s}{\rho g} = \frac{P_{\text{atm}}}{\rho g} - \frac{V_s^2}{2g} - H_{T,\text{ST-S}}.$$

Then,

$$\text{NPSHA} = \frac{P_{\text{atm}}}{\rho g} - \frac{V_s^2}{2g} - H_{IT} + \frac{V_s^2}{2g} - \frac{P_{\text{vapor}}}{\rho g}$$

$$\text{NPSHA} = \frac{P_{\text{atm}} - P_{\text{vapor}}}{\rho g} - H_{T,\text{ST-S}}$$

$$\text{NPSHA} = \frac{P_{\text{atm}} - P_{\text{vapor}}}{SG \times \rho_{\text{water}} g} - H_{T,\text{ST-S}}.$$

It was found that the total head loss in the suction pipe to P-1 was $H_{T,\text{ST-S}} = 13.23$ ft. At 68°F, $P_{\text{vapor}} = 0.0290$ psia. $P_{\text{atm}} = 14.7$ psia.
So,

$$\text{NPSHA} = \frac{(14.7 - 0.0290)\,\text{lbf/in.}^2}{(0.83)\left(62.3\,\text{lb/ft}^3\right)\left(32.2\,\text{ft/s}^2\right)} \times \frac{32.2\,(\text{lb ft})/\text{s}^2}{1\,\text{lbf}} \times \left(\frac{12\,\text{in.}}{1\,\text{ft}}\right)^2 - 13.23\,\text{ft}$$

$$\text{NPSHA} = 28\,\text{ft}.$$

Cavitation will not occur if NPSHA > NPSHR. From the performance plot for this pump (2 in. × 2 in. × $5^1/_4$ in.) operating at 50 gpm,

$$\text{NPSHR} \approx 6 \text{ ft.}$$

Cavitation will not occur, even if the farthest storage tank is nearly empty. Based on this analysis, it is reasonable to conclude that cavitation will not be an issue for the other pumps due to the much shorter suction pipe lengths and smaller number of piping accessories installed in the lines.

Drawings

The layout of the piping system was prepared over the course of the analysis. The final layout, complete with piping specialties, is shown below.

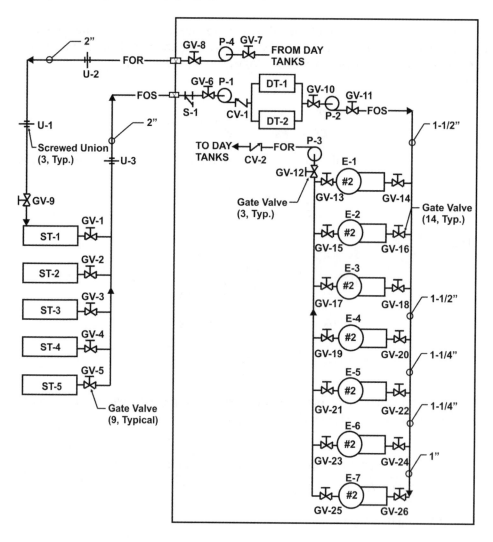

A drawing detail of the piping connections to the day tanks will be needed to facilitate contractor installation. The manufacturer of the tank (Clemmer Steelcraft Technologies) should be contacted to determine if all the required connections are possible.

Equipment Schedules

Shown below are the schedules of the major equipment that will be specified for the fuel oil system. Minor equipment such as valves and unions will not be specified beyond the contract drawings shown above.

						Engine Schedule		
Tag	Manufacturer and Model Number	Maximum Output (kW)	Cylinder	Bore & Stroke (ft)	Maximum SFOC (lb/(bhp h))	Major Dimensions (ft)		
						C	W	H
E-1 to E-7	MAN diesel, MAN-9L32/ 44CR, or equal	4889	9	1.05 and 1.44	0.30	40	9	17

					Fuel Oil Tank Schedule	
Tag	Manufacturer and Model Number	Volume (gal)	Gage	Weight (lb)	Dimensions (in.)	
					Diameter	Length
ST-1 to ST-5	Clemmer steelcraft, AGDW1581, or equal	52835	Double wall	56000	158	630
DT-1, DT-2	Clemmer steelcraft, AGDW1573, or equal	9246	Single wall	7330	99	281

Pump Schedule

Tag	Manufacturer and Model Number	Type	Construction	Fluid		Electrical			
				Flow Rate (gpm)	Working Fluid	Head Loss (ft)	Motor Size (hp)	Motor Speed (rpm)	Volt/pH/Hz
P-1, P-4	Bell & Gossett, Series 60, 612T, or equal	Centrifugal, inline-mounted	Bronze fitted 2" × 2" × 5¼" casing, 5.25" φ	50	No. 2 fuel oil	25	1	1750	208/3/60
P-2	Bell & Gossett, Series 60, 601T, or equal	Centrifugal, inline-mounted	Bronze fitted 1" × 1" × 5¼" casing, 4.38" φ	25	No. 2 fuel oil	7	¼	1750	208/3/60
P-3	Bell & Gossett, Series 60, 605T, or equal	Centrifugal, inline-mounted	Bronze fitted 1¼" × 1¼" × 5¼" CASING, 4.84" φ	25	No. 2 fuel oil	17	⅓	1750	208/3/60

Conclusions

A fuel oil piping system has been designed to transport #2 fuel oil to several power-generating engines. The design of the system was based on the maximum amount of fuel that would be needed to meet the maximum/peak power load of a small community.

A conservative approach was taken in the design analysis. For example, pumps were sized by considering empty or full fuel tanks in an attempt to remove the benefits of hydrostatic head that would drive flow through the pipes. Slightly larger fuel oil storage tanks, pumps, and engines were selected to ensure that the client had some flexibility to cover unexpected surges in power or flow rates. The client may consider the use of a bypass line, complete with pumps of similar performance ratings as P-1 and P-2 (label them P-1A and P-2A), to pump the fuel during times of maintenance or failure of P-1 and P-2. Also, the friction loss chart for a closed Schedule 40 piping system was used. Use of the friction loss chart for open Schedule 40 piping systems would produce larger pipe sizes and slightly lower friction losses. This decision was made to keep the cost of pipe material low given the low impact on pump sizes through use of the smaller pipes.

No allocations have been made for flow control of the pumps. The client may wish to consider speed regulation through the use of a variable frequency drive on the motor or bypass regulation with the use of a bypass line. Discharge throttling is not recommended due to the large system head loss that may occur. Some vaporization of the oil may occur on the discharge side due to the high velocities generated during the throttling process, which may pose a safety hazard.

No allocation has been made to prime the pump should the storage tanks become empty. Complete draining of the storage tanks and pipelines should be avoided. Level switches may be installed or purchased with the storage tanks to form a part of a control or low-level warning system. Vacuum pumping may be a suitable method to prime the pump by removing air from the suction line, if needed. A gate valve (not to be used for flow control), downstream of the check valve on the discharge line of P-1 would be needed. The valves on the discharge lines of the pumps should be closed before initiating vacuum pumping.

While the design and the contract documents (drawings, schedules, etc.) provide sufficient information to initiate construction, the contractor and architect may need to communicate with the mechanical engineer to finalize some points. For example, the size and layout of the plant building still needs to be determined and the engineer will need to approve the selection and installation plan for piping and specialties such as the valves and unions. These should comply with applicable standards such as NFPA 30 and 31. Details regarding the fuel day tanks (DT-1 and DT-2) will need to be determined. In particular, the piping connections will need to be determined once the final tanks are selected.

Though outside of the scope of the present project, control systems will be needed. For example, high-level (spill protection) and low-level switches (empty tank) will be needed at the day tanks. These switches would be routed to a controller that operates the P-1 motor.

Other items outside of the scope of this project are a fuel oil heating system and HVAC (heating, ventilation, and air-conditioning system) for the plant building. The client is advised to seek consultation for the design of a heating system to prevent freezing of the fuel in the pipes. Also, applicable building and ASHRAE (American Society of Heating, Refrigeration, and Air-Conditioning Engineers) codes require that sufficient ventilation of exhaust gases from the engines and air conditioning of the plant building occur to meet safety and human health standards.

Though the efficiency of some of the engines are lowered due to operation at lower capacities than the maximum possible engine capacity, the reduced maintenance, operations, and initial installation and purchase costs may outweigh the lost savings in increased engine efficiency. In this case, a reasonable compromise was made between system efficiency and long-term costs.

Specifications

This section is included to specify equipment or piping specialties that may have special requirements.

1. 2″ fuel strainer

The 2″ fuel strainer shall have a capacity of 50 gpm and have a 100 mesh element.

MAN Diesel & Turbo, MAN-9L32/44CR Cut Sheet

MAN L32/44CR

Bore 320 mm, Stroke 440 mm					
Speed	r/min	750		720	
Frequency	Hz	50		60	
mep	bar	25.3		26.4	
Piston speed	m/s	11.0		10.6	
		Eng. kW	Gen. kW	Eng. kW	Gen. kW
6L32/44CR		3,360	3,259	3,360	3,259
7L32/44CR		3,920	3,802	3,920	3,802
8L32/44CR		4,480	4,346	4,480	4,346
9L32/44CR		5,040	4,889	5,040	4,889
10L32/44CR		5,600	5,432	5,600	5,432

Consumption					
P			100%		
Heat Rate (WB 1998)			7,494 kJ/kWh$_m$ 7,726 kJ/kWh$_e$		
Heat Rate (WB 2007/2008)			7,643 kJ/kWh$_m$ 7,879 kJ/kWh$_e$		
Lube oil consumption					
Cyl. No.		6	7	8	9	10
kg/h		2.0	2.4	2.7	3.0	3.4

Dimensions						
Cyl. No.		6	7	8	9	10
A	mm	6,410	6,940	7,470	8,000	8,530
B	mm	3,515	3,575	4,328	4,328	4,328
C	mm	10,378	11,268	11,798	12,328	12,858
W	mm	2,490	2,490	2,676	2,676	2,676
H	mm	4,768	4,768	4,975	4,975	4,975
Dry mass	t	71	78	84	91	97

Nominal generator efficiencies: 97%

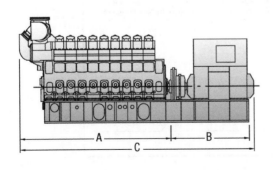
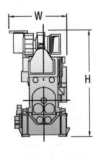

MAN Diesel & Turbo

Source: MAN Diesel & Turbo (reprinted with permission)

Excerpts from the NFPA 30—Flammable and Combustible Liquids Code, 2008 Edition

**Table 22.4.1.1(a) Location of Aboveground Storage Tanks Storing Stable Liquids —
Internal Pressure Not to Exceed a Gauge Pressure of 2.5 psi (17 kPa)**

		Minimum Distance (ft)	
Type of Tank	**Protection**	**From Property Line That Is or Can Be Built Upon, Including the Opposite Side of a Public Way**[a]	**From Nearest Side of Any Public Way or from Nearest Important Building on the Same Property**[a]
Floating roof	Protection for exposures[b]	½ × diameter of tank	⅙ × diameter of tank
	None	Diameter of tank but need not exceed 175 ft	⅙ × diameter of tank
Vertical with weak roof-to-shell seam	Approved foam or inerting system[c] on tanks not exceeding 150 ft in diameter[d]	½ × diameter of tank	⅙ × diameter of tank
	Protection for exposures[b]	Diameter of tank	⅓ × diameter of tank
	None	2 × diameter of tank but need not exceed 350 ft	⅓ × diameter of tank
Horizontal and vertical tanks with emergency relief venting to limit pressures to 2.5 psi (gauge pressure of 17 kPa)	Approved inerting system[b] on the tank or approved foam system on vertical tanks	½ × value in Table 22.4.1.1(b)	½ × value in Table 22.4.1.1(b)
	Protection for exposures[b]	Value in Table 22.4.1.1(b)	Value in Table 22.4.1.1(b)
	None	2 × value in Table 22.4.1.1(b)	Value in Table 22.4.1.1(b)
Protected aboveground tank	None	½ × value in Table 22.4.1.1(b)	½ × value in Table 22.4.1.1(b)

For SI units, 1 ft = 0.3 m.
[a]The minimum distance cannot be less than 5 ft (1.5 m).
[b]See definition 3.3.42, Protection for Exposures.
[c]See NFPA 69, *Standard on Explosion Prevention Systems.*
[d]For tanks over 150 ft (45 m) in diameter, use "Protection for Exposures" or "None," as applicable.

Table 22.4.1.1(b) Reference Table for Use with Tables 22.4.1.1(a), 22.4.1.3, and 22.4.1.5

	Minimum Distance (ft)	
Tank Capacity (gal)	**From Property Line that Is or Can Be Built Upon, Including the Opposite Side of a Public Way**	**From Nearest Side of Any Public Way or from Nearest Important Building on the Same Property**
275 or less	5	5
276 to 750	10	5
751 to 12,000	15	5
12,001 to 30,000	20	5
30,001 to 50,000	30	10
50,001 to 100,000	50	15
100,001 to 500,000	80	25
500,001 to 1,000,000	100	35
1,000,001 to 2,000,000	135	45
2,000,001 to 3,000,000	165	55
3,000,001 or more	175	60

For SI units, 1 ft = 0.3 m; 1 gal = 3.8 L.

2008 Edition

Source: MAN Diesel & Turbo (reprinted with permission)

22.4.1.3 Tanks storing Class I, Class II, or Class IIIA stable liquids and operating at pressures that exceed a gauge pressure of 2.5 psi (17 kPa), or are equipped with emergency venting that will permit pressures to exceed a gauge pressure of 2.5 psi (17 kPa), shall be located in accordance with Table 22.4.1.3 and Table 22.4.1.1(b).

22.4.1.4 Tanks storing liquids with boil-over characteristics shall be located in accordance with Table 22.4.1.4. Liquids with boil-over characteristics shall not be stored in fixed roof tanks larger than 150 ft (45 m) in diameter, unless an approved inerting system is provided on the tank.

22.4.1.5 Tanks storing unstable liquids shall be located in accordance with Table 22.4.1.5 and Table 22.4.1.1(b).

22.4.1.6 Tanks storing Class IIIB stable liquids shall be located in accordance with Table 22.4.1.6.

Exception: If located within the same diked area as, or within the drainage path of, a tank storing a Class I or Class II liquid, the tank storing Class IIIB liquid shall be located in accordance with 22.4.1.1.

22.4.1.7 Where two tank properties of diverse ownership have a common boundary, the authority having jurisdiction shall be permitted, with the written consent of the owners of the two properties, to substitute the distances provided in 22.4.2 for the minimum distances set forth in 22.4.1.1.

22.4.1.8 Where end failure of a horizontal pressure tank or vessel can expose property, the tank or vessel shall be placed with its longitudinal axis parallel to the nearest important exposure.

22.4.2 Shell-to-Shell Spacing of Adjacent Aboveground Storage Tanks.

22.4.2.1 Tanks storing Class I, Class II, or Class III stable liquids shall be separated by the distances given in Table 22.4.2.1.

22.4.2.1.1 Tanks that store crude petroleum, have individual capacities not exceeding 3000 bbl (126,000 gal or 480 m^3), and are located at production facilities in isolated locations shall not be required to be separated by more than 3 ft (0.9 m).

Table 22.4.1.3 Location of Aboveground Storage Tanks Storing Stable Liquids — Internal Pressure Permitted to Exceed a Gauge Pressure of 2.5 psi (17 kPa)

		Minimum Distance (ft)	
Type of Tank	**Protection**	**From Property Line that Is or Can Be Built Upon, Including the Opposite Side of a Public Way**	**From Nearest Side of Any Public Way or from Nearest Important Building on the Same Property**
Any type	Protection for exposures*	1½ × value in Table 22.4.1.1(b) but not less than 25 ft	1½ × value in Table 22.4.1.1(b) but not less than 25 ft
	None	3 × value in Table 22.4.1.1(b) but not be less than 50 ft	1½ × value in Table 22.4.1.1(b) but not less than 25 ft

For SI units, 1 ft = 0.3 m.
*See definition 3.3.42, Protection for Exposures.

Table 22.4.1.4 Location of Aboveground Storage Tanks Storing Boil-Over Liquids

		Minimum Distance (ft)	
Type of Tank	**Protection**	**From Property Line that Is or Can Be Built Upon, Including the Opposite Side of a Public Way^a**	**From Nearest Side of Any Public Way or from Nearest Important Building on the Same Property^a**
Floating roof	Protection for exposures^b	½ × diameter of tank	⅙ × diameter of tank
	None	Diameter of tank	⅙ × diameter of tank
Fixed roof	Approved foam or inerting system^c	Diameter of tank	⅓ × diameter of tank
	Protection for exposures^b	2 × diameter of tank	⅔ × diameter of tank
	None	4 × diameter of tank but need not exceed 350 ft	⅔ × diameter of tank

For SI units, 1 ft = 0.3 m.
^{a}The minimum distance cannot be less than 5 ft.
^{b}See definition 3.3.42, Protection for Exposures.
^{c}See NFPA 69, *Standard on Explosion Prevention Systems.*

 2008 Edition

Source: National Fire Protection Association (reprinted with permission)

Appendix E

Applicable Standards and Codes

A **standard** is a set of technical definitions and guidelines that function as instructions for designers, manufacturers, operators, or users of equipment. Standards do not have the force of law, and are voluntary guidelines. A standard becomes a **code** when it has been adopted by one or more government agencies and is enforced by law, or when it has been incorporated into a business contract.

The following organizations provide codes and standards that may be useful in the design, installation, and commissioning (process by which a system, equipment, facility or plant, which has been installed or is near completion, is tested to verify that it functions according to its design objectives and/or specifications) of thermo-fluids systems and equipment. Most of these organizations provide a variety of different types of codes and/or standards that will be useful for the design, installation, and testing of thermo-fluids systems and equipment.

Air Conditioning Contractors of America (ACCA)
Air-Conditioning, Heating and Refrigeration Institute (AHRI) – Formerly (ARI)
American Petroleum Institute (API)
American Society of Heating, Refrigeration and Air-Conditioning Engineers (ASHRAE)
American Society for Testing and Materials (ASTM)
American Society of Mechanical Engineers (ASME)
American Water Works Association (AWWA)
American National Standards Institute (ANSI)
Canadian Standards Association (CSA)
Gas Processors Association (GPA)
Institute of Boiler and Radiator Manufacturers (I=B=R)
International Association of Plumbing and Mechanical Officials (IAPMO)

Introduction to Thermo-Fluids Systems Design, First Edition. André G. McDonald and Hugh L. Magande.
© 2012 André G. McDonald and Hugh L. Magande. Published 2012 by John Wiley & Sons, Ltd.

International Organization for Standardization (ISO)

Municipal Building and Fire Codes (e.g., Alberta Building and Fire Codes, Building Code of the City of New York)

National Association of Corrosion Engineers (NACE)

National Fire Protection Association (NFPA)

Steel Boiler Institute (SBI)

Sheet Metal and Air Conditioning Contractors National Association (SMACNA)

Uniform Building Code Standards

Uniform Mechanical Code (by IAPMO)

Uniform Plumbing Code (by IAPMO)

Appendix F

Equipment Manufacturers

Currently, there are many manufacturers who provide equipment specified for use in thermo-fluids systems. Below is an abridged list of manufacturers whose equipment may be specified in a system design. Though organized in the groups shown, some manufacturers may provide other equipment and services. The onus rests on the design engineer to check product lines of the manufacturers and suppliers listed below and conduct a thorough search to find other manufacturers and suppliers to provide equipment that meets the required performance targets.

Air Distribution and Heating, Ventilating, and Air-Conditioning
Trane (Ingersoll Rand); York (Johnson Controls); McQuay International (Daikin Industries); Carrier Corporation (United Technologies); Engineered Air (EngA); EH Price, Ltd.; Mortex Products, Inc.; Goodman Manufacturing; FHP Manufacturing (Bosch Thermotechnology); First Company Products; Rheem Manufacturing Company

Pumps and Piping
ITT Bell & Gossett; ITT Goulds Pumps; Taco, Inc.; Peerless Pump; Sulzer Pumps; ACME Pumps; Grundfos Pumps Corporation

Heat Exchangers
Unico System, Inc.; ITT Standard; Armstrong, Ltd.; Sulzer Metco, Inc.; Mortex Products, Inc.

Boilers and Water Heaters
Rinnai America Corporation; Cleaver Brooks, Inc.; Smith Cast Iron Boilers; Lennox; Amtrol, Inc.; Tramont (fuel tanks); American Water Heaters; Bosch Thermotechnology; Navien America, Inc.; A. O. Smith Corporation; Rheem Manufacturing Company; Takagi

Condensing Boilers
Rinnai America Corporation; Triangle Tube; Weil-McLain; NY Thermal, Inc.; Viessmann Manufacturing Company, Inc.; Buderus (Bosch Thermotechnology); Lochinvar

Introduction to Thermo-Fluids Systems Design, First Edition. André G. McDonald and Hugh L. Magande.
© 2012 André G. McDonald and Hugh L. Magande. Published 2012 by John Wiley & Sons, Ltd.

Appendix G

General Design Checklists

The training provided in this textbook has focused heavily on the design, development, and analysis of specific systems and components. Codes, standards, industry rules-of-thumb, client requirements, and engineering intuition have been incorporated in the design analysis. The design checklists of this Appendix serves to provide additional guidance with regard to other items that should be considered in the design analysis and included in the contract documents, such as the drawings and specifications. While the checklists are not extensive, they should be useful to both junior design engineers as well as experienced practicing engineers.

G.1 Air and Exhaust Duct Systems

1. Have appropriate codes and standards from ASHRAE (American Society of Heating, Refrigerating, and Air-Conditioning Engineers), AHRI (Air-Conditioning, Heating, and Refrigeration Institute), NFPA (National Fire Protection Association), and the local jurisdiction been considered?
2. Have adequate balancing dampers been provided at outlets (diffusers, terminal boxes) to restrict noise due to excessive pressure and/or air velocity?
3. Are fire damper and smoke damper locations, type, and flow restrictions indicated?
4. Are access doors at fire dampers, smoke dampers, turning vanes, humidifiers, heating/cooling coils, and other accessories requiring access properly specified and included on the contract drawing sheets or in the general notes?
5. Are flexible connections shown and specified?
6. Is acoustical sound lining required in the duct? Is it properly located and specified?
7. Will the arrangement of the ducts allow the transfer of excessive noise between offices, spaces, and rooms of different function? Do large supply ducts pass over quiet rooms?

Introduction to Thermo-Fluids Systems Design, First Edition. André G. McDonald and Hugh L. Magande.
© 2012 André G. McDonald and Hugh L. Magande. Published 2012 by John Wiley & Sons, Ltd.

8. Is there excessive noise from the intake or exhaust points of fans or other equipment that may be transferred to nearby buildings?

9. Have outdoor air intakes been located sufficiently far from exhaust or relief discharge or plumbing stack effluent? Maintain a minimum 10 ft clearance. For other discharge from contaminated vents and vehicle exhaust, is there sufficient separation distance from air intakes in compliance with code?

10. Have the type of branch takeoffs and duct splits shown on the drawings? Have appropriate detail been provided?

11. Locate exhaust grilles near the floor in operating rooms, storage rooms for flammable and corrosive material.

12. Do not use the space between the slab and finished ceiling above corridors as return air plenums in hospitals, nursing homes, offices, and other facilities.

13. Have fan systems been checked for excessive noise transmission?

14. Is there adequate space for servicing the fans, motors, and other accessories?

15. Is there sufficient straight duct branch length or straightening vanes between the main supply duct and diffusers or terminal boxes?

16. Is balancing required at the fume hood exhaust system? Will orifice plates be needed?

17. Have ventilation systems been provided for equipment rooms and other unconditioned spaces?

18. Is there adequate straight duct upstream of terminal units such as variable-air-volume (VAV) boxes? There should be a minimum of 3 ft of straight duct upstream of all terminal units.

19. For process exhaust systems, does the duct pitch to low points and/or drains?

20. For process exhaust systems, has the correct duct material been selected and specified? Is it stainless steel, Halar coated stainless steel, fiber-reinforced polymer (FRP)?

21. For process exhaust systems with exhaust duct work in unconditioned spaces, will duct insulation or heat tracing be required to prevent water vapor condensation outside or inside the duct?

22. Are process exhaust fans on emergency power as required by applicable code?

23. Process exhaust ductwork cannot penetrate fire-rated wall constructions. Fire dampers are not generally desired. If wall penetration is unavoidable, the ductwork must be enclosed in a fire-rated enclosure until it exits the building. Sprinkler protection may be installed inside the ductwork, if approved by the authorities having jurisdiction.

24. Has coordination with the other trades (architectural, structural, electrical, chemical) been conducted?

G.2 Liquid Piping Systems

1. Have appropriate codes and standards been considered?

2. Are there provisions for piping expansion and contraction, anchors and supports, guides, and flanged joints? Have anchor locations and forces been coordinated with the Structural Engineer?
3. Are balancing valves required on parallel piping loops or systems?
4. Is sufficient space provided for the pitching of pipes?
5. Is there adequate straight pipe upstream and downstream of installed flow meters?
6. Do not route horizontal piping in solid masonry walls or in narrow stud partitions. For piping that penetrates walls, have a link seal, pipe sleeve, and/or flexible piping been used?
7. There should not be piping in electrical switchgear, transformer, motor control center, and emergency generator rooms. If this is unavoidable, have drain troughs or enclosures been provided?
8. Is cathodic protection required for buried piping?
9. Is heat tracing required for piping?
10. For based-mounted pumps, have vibration isolators been specified? If not, is the concrete pad support at least $1^1/_2$ to 3 times the weight of the pump? This should be coordinated with the Structural Engineer.
11. Have major equipment such as pumps been scheduled?
12. Is a bypass line required around the pumps to facilitate maintenance? Are standby pumps required?
13. For open-loop piping systems, has NPSH and cavitation been considered?

G.3 Heat Exchangers, Boilers, and Water Heaters

1. Are there drip pans on cooling coil banks? Have they been piped to a floor drain and has a detail been provided?
2. Are there combustion air intakes for boilers and water heaters? Have the vents, exhaust gas stacks, breeching, and chimneys been shown, specified, and detailed? Have their termination heights been specified?
3. For expansion tanks, has the appropriate size, location, space allocation, support, makeup water pressure, and makeup water location been adequately coordinated with the Plumbing Engineer?
4. Have the owner redundancy requirements been satisfied? Will multiple pieces of equipment be required to prevent system shutdown in the event of equipment failure or maintenance?
5. Has a low load requirement been evaluated? Will the equipment be capable of operating at the low load condition?
6. Does the boiler layout in the design have enough expansion and flexibility in the connection piping to prevent excess stress at the boiler nozzle?
7. Are bypass lines required around the boilers or water heaters to accommodate maintenance and servicing?

8. Is there sufficient access to components in the boiler or water heater that may need servicing?
9. Has a check valve been installed on the boiler or water heater discharge pipeline to prevent backflow of cooled water into the unit?
10. Are water softeners required on the makeup line? Are they shown on the drawings?
11. Is pressure regulation needed?
12. Will the minimum required circulation through the boiler or water heater be maintained?
13. For water distribution systems, will they be reverse-return systems? If not, has proper balancing been considered? Will balancing valves or orifice plates be required?

Index

Affinity laws, 98
Air distribution systems
  low velocity, 13
  small duct high-velocity, 54
Air duct, 9–10, 12–14, 16, 18, 20–22, 29–30,
    54–55, 58, 211, 262
Air handling unit, 67
Air purger, 216
Air separators, 112
Air/fuel ratio, 306
Allowable back pressure, 326
Aquastat, 218
aspect ratio, 15
Atomization, 105, 312
Availability, 299, 319

Back work ratio, 312
Balancing valves, 231
Baseboard heater, 216
Baseboard heaters, 233
Bell and Gosset, 89
Binary cycle, 333
Blasius correlation, 9
Boiler, 138, 209, 210, 216, 245, 266, 323
  annual fuel utilization efficiency, 226
  drum, 323
  dual use, 236
  feedwater, 218
  heating capacity, 220
  heating capacity ratings, 224
  hose bib, 218
  low-pressure, 217
  medium- and high-pressure, 218
  oversizing factor, 222
  pinch point (PP), 324
  pipe cock, 218
  sizing, 220
  stack, 269, 317, 322, 326, 335
  thermal size, 222
  waste heat recovery, 322
Boiling, 146
Brayton cycle, 265, 305, 332
Burner, 218

Capacity ratio, 151
Carnot cycle, 266
Cavitation, 93
Chimney, 219, 226
Churchill and Bernstein equation, 161
Churchill's equation, 9
Circular equivalent method, 18
Circulator, 216, 231, 257
Colebrook equation, 8
Combined-cycle plant, 322
Compressor, 44, 305
Condensation, 146
Condenser, 103, 104, 136, 207, 266, 278, 287,
    335
Conduction, 127, 130
Contact feedwater heater. *See* Feedwater
    heater
Convection, 130

Introduction to Thermo-Fluids Systems Design, First Edition. André G. McDonald and Hugh L. Magande.
© 2012 André G. McDonald and Hugh L. Magande. Published 2012 by John Wiley & Sons, Ltd.

Corrosion, 312
Creeping, 312

Dampers, 29, 34, 65
Darcy friction factor, 7
Diffuser boots, 66
Dip tube, 219
Dittus–Boelter equation, 141
Diversion fittings, 228
Drain cooler, 278
Drain cooler approach (*DCA*), 279
Dropped ceiling, 25, 58

Economizer, 323
Effectiveness-number of transfer units
   (ε-NTU) method, 147
Energy equation, 7
Equal friction method, 20
Equivalent lengths, 20
Erosion limit, 107, 193, 247, 382
Evaporator, 207, 210, 323
Exhaust end loss, 280
Expansion joints, 104
Expansion tank, 111, 216, 257
Expansion-line-end point (ELEP), 282

Fan, 25, 44, 58
  direct-drive, 44
  belt drive, 44
  forward-curved, 46
  backward-curved, 46
  laws, 51
Feedwater heater, 267
  closed, 269
  open, 267
Finished ceiling, 25, 58
Fins, 157, 175
  surface effectiveness, 175
Fittings, 73
Flue gas, 209, 219, 335
Fouling, 137
Friction loss, 12
Fuel, 218, 224, 381
  energy content. *See* heating value
  heating value, 226
  higher heating value, 226
  lower heating value, 227

Full-load operation, 312
Furnace, 226

Gas cycle. *See* Thermodynamic cycle
Gnielinski's equation, 141

Haaland's equation, 9
Head losses, 6
Heat and mass balance diagram, 284, 288,
   297, 318, 330, 334
Heat capacity, 148
Heat exchanger
  analysis, 142
  area density, 129
  baffles, 129
  balanced, 144
  compact, 129
  counter flow, 127
  cross-flow, 129
  design, 147
  duty, 164, 181
  effectiveness, 148
  functional diagram, 213
  headers, 129, 164, 207
  parallel flow, 127
  performance, 147
  shell-and-tube, 129
Heat Exchanger
  flow diagram, 214, 326
Heat transfer coefficient, 138
Heating coil, 164, 171, 174, 181, 188, 190, 207,
  209
Hot water heating system. *See* Hydronic
  heating system
Hydraulic diameter, 5
Hydronic heating system, 216
  one-pipe series loop, 227, 239, 244
  split series loop, 228
  two-pipe direct return, 229
  two-pipe reverse return, 231

Intercooler, 310
International Standards Organization (ISO),
  312
Irreversibility, 300

j-factor, 157

Laminar flow, 6
  fully-developed, 139
LMTD method, 147
Loss coefficient, 19

Mollier diagram, 270
Moody chart, 8

National Electric Manufacturers Association
    (NEMA), 312
Net generator output, 286
Net heat rate, 286, 288, 297, 318, 321
Net positive suction head (NPSH), 93
Noise criteria, 55
Number of transfer units (NTU), 151
Nusselt number, 138

Orifice plates, 231
Overall heat transfer coefficient, 135, 138,
    175, 194, 209, 335

Part-load conditions, 312
Petukhov equation, 9
Pipe
  hangers, 77
  installation, 77
  materials, 75
  plastic, 76
  sizing, 77
  velocities, 76
Piping system, 73
  closed-loop, 111
  open-loop, 103
Plasma spraying, 213
Plenum, 22
Power plant system, 265
Pressure loss, 23
Pump
  best efficiency point, 87
  brake horsepower, 85
  capacity, 83
  dynamic, 83
  efficiency, 85
  equivalence, 98
  free delivery, 86
  gas, 83
  master selection chart, 91

  net head, 85
  performance curves, 86
  positive-displacement, 83
  set, 75
  shut-off head, 86
  system curve, 87
  water horsepower, 85

Rankine cycle, 266, 267, 269, 287
Recovery rate, 222
Regeneration, 267
Regenerator, 310
Regenerators. *See* Feedwater heater
Reheater, 269
Relative humidity, 9, 312, 317, 329
Relative roughness, 8
Resistance network, 134
Reynolds number, 5
Rinnai, 209, 220, 221, 241, 413

Second-law analysis, 298
Second-law efficiency, 300, 301
Steam blanketing, 324
Steam generator, 265
Strainer, 103
Superheater, 209, 323, 325, 326, 327, 328, 335
Surface heaters. *See* Feedwater heater
Swamee-Jain formulae, 8

Taco Pumps, 89
Temperature correction factor, 235
Terminal temperature difference (*TTD*), 279
Thermal resistance, 130
Thermodynamic cycle, 265
Transitional flow, 6
Tube bank
  bare, 157
  in line, 159
  pressure drop, 162
  staggered, 160
Tube-pitch ratio, 159
Turbine
  impulse, 280
  internal efficiency, 270
  reaction, 280
  stages, 280
  steam, 280

Turbogenerator net output. *See* Net
    generator output
Turbulent flow, 6

Unico, Inc., 203
Usable storage capacity, 223
Used-energy-end point (UEEP), 282

Valves, 73
Vapor cycle. *See* Thermodynamic cycle
Vapor pressure, 92
Variable frequency drive, 109

Velocity correction factor,
    235
Vent, 219

Water hammer, 112, 324
Water heater, 209, 216, 220, 236, 258
    tankless, 219

Zone, 227
    condensing, 278
    desuperheating, 278
    drain cooling, 278